The ATL-98 Carvair

The ATL-98 Carvair

A Comprehensive History of the Aircraft and All 21 Airframes

WILLIAM PATRICK DEAN

Foreword by MICHAEL O'CALLAGHAN

McFarland & Company, Inc., Publishers

Jefferson, North Carolina

The present work is a reprint of the illustrated case bound edition of The ATL-98 Carvair: A Comprehensive History of the Aircraft and All 21 Airframes *first published in 2008 by McFarland.*

LIBRARY OF CONGRESS CATALOGUING-IN-PUBLICATION DATA

Dean, William Patrick, 1944–
The ATL-98 Carvair : a comprehensive history of the aircraft and all
21 airframes / William Patrick Dean ; foreword by Michael O'Callaghan.
p. cm.
Includes bibliographical references and index.

ISBN 978-1-4766-6280-0 (softcover : acid free paper) ∞
ISBN 978-0-7864-5177-7 (ebook)

1. ATL-98 Carvair (Transport plane)— History.
2. ATL (Corporation)— History. I. Title.
TL686.A785D33   2015   629.133'343 — dc22   2008023266

BRITISH LIBRARY CATALOGUING DATA ARE AVAILABLE

On the cover: A British Air Ferries Carvair,
1960s, Southend, England

Printed in the United States of America

*McFarland & Company, Inc., Publishers
Box 611, Jefferson, North Carolina 28640
www.mcfarlandpub.com*

Acknowledgments

The Carvair began as an idea in the late 1950s. Its creator, Sir Freddie Laker, is no longer with us, nor is his BAF successor, T.D. "Mike" Keegan. Many of the other significant personalities throughout the Carvair story are also gone. I am privileged to have met so many fascinating personalities over the years who told me their stories and of their fondness for the Carvair.

Special thanks to Tony Merton Jones, Propliner Magazine, who graciously accepted my request to review the manuscript. I am indebted to Ian Callier (ATC Southend), Guy Craven (ATEL photographer), Richard Goring (ATEL Stores), Paul Howard, Sean Keating, Brian Kerry (ATEL chief aerodynamicist), Ruth May (Falcon Airways), Richard Vandervord (Customs Southend), and Michael Zoeller, all of whom provided their personal logs, photos, files and continually assisted in gathering contemporary documentation and fact verification. Many thanks to everyone at Aviation Traders, BAF and British World Airways, who provided data without offering their name. Thanks to Air Britain (AB) for allowing me to serve as the Carvair Specialist and the encouragement of all the staff, and to the American Aviation Historical Society (AAHS), which asked me to write a story on the Carvair years ago.

The majority of memorabilia and photos reproduced came from my collection of 40 years. Credit is given to the original photographer or source. In the 1960s when I began collecting material this volume was not foreseen. Therefore over the years many names were not recorded or have been lost. Apologies are extended to anyone who helped in the preparation of this work or provided material and has been inadvertently not recognized or whose name does not appear in the following alphabetical list:

J.S. Algranti, Ad Jan Altevogt, Michael Anciaux, Norman Atkins, Maurice Austin (CAHS), Robert Austin, Simon Barker (SR Technics), Joseph Baugher, Alan Bennett (ATEL), Clifford Berrett (ATEL), Peter Berry (AB), John Blatherwick, Thomas Blickle, James Blumenthal, Will Blunt, J.W. Bossenbroek, Roger Braga, Keith Burton, Liam Byrne, Ian Carmichael, Guy Cassiman, Tony Clarke, Mike Clayton, Frank Conway, Ruth Cook (AirTran), Paul Crawley, Dave Cummings, Alfred Damen, Robert Dedman (Interocean), Herman Dekker, Joe Dible (Aer Lingus), Tony Doyle, Mike Draper (AB), Murray Dreyer (Pacific Aerospace), George Dyess (Full Air), John Ebel, Ron Edwards, Keith Elliott, Ken Elliott, William Ernst, A.P. Fairchild (Interocean, PAL, PAE, HPA), Bob Farnell, Malcolm Fillmore (AB), Patrick Fitches, John Flanagan, Ray Flynn, Alan Fraser (Ansett), Fredrick Freeman Jr., Ola Furulund, Richard Gaudet (EPA), Andrew Gider (Canadian Airlines), Kenneth Gilmartin, Simon Glancey, Kenny Goo (Century Aviation), Geoffrey Goodall, Dennis Goodin, Keith Gordon (SINGAS), Jennifer Gradidge, Kjell Granlund, David Greer (Berlin Airlift Association), Eddy Gual, Max Harrison, David Hartman, Ian Haskell, Paul Hawkins (HawkAir), Dave Heaney, Leif Helstrom, D. Hollander (Southend Airport Authority), Ian Howse (BUAF, Ansett), Kerry Jones (FBO Griffin), Terry Judge, Derek King (AB), Leo J. Kohn, Dennis Leese, Ruud Leeuw, Terry Leighton (ATEL), Chris Leoni (P&W), Clive Lynch, Ralf Manteufel, Eduard Marmet, Peter Marshall, Peter Marson (AB), Bernard Martin, Brian Martin, J.J. Masterson, Ian Matthews (Aerothrust), Bob McSwiggan (Custom Air

Service), Bruce McSwiggan (Academy Airlines), Mark McSwiggan (Academy Airlines), Brian Mees (BAF), Mark Meltzer, Richard Micklefield (ATL), Larry Milberry (CANAV Books), Stephen Miller, Mike Morgan (Warbelow's Air), Frank Moss (Hondu Carib), Coert Munk, Bob Murdock (Century Aviation), Leo Murray (Irish Aviation Archive), Howard Nash (AB), Bill Nelson (Aer Lingus), Howard Nessen, Odd Nilsen (NAHS), Fred Niven (Historian Ansett Group), Michael O'Callaghan (Aer Lingus), Arthur Pearcy, Stephen Piercy (AB), Mike Pitman, Paul Rakisits (Interocean, SEAAT, PAE), Han de Ridder, Gordon Reid, Don Robesky (HPA), Matthew Rodina, Tony Rogers (ATEL), Marisa Ruiz-Ayucar (Iberia), Lawrence Safir, G. Salerno, Richard Sartini (AirTran), Nicky Scherrer, Robbie Shaw (ATC Gatwick), L.B. Sides, John Simms, Steve Siviter, Douglas E. Slowiak, R.N. Smith, Charles Stewart, J.J. Sullivan (Aer Lingus), Lynn Taylor (British World Airways), Henry Tenby, Roger Thiedeman (Airways), John Thomas (ATL), Kenneth Tilley, Rob Tracz, David Truman, Randy Tucker (PAE), Denny Turner (PAE), Ferand Van De Plas, Aad Van De Voet, Ad Vercruijsse, Philip Warnault (Le Touquet Airport), Dacre Watson, John Wegg (Airways), Carl Gustaf Wesslen, James R. Willis (Eastern Airlines), Roger Winfield (Monarch Airlines), Lindsay Wise (Ansett, DCA Australia), Charles Woodley, Paul Zogg.

Table of Contents

Foreword by Michael O'Callaghan

Most historians view the Carvair as a moment in aviation history. Those who read this work will find that Patrick Dean has done extensive research to relate the story of an amazing low production aircraft. This accounting provided details that I did not know and refreshed my memory of those that had receded into the mist of time. It is an objective review of the aircraft and all those who envisioned, designed, built, flew, repaired and loved it.

I was introduced to the Carvair when I applied for a position on the Aer Lingus cadet pilot scheme. During the interview, I was asked to list the aircraft currently flown by the company. I replied that Aer Lingus had Douglas DC-3s, Fokker F27s, Vickers Viscount V808s and Boeing 720s. It was noted that I had left one type out. Somewhat taken aback I replied that I thought the Bristol Wayfarer had been taken out of service and sold. The interviewer agreed then informed me the company had just issued a press release announcing the purchase of Carvairs for car-ferry service to the United Kingdom and France.

At the end of my time in flying school I returned to Dublin to fly F27s noting the new Carvairs parked on the southern edge of the main ramp. They lived apart from the main center of operations. The reputation of "unreliability" became well known to us as we watched them taxi out only to return with a magneto drop or other ailment. It was said that if you did depart on schedule you could not be sure to return on schedule, or on occasion, even that day.

The 16 most junior first officers were converted to the Carvair when Aer Lingus F27 operations were phased out. None of us relished the prospect of flying one, but after completing the conversion courses we found that despite its reputation it was in fact not as unreliable as said to be. It was decidedly different to operate from the turboprops that we had flown. It reintroduced us to vacuum operated horizons and directional gyros that had to be set before takeoff and regularly during the flight. There were engine run–ups and magneto checks, not to mention the multitude of levers on the pedestal that did not exist on the turbine craft to which we were accustomed. But then, the Carvair was based on a 25-year-old design.

During the conversion course we were all pleasantly surprised how simple and indeed how pleasant the aircraft was to fly. Heavy, yes, but it was very stable, and the precision instrument flying that was demanded by Aer Lingus flying instructors and necessary to operate in the cloudy climate of Ireland and the United Kingdom was easy to achieve. One of the tricks was to increase the engine rpm above the recommended for approach. This increased the airflow over the tail, giving a greater, or quicker, response from the elevators. It also provided more power in the event of a missed approach as you could go straight to METO power if required and eliminate one set of levers from the task of initiating the missed approach.

After the conversion course we were assigned to the regular schedule of the car-ferry followed by freighter operations that even in those days either started very early in the morning or finished very late at night. The car-ferry operated to schedule, late at times, but not so much so as to cause passenger worry. Freighters could be late for many reasons, but of course

1

the freight did not care, unless the newspapers and mail arrived too late from Manchester to be distributed around the country. Then, questions were asked.

Despite the legendary unreliability, I only had one engine failure, and that was on a test flight prior to return to service after overhaul. We adapted and had fewer mechanical problems but reliability could never approach that of turboprops. Even with high load factors, Aer Lingus management was never enthusiastic about the Carvair operation. The service was terminated in the autumn of 1966 and I flew the last freighter out of Manchester on the evening of 30 October. Henceforth, modified Viscounts operated cargo service. The Carvairs were retired to the hangar ramp until sold to Eastern Provincial in Newfoundland. The Captains were redeployed to Viscounts and the junior first officers were sent on a conversion course for the Boeing 720/707, never again to fly propeller aircraft with Aer Lingus.

It was the end of an era at Aer Lingus and within a few years even the Viscounts were replaced by Boeing 737s. The Carvair receded into folklore within the company but EI-AMP managed to add to this store with a major contretemps on its acceptance flight for its new owner. It was to prove a portent for the future as its life came to a sudden end in a crash landing in Newfoundland.

Conversion to the transatlantic operation was not the last time I saw the Carvairs. In May 1970 I arrived in Gander because of a bomb scare. As it was raining, after the aircraft was declared clear by the Royal Canadian Mounted Police, the passengers were required to complete their baggage identification in the shelter of a hangar. Inside, parked side by side, were the two remaining Carvairs in Eastern Provincial colors. The distinctive smell of aviation gasoline, oil and hydraulic fluid brought back the memories of the past. I was glad to see the drips from the usual places that meant they were not dormant but maybe just resting between flights.

Despite the short career with Aer Lingus the Carvair served successfully with BUAF, latterly BAF, in England and with other small European airlines. It served with distinction in Australia, but unfortunately, economics began to tell against that operation in the front line of aviation. The expansion of motorways in Europe coupled with improvements in faster cross-channel sea-ferry service added to the difficulties facing the airborne car transport business. The Carvair could not cope, continuing only in pure freighter operations in Europe but even that succumbed to turboprops and new generation turbofan freighter aircraft. It quietly served throughout the world in freight transportation, specializing in bulky, outsized cargo. The rugged basic design enabled it to work in areas that were, and indeed still are, unsuitable to more modern aircraft.

This book is exceptional in bringing to life the history of an aircraft that attracted interest wherever it has been. At Aer Lingus the Carvair was said in lighter moments on the flight deck to have been the inspiration to Boeing for the B747. I am not sure, but the field of aircraft design is small, so maybe there is a grain of truth in it! Today the quickest way to describe a Carvair to the younger generation of aviation enthusiasts is ask them to think of a B747F with four propellers and piston engines. No matter how you wish to describe it, the Carvair stands as a monument to American engineering wedded to British ingenuity. It fulfils a small place in modern aviation history some 60 years after the original Douglas DC-4 was conceived and 40 years after Aviation Traders introduced an improved airborne car-ferry to the English and Irish tourist.

Michael O'Callaghan was born in Dublin in 1945. He joined Aer Lingus in 1962 and flew as first officer on the F27, Viscount and Carvair, and engineer and first officer on the Boeing 720 and 707. He was promoted to 737 captain in 1976 and then turned to the 747–100. In 1996 he converted to the A330 and in 2003 retired as the last active pilot at Aer Lingus to have flown the Carvair.

Preface

The goal of producing a book on the Carvair began as a desire to track the 21 airframes from C-54 war production at Douglas through ATL-98 conversion in England to breakup or loss. In compiling data I became acquainted with the most unique and interesting individuals and characters. It became evident that the history of the Carvair is more than just dates and facts. Tracking these unique aircraft for 40 years became an incredible learning process. It would be hard to imagine any fleet of aircraft more woven into world politics, operating in more remote areas, or transporting more diverse cargo.

My interest in aircraft began as a child. I later served in the United States Air Force and spent 36 years with United Airlines. As any historian or writer can tell you it would be impossible to record every event on an aircraft type in chronological order, especially on a special purpose aircraft developed for a niche market then dispersed throughout the world. The order of text is arranged in such a way to offer the human-interest side of the fleet history. It is not possible to tell every anecdote about each aircraft, therefore, selected incidents are woven into the dated history to relate an overall experience. A wise man once told me that it is very risky offering a work as a definitive source. Although every effort has been made to assure accuracy and completeness, we as humans are not perfect. Corrections and additional information are always welcome. It is my desire to tell the Carvair story by introducing the reader to the individual aircraft and many personalities that cover 70 years since the idea of a four-engine transport was conceived.

The ATL-98 Carvair is a unique example of a short time need for a single purpose aircraft. Although it was envisioned and designed for that special purpose and market, it is the end result of many unrelated events, corporate consolidations, and engineering designs that evolved to create a unique flying machine.

I first saw the Carvair in the 1960s because of the unrelated fact that I was serving in the USAF and stationed at RAF Mildenhall, England. I had previously seen a photo of the Carvair, which first struck my interest. As a mechanic on the C-124 Globemaster II, I was quite intrigued with another front-loading Douglas derivative. While on a weekend exploration of the Channel Coast of England I visited Southend. When I actually saw the Carvair for the first time, I was immediately fascinated by the design. As a result, for more than twenty years, I collected photos and articles and tucked them away in a file.

Long after leaving the Air Force and joining United Airlines I was assigned to Hawaii. On a morning in 1984 I was standing in flight operations when a Carvair taxied by. I was reminded of my past fascination and ran across the ramp to the parking spot to get a better look. Little did I know that my interest combined with a series of events and discoveries would result in an attempt to compile a historical accounting of this novel aircraft. On an early morning in 1990 I was again surprised to see another Carvair landing at Maui that I had not previously seen. My fascination was renewed when I discovered it was one of a pair that had been imported from New Zealand.

I discovered in 1994 that three Carvairs were at two airfields near Atlanta, where I grew up. In 1996, at the urging of the American Aviation Historical Society (AAHS), I reviewed my files for a possible article on the Carvair. I realized I had enough material for a series of articles or possibly a book. The article was never written, and I decided to set out on putting together the fleet history.

While searching for additional material in January 1997 I contacted Ruth May, who in 1979 purchased Carvair five in England and brought it to the United States. We made arrangements to meet three months in the future when she planned to visit Honolulu. I met her at the airport in the late afternoon of 03 April and we went to her hotel for the interview. Since there is a six-hour time zone difference it was already 04 April in Georgia, where the aircraft was currently based. At precisely the same time we were discussing the years May owned the fifth Carvair, N83FA, it crashed at Griffin, Georgia. I had been on the aircraft only weeks before when I visited Griffin. The coincidence that we set a date to meet three months in advance and it was the day Carvair five was destroyed was overwhelming. While continuing research I discovered a number of other events that reinforced my desire to complete the project.

I began my career with United Airlines on the same day that BUAF became BAF. This was before I knew that William Patterson, the first president and father of United Airlines, was the force behind the development of the DC-4. I now find my interest in this aircraft special since I worked for the company Patterson built and the DC-4 was his vision of a four-engine transport, which is also the basis for the Carvair. Patterson unknowingly provided a suitable airframe years before the idea of the Carvair was conceived.

I also discovered a number of other events took place on my date of birth. Freddie Laker's Air Charter made its first vehicle ferry flight in a Bristol (G-ANMF) to Calais on 28 August 1954. The first Carvair, G-ANYB was seriously damaged by forklift on 28 August 1961. Carvair five, N9757F, the first Carvair sold in America, applied for U.S. registration on 28 August 1962. On 28 August 1975 the original registration of the last Carvair built was dropped from the Australian register.

When I was a boy in Georgia I saw a squadron of B-36s fly over in formation on a training deployment. It was the most unbelievable sight and sound. From that time on I wanted to be assigned to the B-36. But by the time I was old enough to be in the Air Force the B-36 was redundant. However, I found myself in a C-124 squadron in England, where I was introduced to the Carvair. It can be said without doubt that seeing the B-36 was the first of many events that brought me to the Carvair. It wasn't until years later that I learned the first B-36 (44–92004) made its maiden flight on 28 August 1947.

I was at Griffin, Georgia, in May 1977 photographing the progress on the re-build of Carvair 20. The FAA inspector was there and the logbook was lying on a table outside operations. I had to look twice to make sure my eyes were not deceiving me. The cover of the logbook was a photograph I had taken of Carvair 21 in Hawaii years earlier. It was not a published or a circulated photo and yet somehow it was there. I was unable to determine how it had gotten there.

I have always been drawn to the Carvair since my first sighting in the 1960s at Southend. However, after the crash of N83FA in April 1997, I was no longer an outsider writing the story but had become a part of it. I have attempted to relate events that I believe contributed to the existence of the Carvair. The accounting begins with an overview of corporate histories, personalities and profiles. It is followed with engineering and design then completed with the individual history and stories of each of the 21 aircraft beginning with DC-4 production. Appendices of fleet data, cross-references of registrations and serials, aircraft names, individual dated history, and lists of owners and operators are near the back of the book, followed by a glossary, chapter notes, a bibliography and an index.

Introduction

The ATL-98 Carvair is one of the world's most unique aircraft. It is a true hybrid considering the use of an existing airframe for conversion to a new design of special purpose aircraft. Its existence is as much chance circumstance as planning. It is the end result of a number of unrelated forces, events, and decisions by individuals that unknowingly contributed to the Carvair concept and creation. There are other large "Guppy" and hybrid transport aircraft built both before and after the Carvair. They are the conversions of the Douglas C-74 to C-124, Convair B-36 to XC-99, Boeing C-97 to Super Guppy, Airbus A300 to Beluga, CL-44 to Conroy Guppy and most recently the Boeing 747 to LCF. Each are unique to a specialty but have not achieved such multifunction, recognition, or acceptance throughout the world as the Carvair.

The Douglas C-124 Globemaster II is actually a modification of a previous type, the C-74 Globemaster I. The C-74 was conceived in 1942 to fulfill the military need for a long-range heavy transport. It was built at Long Beach and based on technology Douglas had developed from the DC-4. The first C-74 was delivered to the Air Force late in 1945. A commercial version was developed for postwar airline service and designated DC-7. Pan Am ordered 26, which were later cancelled. As a direct result of C-74 service in the Berlin Airlift the decision was made to build a larger version by returning the fifth aircraft to Douglas for conversion to the YC-124. The fuselage height was expanded to a double deck using the same wing, empennage and gear. Straight-in loading was achieved with double clamshell doors in the nose below the cockpit. Although different in design, the concept of front-end loading and raised cockpit are similar to the Carvair. The C-124 became the hybrid and saw only military service of less than 30 years.

In May of 1942 Consolidated (later Convair) considered a large cargo aircraft utilizing the wings, empennage, and landing gear of the B-36. The design was to be a commercial aircraft 13.5 feet wide and 173 feet long. Ultimately one unpressurized version was built for the Air Force with a fuselage width of 20.6 feet and 182.5 feet long. The advanced nose loading design which never got past the drawing board, appeared to be a cross between the C-124 and the Carvair. The fact that it was a new fuselage utilizing B-36 wings, engines, landing gear, and empennage made it a hybrid.

Another significant and more bizarre hybrid is the Super Guppy modification of the Boeing Stratocruiser. The XC-97 was ordered by the military in early 1943. The only competition for C-97 was the Douglas C-74 that had the same side loading problems as the C-54. The C-97 was designed with rear clamshell doors for straight-in loading. It was actually a hybrid B-29 with a greatly enlarged double deck fuselage using the same wings and empennage. The C-97 vertical fin was later increased in height for reason similar to the Carvair extended tail. The Stratocruiser saw commercial service with the big overseas carriers but came at the end of the propeller age and was soon outclassed by jets.

Aerospace engineer Jack Conroy recognized the need for an oversized cargo aircraft to transport rockets and space hardware, which resulted in the creation of the hybrid Super Guppy. Aero Spacelines, a cargo carrier based in Santa Barbara, California, specialized in transporting space hardware for NASA. In 1961 Conroy began converting a surplus Boeing Stratocruiser for the company by expanding the fuselage to a size that resembled a blimp using the same wings and tail. Originally the Guppy had a swing tail for straight-in loading then changed to a swing nose on later conversions. It became an extremely limited special purpose hybrid which did not interface with the public. It was eventually acquired by Airbus to transport aircraft parts, but the aging airframe created a need for a suitable replacement.

Airbus officials decided in 1991 to develop a replacement Guppy from an existing type. The Beluga was developed from the A300 flying the first time in 1994. The aircraft with a blimp-like fuselage features straight-in loading with a large swing-up cargo door above the cockpit. Although a hybrid, the A300B4–600ST did not utilize a pre–existing fuselage for conversion. Instead a new aircraft was built by mating a new fuselage to an existing design of wing and tail components from the proven A300 aircraft. The Beluga, like other transport hybrids, is strictly all cargo with no interface with the public.

Jack Conroy created another special purpose "Guppy" aircraft by using the existing CL-44 as a base airframe. The sixteenth CL-44 built was modified in America by Conroy Aircraft to produce a large capacity transport that first flew on 26 November 1969.

Although only one aircraft was produced, it is significant in relating the Carvair story. The fuselage of the CL-44 was expanded, almost doubling the capacity to 13,000 cubic feet. The swing tail feature of the CL-44 was retained to utilize straight-in loading. This one-of-a-kind aircraft was named the Conroy Monster. Jack Conroy designed an even bigger conversion referred to as the Conroy Colossus. He also conceived a six-engine version that would have been the world's largest aircraft. The aircraft was never developed because of a lack of available airframes.

The single Conroy Guppy built became a member of the Transmeridian fleet during the period T. D. (Mike) Keegan held majority financial control and was chairman of Transmeridian-BAF. The aircraft was operated along side the British Air Ferries Carvairs. Both types were featured in advertisements comparing and emphasizing their ability to transport oversize loads.

The Carvair, although not as large as some of the other hybrid aircraft, still qualifies as a Guppy or Mini Jumbo. With exception of the Conroy CL-44 Guppy, all the others served only in military or aerospace industry while the Carvair acquired a public image. It has served a much more diverse role and stands out from the others by transporting cargo and passengers. The Carvair is indirectly the result of a vision for a four-engine transport (DC-4) by a man who possibly never saw the conversion.

The need for oversize Guppy aircraft continues as seen in the latest design of B-747 LCF. Boeing, like Airbus, realized the need to transport complete airframes between manufacturing and assembly facilities. Boeing took the concept one step further than the Airbus Beluga when modifying an existing type. The 747 fuselage has been increased, creating a capacity three times the normal airframe with a swing tail for easy loading. The first of these monsters flew in 2006.

To understand why the Carvair is a mutation of multiple ideas we have to look back some 26 years before it first flew. The need for a large transport prior to World War II was not considered urgent. One man envisioned the era of large 54,000 pound, 175 mph, 40 passenger transports. In 1935 William Patterson, President of United Airlines from 1934 to 1966, envisioned such an aircraft. As a banker he was a most unlikely person to have fathered such an idea. He became interested in air travel while working on the merger between Pacific Air Transport and Boeing Air Transport that formed United Airlines.

William Patterson discussed with Donald Douglas the need for a transport aircraft with a 2200-mile range. Neither company had the resources to commit to such a project. Patterson was so convinced of the need that he encouraged American, Eastern, Pan Am, and TWA to join United in this venture. Each carrier put up $100,000 toward the $3 million development cost for the new aircraft. Patterson reviewed his ideas with G.T. Mead of United Aircraft and Jerome Hunsaker of M.I.T., who did the basic drawings. Hunsaker started one of the first programs devoted to aeronautical engineering

In 1936 design studies began for the large tri-finned DC-4E that first flew 07 June 1938. Pan Am and TWA both withdrew before the aircraft was completed and opted for the Boeing 307. The new DC-4E was loaned to United from May to August 1939. Although the new aircraft was painted in the United livery and made many proving flights over the carriers' route systems, it had many performance and maintenance problems. The remaining sponsoring carriers agreed with Douglas to suspend the program in favor of a smaller, less complex design designated DC-4. The DC-4E was returned to Douglas and eventually sold to Japan for Greater Japan Airlines.

United, American, and Eastern went back to Douglas in 1939 for the new design. The design team, headed by A.E. Raymond and E.F. Burton, designed a completely new model designated the DC-4. No prototype was built. Instead, the first aircraft (c/n 3050) was flown by John F. Martin on 14 February 1942. Production was placed in jeopardy by the war in Europe and the need for combat aircraft but Douglas was determined to keep the project going. After the Japanese bombing of Pearl Harbor the military recognized a need for long-range transports for worldwide operations. Since the DC-4 development program had gone well with few problems and was immediately available, it was put into mass production as the C-54 and pressed into military service.

If not for the vision and persistence of William Patterson the DC-4 probably would not have been developed. Chances are that any other large transport design evolved from the war would not have formed a practical base for the Carvair. Even more unique is the fact that five of the C-54s later converted to Carvairs were originally acquired by United, Eastern, and American after the war. These three carriers were financial backers of the development of the DC-4. Since the DC-4 first flew 20 years prior to the Carvair it appears unrelated yet is a basic factor in the existence of the Carvair.

If the DC-4 had not been in production at the outset of World War II the C-54 version possibly would not have been built. There certainly would not have been a large surplus of C-54 aircraft and spares after the war, a primary factor in selecting it for Carvair conversion.

Another series of events that gave reason for the Carvair development began at the end of World War II. Silver City Airways was formed in 1946 as an in-house airline by the Zinc Corporation to support mining activities in Australia. In 1948 Air Commodore G.J. (Taffy) Powell and British Aviation Services purchased the company. In an effort to increase revenue, Commodore Powell conceived the idea of transporting automobiles. Powell was met with strong opposition but persevered. To prove his idea he transported his personal automobile to Le Touquet in July 1948. A large group gathered at the airport to observe the event.

While promoting the proposed service Powell was approached by Sammy Norman, who was about to fly back in his private plane. Since his Bentley was to be transported by boat he asked if it could be accommodated. They agreed on 30,000 francs and the two automobiles were flown back to England as Sammy Norman followed in his private plane. Norman became a frequent customer until he passed away in 1960. He never saw the Carvair yet played a role in the development of the car-ferry service by being the first car-ferry passenger.

All other oversize and specialty transport aircraft have not had the direct impact on the public at large in a personal way as the Carvair. In researching the history everyone interviewed reminisced about the Carvair with affection and nostalgia. These people included flight crews, mechanics, engineers, owners, and passengers.

The Carvair is a result of unrelated ideas, decisions, visions, and dreams of individuals who never planned or saw the need for the aircraft. Some passed away long before Freddie Laker, while sitting in the bath, had his vision. Yet if William Patterson had never seen the need for a four-engine transport and if Donald Douglas had not designed it with his specifications there would not have been a DC-4. If Commodore Taffy Powell had not believed in the car-ferry using the Bristol, Freddie Laker would not have envisioned a cost-effective long-range replacement and there would not have been a Carvair.

PART I

Getting Off the Ground

1

Corporate History

Silver City

Car-ferry service can be traced back to Silver City Airways, which was registered 25 November 1946 and formed to support mining activities in Australia. Because of limited and frequently booked commercial flights between the U.K. and Australia a growing need existed for transporting company personnel. Silver City was tied to British Aviation Services (BAS) after it was formed in 1947 along with Manx Airlines. BAS was later changed to Britavia in 1954, and shared the same managing director, Air Commodore G.J. Taffy Powell.

Griffith J. "Taffy" Powell is well established in British aviation history having broken two Trans-Atlantic records and becoming the first British pilot to hold both land and marine aircraft license. He successfully set up and commanded the Air Ferry Command to move military aircraft across the North Atlantic from U.S. factories to the war in Europe. After the war he decided not to return to BOAC and accepted the position of the first Managing Director of British Air Services (BAS). He maintained the position of Managing Director of Silver City until his retirement in 1957 because of poor health. He was replaced by W.C. (Bill) Franklin and C.M. Fox as joint directors.

The name for the airline came from one of the Zinc Corporations major operations. The area of Broken Hill, New South Wales was known as "The Silver City." Since the primary function of the newly formed in house airline was to support this project, the name was adopted and service began under the direction of British Aviation Services. BAS became the owner of Silver City Airways along with Manx Airlines in 1947. Lancashire Aircraft Corporation and Dragon Airways was acquired in 1957 to form the northern division of Silver City Airways. In 1958 Air Kruise was added further expanding routes and increasing the fleet size.

Silver City initiated service with the Lancastrian, later adding DC-3s and eventually the Bristol 170, which is the forerunner of the Bristol Freighter Mk.32. During this time the airlines of the world were becoming more efficient yet Silver City's revenues were slipping. Leslie McCraken suggested to managing director Air Commodore (Taffy) Powell that a niche market existed in transporting of automobiles by air and it could be a new source of revenue. The cost would be considerably more than by sea, since the distance between the two cross channel airfields was 47 miles. However, the fuel tanks of the automobiles would not have to be drained, which would be another plus.

Ferrying cars was not a new idea since it had been tried in America in 1927. Also in July 1947 a newly married couple in England chartered an Aerovan to fly their car from Croydon to Le Touquet. Although novel, the idea of scheduled service did hold promise.

Powell continued to pursue his idea against very restrictive regulation believing that people were willing to pay for fast service. As a publicity ploy on 06 July 1948 Taffy Powell flew his personal Armstrong-Siddeley automobile across the channel from Lympne in Kent to Le Touquet then on to Paris (Le Bourget) and stayed overnight. The next morning he returned to Le

Touquet where he met Sammy Norman. The Bristol 170 G-AGVC, which Powell had fitted with nose doors, was commanded by Captain Storm-Clark with co-pilot P.E. "Jerry" Rosser. Silver City Airways officials were so pleased with the eventual market success of this first flight that the story was used in advertising for many years. Flight schedules and brochures carried a photo of Rosser stating he had flown 6000 more crossings since the first car-ferry flight.

A large crowd led by Mr. M. Sainsard, the Le Touquet airport manager, met the first flight. During the festivities, Le Touquet resident Mr. Sammy Norman approached Commodore Powell. He asked if the aircraft could accommodate his Bentley, which was scheduled to return by boat. Little did they know this was the dawn of the car-ferry era. The Silver City Bristol returned to Lympne with the two automobiles. Sammy Norman's Bentley became the first revenue automobile at a fare of £22 ($61.62) and the car-ferry was born. Sammy Norman also had a home at Southend and was regular customer until his death in 1960.

In June 1948 Silver City began advertising a fare of £15 ($42.00) for one car and four passengers from Lympne to Le Touquet. This was considerably more than the sea-ferry at £8 10s ($22.50). Regular car-ferry service began on 13 July 1948 with Bristol Freighter G-AGVC. During the three months after the first revenue flight Silver City transported 180 cars. The fare was then set at £7 15s ($20.00) per car. It became an immediate success and the next year 2,600 cars and 7,900 passengers were transported. By 1950 the fare was £32 ($89.63) for larger cars and £27 ($75.62) for cars under 14 feet.

Powell's perseverance paid off with Silver City Airways starting the first scheduled car-ferry service in the world. The route from Lympne to Le Touquet was selected because of the short distance across the channel. Lympne is actually directly across the Straits of Dover from Calais. The success of the service was so great that Silver City built its own airport further south at Lydd. The airport was appropriately named Ferryfield. The carrier virtually controlled the car-ferry market by owning its own airport. To further exercise the hold on the market Silver City acquired the rights to all Bristol MK.32 aircraft produced, which eventually contributed to the carriers demise. Because of too many aircraft for the existing market it could not purchase anymore allowing other airlines to acquire MK.32s. At its peak Silver City was operating 18 Bristols to six continental destinations. It expanded car-ferry routes and continued service until merged into British United Airways in 1962.

Silver City's success in the car-ferry market did not go unnoticed. Freddie Laker, owner of Air Charter Limited, maintained a base at Southend along with Aviation Traders Engineering Limited (ATEL), which did contract work for other air operators. His ATL division was involved in aircraft sales and operating Bristols on contract work from the base at Southend.

Silver City eventually dropped the Southend-Ostend car-ferry route and suspended flying the Southend-Calais route citing it was not cost effective. The higher cost was related to Silver City having to ferry empty aircraft in for positioning, prompting the carrier to abandon Southend in 1952. It did not have a large base at Southend with maintenance like Air Charter. The decision by Silver City management to drop Southend service has been considered a fatal mistake since it gave Freddie Laker the opportunity he was seeking. He immediately applied for the routes for Air Charter and succeeded with the support of Benard Collins, Commandant of Southend airport. In spite of strong objection by Silver City officials the routes were approved creating competition in the car-ferry market.

Silver City set up a French company Société Commerciale Aérienne du Littoral (SCAL) to handle operations at Le Touquet but by 1961 Silver City losses were not sustainable. In an effort to raise cash the carrier sold three MK.32 Bristols to French carrier Compagnie (or, Cie) Air Transport (CAT) for £192,300 ($538,632). This gave CAT a 25 percent stake in the British/French car-ferry market and the two carriers shared a common logo on the aircraft. CAT later became a Carvair operator with marketing agreements with British United.

Air Charter

Air Charter introduced car-ferry service in 1954 between Southend and Calais and Laker began referring to the service as Air Bridge becoming known as Channel Air Bridge in 1955. The service was extended to Ostend Belgium in early 1955. On Monday 01 October 1956 the city of Rotterdam opened its new airport and Laker seized the opportunity to introduce international service to another country. Three Bristol Superfreighters operated that morning in typical Laker grand style with the first aircraft, G-ANVS, departing Southend at eleven o'clock commanded by Aviation Traders Chief Test Pilot, Captain L.P. Stuart-Smith.

In 1957 Belgian authorities were becoming aware of Laker's expanding operation at Ostend with concern. Air Charter and SABENA entered into a Southend — Ostend joint agreement resulting in three Bristol Superfreighters being painted in SABENA livery. By 1959 there were 27 Bristol Freighters in car-ferry service. The success of Air Charter caused it to be bought out by Airwork in January 1959. The operation became British United Airways and the car-ferry division was split off in February as Channel Air Bridge.

In 1962 when the two competing car-ferry companies were merged the car-ferry operations were combined at Ferryfield (Lydd) and Southend. The combined car-ferry fleet after the merger consisted of 25 Bristols and three Carvairs. The first visible indication of the merger took place on 29 May 1962 when Silver City Mk.21 G-AIMH arrived at Southend to operate the Southend — Calais/Ostend Channel Air Bridge route. The aircraft of both fleets were repainted beginning in March 1963 with the British United livery of navy stripe and red lettering. At the time of the consolidation the projected requirement was set at 28 Carvairs. The combined companies became the last British car-ferry operator and never fulfilled the requirement of 28 Carvairs.

In 1957 Belgian authorities became concerned with Freddie Laker's expanding operation at Ostend Belgium. Air Charter and SABENA entered into a joint agreement resulting in three Bristol Superfreighters being painted in SABENA livery. After repainting, G-APAV was seen regularly at Southend (courtesy Transportraits/Richard Goring).

The Silver City name was revived for a short time in 1973 by Air Holdings. Invicta Airways failed to make payments on five leased Vanguards belonging to Air Holdings. The aircraft were repossessed causing Invicta to cease operations. Air Holdings resurrected Silver City to operate three of the Vanguards on German based cattle charters from Norwich England. The venture was not considered a success prompting Air Holdings to find a buyer for the Vanguards in October, shutting down in November, and retiring the Silver City name at the end of 1973.

Air Holdings

Channel Air Bridge Carvair operations is a product of corporate consolidation and a need for a suitable replacement for the Bristol Freighter. This combined with Freddie Laker's competitive drive led to the concept and development of the ATL-98. The merger with Silver City Airways was completed in 1963 after the Carvair was introduced. The ultimate surviving carferry company of the merger was named British United Air Ferries which was announced 16 July 1962. Channel Air Bridge and British United Air Ferries have a history that can be traced back through many acquisitions and companies of the Air Holdings group. Additionally, Aviation Traders acquired by Air Holdings in 1958 developed and manufactured the Carvair.

Although the car-ferry era and the evolution that generated the Carvair most likely would have occurred in some fashion, the development was guided and nurtured somewhat by default as many companies eventually became part of Air Holdings. Old established shipping, transportation and financial institutions with formidable histories financed the group. Air Holdings financial control consisted of British and Commonwealth Shipping, 29 percent; Furness, Withy and Company, 18 percent; Blue Star Line, 16 percent; Peninsular and Orient, known as P. and O., 12 percent; Guinness Mahon, 8 percent; Whitehall Securities, 8 percent; Hunting, 5 percent; and Eagle Star, 3 percent.

Nigel Norman and Alan Muntz formed Airwork, which is the origin of Air Holdings, in 1926. The two believed there was a need for a single company to provide services for private aviation. After a slow start a base for private aviation was established at Heston in 1928. The success attracted Whitehall Securities in 1929 to become a financial backer. Airwork continued to grow and was instrumental in several overseas ventures creating new air carriers. The Guiness family became investors during this period. Because of the war the government took control in 1938 expanding Airwork. The engineering subsidiary employed more than 3500 and the flight school another 1500.

After the war Airwork renewed its commitment to commercial aviation by obtaining new Vikings and securing inclusive tour contracts. These contracts organized by George Wenger later became Whitehall Travel. Airwork continued to expand internationally by taking a financial position in Straits Air Freight Express (SAFE) in New Zealand. SAFE was founded in 1950 with the U.K. backing of Tradair. Some years later the Carvair was reviewed for service with SAFE.

Air Holdings enjoyed a windfall in 1951 with the return of a conservative British government. Under the administration shipping companies were encouraged to become involved in air transport. The independents were urged to diversify into schedule service, all freight service, inclusive tours, and trooping flights. The Blue Star Line and Furness Withy provided Airwork with the financing to expand. Eventually the competition and overlapping services created chaos throughout the industry. The only answer to stabilize the situation was consolidation. By 1958 Airwork was experiencing considerable financial problems prompting the shareholders to put forth a plan to bring in strong established independents to gain skilled new management.

During this time a behind the scenes mass effort was being undertaken to merge Airwork, Jersey Airlines, Hunting-Clan, Transair, Mortons, Bristow Helicopters, Aviation Traders, Air

Charter, Channel Air Bridge and a number of other subsidiaries and related companies into one group. Since Laker was founder of several of the companies, during these mergers he held out for car-ferry autonomy and was successful in keeping it separate as Channel Air Bridge.

To achieve the consolidation Airwork purchased Transair from Gerald Freeman late in 1957 to gain newspaper transport contracts and Aviation Traders, Air Charter, and Channel Air Bridge from Freddie Laker in 1959. Morton-Olley, originally established by Gordon Olley, was also purchased from Captain T.W. "Sammy" Morton in 1959. Airwork was then merged with Hunting-Clan in 1960 to become British United Airways (BUA) on 27 June. Airwork co-founder Alan Muntz retired and financial control was taken by British and Commonwealth since it controlled Hunting-Clan. Freddie Laker was named executive director of British United Airways, that was made up of companies he founded.

Gerald Freeman had actually negotiated with Laker on behalf of Airwork to purchase Air Charter and Aviation Traders. After the consolidation Laker became involved in a power struggle with Freeman resulting in Freeman resigning. Shortly after that Laker and Myles Wyatt, Chairman of Airwork and responsible for the name British United Airways, had struggles because of their different management styles. Laker had aspirations of succeeding Wyatt but his love of aviation often conflicted with Wyatt's view of making a profit. Even though there was friction Laker remained managing director at British United until after British United purchased VC-10s and brought them on line.

In 1960, during the period of consolidation, Cunard acquired control of Eagle, which was owned by Laker's old rival Harold Bamberg. This might not be of significance in the Carvair story except for the fact that after Mike Keegan acquired BAF and operated it in concert with Transmeridian (TMAC) he eventually sold TMAC to Cunard.

By 1960 prior to consolidation Silver City and Channel Air Bridge were both grossing about the same on the car-ferry service. Silver City continued to reduce fares in order to protect market share and fill excess capacity caused by too large of a fleet. The result was continued losses amounting to £650,000 ($1,820,650) in the first half of 1962 for Silver City. Sir Donald Anderson, president of Peninsular & Orient, a major shareholder of Silver City, devised a plan with Sir Miles Wyatt of Airwork. Consolidation had reached a peak and the uneconomic situation at both British United and Silver City led to another merger.

Anderson was adamant that Laker not be informed of the plan. British Aviation Services (BAS), parent of Silver City Airways, was merged in 1962 completing the Air Holdings Group. After the amalgamation the two airlines were operated separately but traded as British United Air Ferries. Douglas Whybrow was named General Manager of both Silver City and Channel Air Bridge.

Freddie Laker Era — (Aviation Traders, Air Charter, BUA)

Freddie Laker, who is credited with being the father of the Carvair, is a most interesting and colorful personality. He was born in Canterbury Kent Southern England and is reported to have loved aviation since he was a small boy. At age 16 he went to work for Short Brothers Aircraft in Rochester Kent and crossed paths with Gordon Thomas who would later become the Chief Engineer for the Decca Navigator System. In 1940 Laker worked for General Aircraft then during World War II he flew for the Air Transport Auxiliary, ferrying aircraft. After the war he worked part-time for London Aero Services and later founded Aviation Traders Limited to deal in war surplus aircraft and spares.

Shortly after setting up Aviation Traders in October 1947 Laker and his Scottish friend Bobby Sanderson were drinking in the Silver Cross Pub at Whitehall. Laker had done contract work for Sanderson and helped him sell a number of aircraft. The two were discussing BOAC selling off its fleet of 12 Haltons. Sanderson asked Laker why he was not bidding on the Haltons. Laker replied that he only had £4000 ($11,204) and it would require an additional £38,000 ($106,408). Sanderson, wanting to repay a favor replied, If that is all that you need I'll write you a check and you can pay me back when you can.[1]

Laker's maintenance and servicing company, Aviation Traders, had already begun purchasing surplus aircraft and parts from the government and stockpiling them at Southend. ATEL was becoming a successful aircraft maintenance company and parts supplier. The £42,000 ($117,642) purchase of the BOAC Haltons included six hundred tons of parts. The purchase put Laker in position of an unexpected windfall having cornered the market on Halifax spares. Aviation Traders (ATL) was a salvage company and the engineering division (ATEL) was certified for aircraft service and repair.

Shortly after the purchase of the Haltons on 24 June 1948 the Soviets blockaded Berlin. The Americans accepted the challenge and British companies with large aircraft were also pressed into service. The Berlin Airlift was on and Laker was looking at a great opportunity. Aviation Traders however, was not an operating airline but a maintenance and aircraft salvage company. He rented six of the ex–BOAC Haltons to his friend Robert Treen, owner of Bond Air Services. Laker agreed to provide the aircraft and all maintenance to Treen in exchange for half of the freight revenue. To cover himself he kept six of the Haltons for his other operations then rented hangar space at Southend. Almost immediately he began servicing more than 100 aircraft because British registered aircraft were required to return to home base every 300 hours for maintenance. Within a short time ATEL had more than 400 men at Southend and 30 in Hamburg.[2]

During the Berlin Airlift Aviation Traders became quite profitable supplying parts and maintenance to the other companies. The six aircraft leased to Treen became the first civilian carrier to fly the Berlin Airlift. After only 54 weeks of the Airlift Freddie Laker was a millionaire at the age of 27. He had accumulated 200 four-engine aircraft and 6000 surplus engines. He realized that when the Airlift was over many of the companies he was servicing would be out of business. He decided to suspend flying and shut down for at least a year because of the glut of aircraft.

During this time he began scrapping more than 200 of the aircraft he had purchased for £50 ($140.05) each and salvaged the platinum from the spark plugs of 6000 engines. He melted down the aircraft and sold the ingots to manufactures of cookware and toothpaste tubes. During the war cookware was melted down to make airplanes and he was just completing the cycle.[3] Laker purchased six damaged Avro Yorks and several Lancasters in 1950 rebuilding three of them for a newly planned operation. At the same time he secured a contract with Bristol Siddeley for Aviation Traders to build wing sections.

Air Charter Limited was established on 28 May 1946 and was acquired by Laker's rival Harold Rolf Bamberg in December 1951. Laker purchased Air Charter for £20,000 ($56,020) from Bamberg who was in financial difficulties because of the lack of steady charters. After gaining control he revived it in June 1952 by purchasing Surrey Flying Services, which was established in 1921. It was merged into Fairflight, owned by Air Vice Marshal Don Bennett. By doing this Laker inherited the profitable freight contract between Hamburg and Berlin for the continued supplying of Berlin. The acquisition of Bamberg's Air Charter brought an operating company of more than 300 employees under Aviation Traders allowing Laker to bid on government trooping flights.

Laker's Aviation Traders converted the previously rebuilt Avro Yorks to passenger configuration but was not authorized to fly passengers. With the newly acquired operating company Laker concocted a plan to be awarded government trooping contracts to the Middle East, Far East and Pacific in order to get into the passenger business. He determined that troops, which were at that time being transported by ship, took one month each way in transit. This resulted in every soldier doing nothing for two months of his or her duty rotation. Laker proposed sending them by air, which took less than three days, and it was less costly. At the time trooping flights were deplorable. Families and children were crowed into sparsely equipped noisy cold surplus transports. Laker installed sound proofing, airline type seats, and even supplied napies for military dependant children. His changes were so convincing that he eventually received all British trooping contracts. In addition he was granted a seven-year contract to transport nuclear material to Australia.

Laker was so pleased with the granting of the trooping contracts to Air Charter that he flaunted the reliability of his operation. In April 1953 he placed advertisements stating that Air Charter Limited had been entrusted by Her Majesty's Government to provide air transportation for British Forces. The route from London to Fiji was 13,319 miles, staging through Malta, Nicosia, Bahrain, Karachi, Delhi, Calcutta, Rangoon, Singapore, Makassar, Darwin, Townsville, Tontouta (Noumea), and Fiji, becoming the longest air route that had ever been undertaken by a Private Air Operator. The copy further stated that the mechanical efficiency of the aircraft were backed by maintenance facilities of Aviation Traders Engineering.[4]

Channel Air Bridge

As a result of the trooping service Air Charter was granted an unrestricted passenger certificate in 1954. Air Charter operated its first car-ferry proving flight on 28 August 1954. The Bristol Mk.31 G-ANMF flew from Southend to Calais under the command of Captain Flannigan. Laker exploited the press on 01 September by flying the Mayor of Southend and his Rolls-Royce to Calais where his party drove to the official reception with the Mayor of Calais. Afterwards he returned to the airport, the Rolls was loaded, and the party returned to Southend. Air Charter began cross Channel service from Southend to Calais with three Bristol MK.31 Freighters to test the market. The service continued until the close of the season at the end of October. Air Charter was now in the car-ferry business challenging Silver City Airways. The fare was established at £7 15s ($20.00). Laker had approached Douglas Whybrow on several occasions to come to work for him and manage Air Bridge. Whybrow finally accepted in 1955 after resigning his military commission.

Jersey Airlines was formed in 1956 and initiated Channel Island car-ferry service. When the new Rotterdam airport opened in October 1956 Air Charter added twice daily service, which takes the Bristol 75 minutes for the 186-mile flight. Ostend Belgium was added with the three-car Bristol Superfreighter Mk.32. The car-ferry division became a separate company known as Channel Air Bridge, a name conceived by Freddie Laker. Silver City called its service Air Ferry, but had referred to it as Air Bridge in advertising. The name Channel Air Bridge was adopted and was first seen painted on a placard inside the Bristol Freighters. Later the name became part of the British United operation. Laker prevailed in the media over his rivals when Channel Air Bridge became the first airline to advertise on the new British independent television channel. He found success in the car hauling business with Air Charter and continued operating as Channel Air Bridge. Two more Bristol Mk.32s were ordered for the 1955 season and the challenge was on between Channel Air Bridge and Silver City Airways.

Although promising, Channel Air Bridge was anything but a success when service began in 1954. The holiday season only lasted for three to four months leaving the aircraft idle during the balance of the year when there were no bookings. By the third year the tiny airline only operated three Air Ferry routes of 73, 94 and 165 statute miles. The carrier began service from Southend to Calais in September 1954. Southend to Ostend was added in October 1955 and Southend to Rotterdam in 1956. Combining the three-destination route structure with the high operating cost Bristol Freighter and fares that are based more on boat fares than international fares left only a slim chance of success.

During this same period Laker became occupied with building what he thought would be a replacement for the DC-3. He was unable to entice anyone to join a partnership consequently he decided to proceed on his own. The Aviation Traders division of his group of companies produced the Accountant ATL-90. It was a 28-passenger aircraft, 60 feet long with an 82 foot wingspan and powered by twin Dart turboprops. It was flown in the summer of 1957 but was not met with success after costing Air Charter and Aviation Traders £650,000 ($1,820,650). Freddie Laker was so disappointed with the results, when Gerald Freeman approached him in 1958, he decided to sell his companies to Airwork as part of the Air Holdings consolidation.[5]

Channel Air Bridge entered into a marketing agreement with SABENA in 1957 and painted a Bristol in SABENA colors. Belgian officials were concerned about Silver City and Channel Air Bridge success into Ostend but the market was considered to small for SABENA to purchase aircraft for the service. Belgian authorities considered withdrawing British route authority. To prevent Silver City from making a deal with the Belgian authorities, Channel Air Bridge offered to operate Bristols in SABENA livery allowing SABENA and Channel Air Bridge to advertise joint service.

Silver City still remained the primary car-ferry operator but market share was declining. The Bristols were becoming tired and were not suited for the deep penetration routes that were planned into mainland Europe. Silver City was not in as good of a financial position as Air Charter and it was believed new aircraft would be needed to continue to compete.

After a slow start Air Charter became quite successful in the car-ferry business with the Bristol Freighter. When Airwork approached Laker about merging Air Charter into the Airwork Group, Channel Air Bridge was a division of Air Charter. Although disenchanted with the airline business, Laker believed the car-ferry should stand alone. When he sold controlling interest in November 1958, Douglas Whybrow and Alan Nickals became directors of the Channel Air Bridge division of British United Airways. Bob Batt and Jack Wiseman became directors at Aviation Traders and Laker was made a director at Airwork and executive director of British United Airways. Bob Batt and Freddie Laker were loyal friends from "Sky Tramping" days. Batt began in aviation as an airframe fitter with the Royal Air Force then joined London Aero Motor Services (LAMS). Laker began as a flight engineer with LAMS. Batt went on to Alpha Airways in Africa. When that carrier went under Batt and several other airmen returned an aircraft leased from Laker back to England. Within nine years after going to work for Laker he was on the board. Laker who went into business with £240 ($672.00) became the head of a £20,000,000 ($56 million) company.

It was agreed to split the operation spinning off Channel Air Bridge as a separate car-ferry airline company. Although the car-ferry had operated under the Channel Air Bridge name for years it did not become official until 25 February 1959. The new Channel Air Bridge color scheme was first applied to Bristol Freighter G-ANVS. By 1960 Channel Air Bridge was making a profit with revenues equal or higher than competitor, Silver City Airways. It was realized that longer routes could improve profitability and expand market share but in order to accomplish this dream a better-suited aircraft would have to be found. The Bristol Freighter was noisy and generally uncomfortable and operates at 2000 feet across the channel. As long as they were used on the short routes the service was tolerable. Considering the Bristol for long haul routes was not an option.

The holiday market is highly seasonal with many aircraft remaining idle during the winter season. In addition high maintenance cost and modifications to the Bristols increase operating cost. Car-ferry airlines were unable to increase fares because of the lower boat ferry fares. As an example of comparison the round trip boat ferry from Dover to Calais was £40 ($112). The Channel Air Bridge fare for a car and two passengers was £56 ($156.85). Sixteen pounds ($45.00) is a considerable amount to consider when the savings in time are only a few hours. Boat traffic did decline in 1960 with 100,000 cars being transported by air and 289,000 by boat.

In order to increase market share and maintain yield, plans were made for longer flights into Europe known as "Deep Penetration" routes. Laker was anxious to expand service and quite aware the Bristol could not be effective on two and three hour segments. Since the car-ferry market required a limited production specific design aircraft he was left with few choices. In order to survive in the car-ferry market the not yet named Carvair replacement had to be found.

As Executive Director of British United Airways, Laker had to make Channel Air Bridge profitable since he demanded it set separate during the Air Holdings consolidation. The Bristol could not handle the projected increase in traffic and deep penetration routes. Further review determined that wing fatigue would shorten the operational life. He also realized that BUAF would have to find a solution. There were few options, purchase an existing aircraft, design a new aircraft, or modify an existing aircraft. Cost was a factor preventing manufactures from producing a new aircraft in limited numbers for a limited market. A new turbine powered aircraft would cost at least £500,000 ($1,400,500). The amortization charges of nearly £80 ($224.80) per hour would be prohibitive for and aircraft of 1000 hours utilization per year. Furthermore the new aircraft must be capable of transporting supplemental passengers during the tourist season and cargo during the off season to increase the aircraft utilization hours.

Laker determined that modification of the existing DC-4 was the only cost effective solution. The amortization cost of a turbine aircraft was greater than the operational cost of a reciprocating engine. With no other options he would have to convert aircraft in house. Aviation Traders was the obvious choice for the project since he built the company and knew it had the ability and it was in the British United family of companies under Air Holdings. Although small, Aviation Traders had considerable experience in aircraft modification. The company converted Handley Page bombers to Haltons for civilian use, also the Avro Tudor to freighters, and produced Laker's ATL-90. Although a market failure it did prove the ability of ATEL.

Aviation Traders director Bob Batt and Chief design engineer A.C. Leftley conducted studies to find a suitable replacement for the Bristols. Channel Air Bridge submitted specifications for an aircraft with a cargo hold that would transport five cars. The cargo compartment would have to be a minimum of 70 feet long with a nine-ton capacity. It would also have to contain a 25-seat cabin, galley, and toilet. Leftley reviewed the Blackburn Beverly, Handley Page Hermes and Hastings, and Argosy AW650. The Argosy was eliminated early on because of being pressurized and having turboprop engines. Eventually the search came back to converting the DC-4/C-54 because of low cost and availability.

Flight International magazine reported in June of 1959 that the aircraft did not have an elegant name. Once again Freddie Laker seized the opportunity to promote the new ATL-98 and Channel Air Bridge. An aircraft-naming contest was promoted for Channel Air Bridge passengers. The airline offered as a prize of a free return journey for a car and two passengers from Southend to Calais, Ostend, or Rotterdam for the winning name selected from entries. Two local gentlemen from Southend entered the winning name of Carvair, which was a contraction for Car-Via-Air.

In 1959, after coming under the British United Airways group, Air Charter planned seven new long range routes to France and beyond from Southend. Applications were filed to fly to Paris (Le Bourget), Dusseldorf, Bremen, Dijon, Strasbourg, and Lyon with the new Carvair. Channel Air Bridge had already determined that the very popular short cross channel flights were only marginally profitable since they were seasonal and limited aircraft utilization.

The longer routes proposed for Air Charter would increase utilization and balance the seasonal fluctuation. This would also allow a profit at a lower load factor. A proposed fare of £150–180 ($420–504) for car and passengers would not need to be seasonal adjusted. The Bristol was very uneconomical and reaching the end of its operational life and the proposed Carvair could transport two more cars than the Bristol. For example the round trip operational cost of the Bristol from Southend to Rotterdam is £175 ($490) and the Carvair is £200 ($560).

The Carvair has a 15 percent higher operating cost offset by a 100 percent higher capacity. If estimates were correct the Carvair could generate twice the revenue as the Bristol at a cost factor only 15 percent higher. The Carvair cruises at 200 mph, which is 45 mph faster than the Bristol. Additionally the Carvair is equipped with a toilet, galley and will be staffed with a Stewardess.

Air Bridge transported 265,000 passengers with the Bristol in 1961 of which 60,000 were supplemental bookings without cars. This further verified the Carvair with increased capacity being the correct choice. When factoring in the no show of some bookings the Carvair could replace two Bristol flights. The longer routes reduced cost increasing profit margin, which supplemented the low margins of the cross channel Bristol flights. The Carvair has an economic range of 400–450 miles, which could maximize the balance between cost and price and justify the plan to operate the Carvair on long haul routes and Bristol on the short routes. Laker also believed that the 60,000 supplemental bookings indicated additional market. To access more market He also ordered 20 BAC 111 aircraft in April 1961.

The Carvair was scheduled to operate the shorter Rotterdam route supplemented by Bristols during the peak months since the primary Channel Air Bridge freight route was Southend — Rotterdam. In fact the carrier transported three times more freight to Holland than BEA. Cargo business was always strong and by the 1970s BAF was third in the league of carriers behind British State giants BOAC and BEA. The cargo transported between Holland and the U.K. in the early 1960s approached nearly 500 tons per month. Air Bridge at that time transported 80 percent with the troublesome Bristol. Laker and others believed the Carvair could capture even more of the cargo market by cheaply transporting new Jaguar cars, diesel engines, washing machines, fresh produce and Dutch flowers and bulbs.

It was planned that by 1963 as the larger capacity Carvairs were delivered, the Bristols would operate only on the Southend to Calais route and completely phase out of car-ferry service by 1965. The first three Carvairs were actually delivered to Channel Air Bridge with small British United lettering on the tail. Seven more Carvairs were ordered for delivery over the next two years and option was taken for eight more aircraft.

By the end of August 1962 all three of the original Channel Air Bridge Carvairs were in service. Their potential was viewed with great optimism. The trio was running an 80 percent load factor on the deep penetration routes in the first season. They transported 787 cars and 2,104 passengers to Basle, 950 cars and 2,362 passengers to Geneva, and 346 cars and 975 passengers to Strasbourg. These longer routes that the Bristols were incapable of operating proved much more profitable, which created an optimistic euphoria.

To introduce the Carvair Laker decided to expand on a marketing idea that had been suggested to him by James Watson who was the Director of Castrol Petroleum. Laker and Watson became friends when they served together in the Air Transport Auxiliary. Laker believed in Castrol petroleum products and used them exclusively in his vehicles and aircraft. Beginning in 1954 departing passenger names and address were sent to Castrol. Upon their return home they found a letter waiting with compliments and the hope the flight was enjoyed. Included was a booklet describing how to keep your car in top operating condition and a lubrication chart for your exact model. The booklet also described the history of British aviation and Castrol. The letter was used to introduce the Carvair and to point out its advantages and superiority to previous car-ferry service. This successful promotion targeted the individual that used the car-ferry service.

Cabin staffing changes were created with the introduction of the Carvair. The bigger aircraft and longer routes required more stewardesses. They began replacing the male cabin staff that previously dominated the Bristol car-ferry flights. After takeoff the flight attendant was required to check the automobiles in the cargo area for any fuel or oil leaks. After this was verified and reported to the Captain the "No Smoking" sign could be turned off.

British United/British United Air Ferries

When British authorities made an effort to strengthen the aviation industry in 1959–60, many airline companies and manufactures were forced into mergers prior to the Civil Aviation Licensing Act. The British United Airways division of Air Holdings was organized to compete directly with BEA. Hunting-Clan, formed in 1945, also came under the new corporate structure through the earlier combined services with Airwork Limited, which was formed in 1928, to operate cargo service to Africa. It was a good fit since Hunting-Clan specialized in livestock transport, primarily pigs, also race horses and zoo animals. Under the new corporate umbrella there were 19 subsidiaries which included Aviation Traders.

When British United became Britain's largest independent in 1960 prior to the first Carvair flight Laker insisted it be allowed to stand alone and operated with complete autonomy. He wanted to keep the product identity of Channel Air Bridge as the car-ferry and not be associated with livestock flights. The image of the new Carvair could not be tarnished. Silver City Airways was the only competitor and it had not ordered replacement equipment for the Bristols. Consequently Freddie Laker argued that the public might have identity problems if the previously known car-ferry name of Channel Air Bridge was changed.

Laker created the Air Bridge name for car-ferry service in 1954. After the introduction of the Carvair he wanted to further capitalize on the brand by naming each aircraft giving them special identities known as "Bridge" names. The first Carvair, G-ANYB, was named "Golden Gate Bridge" and christened on 16 February 1962 by Mme. Daeniker, wife of the Swiss ambassador. This began the tradition that gave eight Carvairs special names; Golden Gate Bridge, Chelsea Bridge, Pont de l'Europe, Pont du Rhin, Maasbrug, Channel Bridge, Menai Bridge, and Pont d'Avignon.

The low yields in 1960 made it obvious that two car-ferry airlines could not continue.

Mme. Daeniker, wife of the Swiss ambassador, christened the first Carvair, G-ANYB, "Golden Gate Bridge" on 16 February 1962. A total of eight Carvairs received "Bridge" names (courtesy Brian Kerry).

Laker made it one of his first priorities, after becoming Managing Director of British United, to devise a plan to control the car-ferry market. When Air Holdings was formed in 1961 as the parent of all the consolidated companies he was instrumental in the purchase of British Aviation Services (BAS) from Commodore Taffy Powell, who had conceived the car-ferry years earlier. The deal for Silver City Airways also included Manx Airlines and British Aviation Engineering services. The major component of the acquisition was Silver City Airways in order to eliminate all competition. It was merged with Channel Air Bridge on 01 January 1963 officially becoming British United Air Ferries (BUAF).

Three Carvairs were delivered to Channel Air Bridge prior to the merger with Silver City Airways, which prompted a name change to British United Air Ferries. Actually the British United Air Ferries (BUAF) name was in use from 16 July 1962 although the merged airlines continued to operate independently until 01 January 1963. The merger forced Laker to drop Channel Air Bridge as a separate company bringing all car-ferry service under British United. He agreed with the British government to never use the initials BUA to avoid confusion with state carrier BEA. He was instructed to always use the full title "British United." Furthermore the titles inside the nose door have "British United" with "Air Bridge" outlined. This was to prevent any misconception and emphasize the "Air Bridge" theme.

It was a commonly shared opinion that Freddie Laker ordered 10 Carvairs to create work for Aviation Traders and to absorb three surplus DC-4s owned by BUA. Ten Carvairs were not immediately needed since many of the deep penetration routes were not yet negotiated. This is further supported by the fact that Carvairs four and five originally scheduled for British United were quickly declined to allow delivery to Intercontinental for Interocean. It was also quite positive when ships six and eight went to Aer Lingus. If there had been a need for 10 Carvairs, British United would not have given up delivery positions so readily.

Aviation Traders conducted wind tunnel test for a DC-6 and-7 version during development of the conversion program. Almost immediately other carriers expressed an interest in the DC-6 Carvair conversion but not DC-4s. The choice to reduce cost by using the DC-4s ultimately limited the Carvair production to only 21 aircraft. If not for the proposed long haul route expansion the Carvair would have not been needed and the Bristol could have continued for sometime. The Bristol life is based on landings, which had almost 10 years left under usage at that time. Silver City previously got into financial trouble by over purchasing the Bristol Mk.32. Laker was aware of this along with the reality that Channel Air Bridge could not possibly use ten Carvairs.

Another Southend based carrier, Channel Airways (East Anglian Flying services), did extensive market research and considered the many problems involved when dealing with foreign governments for routes and marketing. Laker also realized the importance of market research since Channel Air Bridge derived 40 percent of its bookings from the two motoring organizations AA and RAC. Another 35 percent were booked by travel agents attempting to boost revenues by capitalizing on the holiday motorist with the balance was booked directly. Channel Airways marketing officials understood the demands of market and worked closely with the different agencies and organizations. Initially Channel Air Bridge carried out extensive research for routes it applied to operate. As British United Airways slowly evolved after the merger it was noted that the company did not research markets and often appeared amateurish. Laker's Channel Air Bridge did this extensive research although it operated a fleet valued at less than one half million pounds. However, Parent company British United in a massive expansion program ordered a fleet of jets worth 20 million pounds without researching the market. The smallest division of the corporate group did a better job of market analysis than the major division.

British United was formed in 27 June 1960 after the merger of Airwork with Hunting—Clan and others. Eventually all of the associated companies came under Air Holdings, which was formed in 1961.

BUAF secured authority and license to operate the longer "Deep Penetration" Carvair routes to Basle and Strasbourg in 1962. Because of the success of the new aircraft traffic projections were quite optimistic, making it obvious that the facilities at Lydd Kent were inadequate. Channel Air Bridge had 1,963 firm bookings by 16 February 1962 for the Basle and Geneva routes with 800 more on the waiting list. The carrier projected 240 round trips between April and October with a potential of 2,400 cars. French authorities had not approved the Strasbourg route, however 340 firm bookings were on the waiting list.

Traffic for 1962 was quite successful with 1963 projected at 45,000 vehicles, 15,000 tons of freight, and 180,000 passengers. To handle these optimistic projections a new £150,000 ($420,165.00) terminal building was built at Southend. The terminal opened in July 1963 with additional vehicle gates and improved passenger facilities. By August 1963 British United had a combined total of 29 aircraft (Carvairs and Bristols) operating car ferry service. During the first two weeks of August alone the fleet made 4,762 Channel crossings transporting 12,150 cars. It is rather unique that the car-ferry traffic actually peaked in 1963 the year after the merger of Silver City and Channel Air Bridge. That year Channel Air Bridge made 58,000 sector flights transporting 400,000 passengers, 50,000 tons of freight and 100,000 cars. The Carvair became so popular that years after the service no longer existed calls were still received for car-ferry bookings.

In 1963 expansion began at Lydd with the new cargo agents building and Customs long room being constructed. Lydd was viewed as the most important export/import gateway to the European Continent. The main runway, 04/22, was extended 600 feet to a total length of 5000 feet. New radar and approach lights were installed to aid in the Carvair operation. Lydd was considered strategic for the Carvair long haul routes referred to as "Deep Penetration" routes.

2

Car-Ferry Evolution

British Air Ferries

Freddie Laker became disenchanted with the overwhelming corporate structure of Air Holdings and all the sub-companies. After getting the VC-10s into service and taking over the BOAC routes to South America he decided to follow another dream. He resigned in 1965 to form his own inclusive tour airline named Laker Airways, of course. Douglas Whybrow also resigned from BUA in 1965. He had been with Laker since the beginning of Air Bridge with the Bristols and had served as General Manager of both Channel Air Bridge and Silver City after the merger.

The market dynamics were changing for the car-ferry as early as the mid–1960s. BUAF announced that the 1965 season would be operated with five Carvairs in the 55-seat 3-car configuration to meet the changing demand. The carrier was losing money by the end of 1966 causing management to consider selling the Carvairs and reverting back to an all Bristol fleet for car-ferry service. The idea was based on the theory that Carvair engines were unreliable. The biggest problem with the engines is short sector flights. The engines are stressed by full throttle takeoff and climb. After a short cruise they are once again stressed by descent when the throttles are pulled back with minimum boost causing the props to drive the engine. This causes the rings to flutter and lubrication problems accounting for a higher rate of failure. It was later determined some engine shop overhaul practices were as much to blame.

BUAF announced on 23 January 1967 that it would discontinue seven of the Carvair longhaul routes because of a drop in traffic. This decision tends to verify the British United reputation of not researching markets. Because of better roadways and drive on/off sea-ferries the short haul routes were being sustained by cargo not cars. BUAF managing director Mr. Max Stuart-Shaw stated it would continue to operate the 11 cross-channel routes, however some Carvairs would be withdrawn from service. All Carvair service was concentrated at Southend. The Lydd-Basle, Southend-Basle, Southend-Geneva, (Gatwick)-Geneva, Manchester-Rotterdam, Coventry-Calais, and the short-lived Gatwick-Le Touquet service were all discontinued. The Bristols were also retired from the car-ferry service by years end but continued to operate in a limited cargo role.

The average car weighs as much as ten passengers and occupies twice the volume yet yields only ten percent of the revenue. When comparing yield per ton-mile, cars alone lose revenue. Supplemental passengers and cargo are required to break even. BUAF increased and decreased fares to meet the changing market conditions dictated by the roll-on/off cross channel ferries. Although not realized at the time this was the turning point in the demise of the Carvair car-ferry era. From this point on the market declined.

By mid–1967 the Air Holdings British United Airways Empire began to feel economic pressures from several of the group's holdings. BUA alone had 19 subsidiaries including not only car-ferry operations and Aviation Traders but Bristow Helicopters, Sierra Leone Airways, Straits Air Freight Express, a number of travel agencies and engineering companies. An independent study suggested that changes were necessary and that BUAF could be profitable if it stood alone. It suffered from too much corporate control and could only improve if allowed complete independence. This is the same reasoning that Laker had campaigned years before. Air Holdings reorganized and sold off a number of the aviation interest retaining BUA and Aviation Traders. British United Air Ferries (BUAF) emerged as the car-ferry airline British Air Ferries (BAF) on 01 October 1967.

Derek Platt began his aviation career with Imperial Airways in pre-war days coming to British United in 1966. Prior to Platt the aircraft at each division of the British United group had different color schemes. Under his direction in November 1966 a common British United livery of sandstone and blue with stylized bird logo, often referred to as the "Toppled Mushroom" was adpoted.[1] After only a year the BUA letters were changed to BAF with the 1967 cutbacks and reorganization. The old sandstone and blue color scheme was retained and the stripe was extended over the "Toppled Mushroom" and British Air Ferries was applied to the fuselage above the windows. The interior was not changed and remained in the same decor. The seating was changed on three of the Carvairs from the 22 passenger five-car to the optional 55-passenger three-car configuration. The transition to the new color scheme was not uniform. Some ships like G-ASHZ still had not been repainted in the sandstone-blue colors. It skipped that color scheme retaining the dark stripe and the red titles were changed from "British United" to "British Air Ferries," becoming the only aircraft to wear this one-of-a kind livery.

BUA operated only six Carvairs in 1967 with two ships, (G-ANYB, G-AOFW), in storage at Lydd Kent. Carvair two was removed from service four days after the company was changed to BAF and also stored at Lydd. Carvair one, G-ANYB, was still painted in first British United livery with dark cheat line and red titles. The second Carvair, G-ARSD was painted in the sandstone and blue with bird emblem. Carvair 12, G-AOFW, was painted in the old British United livery like number one. BAF began service in 1968 with a fleet of only five active Carvairs.

The British Carvair fleet saw a number of livery changes over the years. The first change was in October 1962 from the red and white Channel Air Bridge livery to the dark cheat-line and red British United titles. The BUA blue and sandstone "Toppled Mushroom" appeared in November 1966. When British United became British Air Ferries (BAF) in October 1967 the BUA titles were changed to BAF. Beginning in 1968 the cheat-line was changed to two-tone blue with black BAF letters. All operational ships owned by BAF remained in the two-tone blue stripe livery until after the company was purchased by Mike Keegan in 1971.

The car-ferry market declined from 29-percent in the 1960s to only five-percent in 1970. When Keegan purchased the company in 1971 he recognized the need to diversify and began to make many changes. One of these decisions was to change the color scheme. The dark blue lower fuselage was the outcome of a discussion between two of the company directors. As they were considering a new color scheme one of them took out a pack of "Cambridge" brand cigarettes. He indicated that the color of the pack was his choice and would match the light blue cheat line, the dilemma was solved. Beginning in 1972 the Carvairs were painted with navy blue bottom with white upper fuselage and wide light blue cheat-line retaining the black BAF letters which was the last color scheme for the British Carvairs.

In 1973 a Carvair was painted in a special BAF livery to haul racecars. In 1975 another was converted to all cargo and stripped to bare metal in an experiment to reduce weight and increase payload. In June 1977 it received the livery of yellow and black sash on the forward fuselage. One other Carvair received the yellow and black colors. This sash livery design is a direct copy of the Maritime Surveillance livery first displayed at Farnborough.

All of the ships retained the "Bridge" names they were christened with until 1969. The change in advertising strategy prompted character names. The first name, Big Joe, was added in the spring of 1969. Two more names, Porky Pete and Fat Annie, were added in 1970. The last four Big John, Big Bill, Big Louie and Fat Albert came in 1972. Big Louie was changed to Plain Jane in 1975 when it was converted to all cargo and stripped of paint to save weight. It would seem that "Plane Jane" and not "Plain Jane" would have been more appropriate. After all the intent was to create catchy character names for public reaction.

The Keegan Era

At the end of the 1971 summer season Air Holdings accepted a purchase proposal from T. D. (Mike) Keegan for British Air Ferries. The company changed ownership on 01 October. British & Commonwealth held a major financial stake in BAF parent company Air Holdings at 29 percent. For this reason it was instrumental in the sale noting the poor performance of BAF and lack of growth in the market. It became a subsidiary of Keegans Stansted based Transmeridian Air Cargo (TMAC), which he controlled with 58-percent of the shares. The balance was held by Charles St George (18 percent) and Merchant Bank (24 percent). The BAF company officers were Chairman T. D. "Mike" Keegan, directors Charles A.B. St. George, L.L. Orr, A.L. Macloed, financial director Alan Judd, and director of sales Dennis R. Day. The company had 300 employees and a fleet of seven Carvairs (-AREK, -ASDC, -ASHZ, -ASKG, -AOFW, -ASKN, -AXAI), two DC-4s (-BANP and -BANO) and one DC-6.

Thomas Dennis Keegan was born in Liverpool in September 1925. At age 17 he worked for a company in Wales assembling Wellington bombers. He joined the RAF at age 18 and became a flight engineer on a Lancaster. He never liked his name so at this time he changed it to Mike.

Keegan was well known at Southend because of his association with Crewsair, which he co-founded in 1948. He also owned the Flarepath Cafe at the airport and is remembered for sleeping there on many occasions. He was also a co-founder of BKS Air Transport. The company was established on 12 October 1951 when four of the Crewsair directors resigned to form their own charter airline and maintenance company. The four were T.D. (Mike) Keegan, James W. Barnby, Cyril J. Stevens and Captain J.P. Falconer. They were all multi-experienced entrepreneurs and elected to accept a DC-3 (G-AIWE) from Crewsair, instead of pay out for their shares. The new company was registered as an engineering operation and set up a base at Southend. On 7 February 1952 the name was changed to BKS Aero Charter using the initials of three of the founders Barnby, Keegan and Stevens. It was operated until June 1970 when it became Northeast Airlines. The Northeast name was retired in 1976 when it became part of the British Airways regional division.

Mike Keegan was also known in British aviation circles from his aircraft sales company, which he previously operated out of the Luton. In 1958–59 Keegan found himself short on cash. He went to the London docks and somehow got a job as a stoker on coal-burner to gain passage to the United States. Upon arrival he checked himself off and managed to purchase a Piper Apache. He ferried it from Boston to Southend and sold it as (G-ARCW). Then he did the same thing again and purchased another Apache and flew it back (G-AREW). Making a handsome profit he resolved his cash flow problems.

Keegan gained control of Stansted based Trans-Meridian Air Cargo in 1967. It was founded by Captain A.H. Benson as Trans-Meridian Flying services on 05 October 1962. Benson established it as a long distance charter carrier based at Luton. The executives included Viscount Long of Wraxall as chairman, A.H. Benson managing director, R. Huggins director, and Captain A.J. Burridge as Chief Pilot. Benson's original operation was primarily transporting ship crews from Rotterdam to Hong Kong. The name was derived on one of these flights when a crewmember commented on how many meridians were crossed. Eventually Transmeridian was referred to as T-Mac (TMAC).

The Transmeridian operation was financed by John Gaul, a wealthy London based real estate broker, who contracted to Keegan to handle his aviation interest. In 1965 Keegan became deeper involved with T-Mac when he leased the carrier a DC-4B he had acquired from United Airlines. John Gaul retired in 1968 creating a struggle between Keegan and Benson to gain control of the company. Keegan eventually won causing Benson and his group to resign and move on to Monarch Airlines. In December 1969 Transmeridian took delivery of its first CL-44 eventually receiving the ninth and final aircraft on 29 January 1972. Keegan was quite successful with T-Mac and became a master of short-term leases for specific routes.

With the acquisition of BAF Keegan along with his sons began their attempt to resurrect the declining car-ferry business. The BAF base was maintained at Southend operating the cross channel flights to Ostend and Rotterdam. The new dark blue paint scheme and change in service was touted in an advertising campaign for the "New" British Air Ferries— BAF. The Carvair was configured for 17, 40 or 65 seats with cargo hold of 69 or 42-feet, as loads demanded. The Cargo service was upgraded to a 24-hour operation in an effort to utilize the aging Carvairs.

Keegan was no stranger to the car-ferry market being involved with BKS Aviation and the Bristol Freighter. The airline also operated DC-3s between Newcastle and London slowly expanding from 1952 to 1964 eventually building a large network. BKS operated from Teesside, Newcastle, and Leeds to London, Glasgow, Belfast, Bergen, Jersey, Ostend, Basle, Dusseldorf, Paris, Biarritz and Bilbao. There was additional car-ferry service between Liverpool and Dublin. BKS also converted specific aircraft for airlifting racehorses for Dublin bloodstock charters. Bases were maintained at Woolsington Airport, Newcastle and Teesside, airport Middleton St George. The maintenance base was at Southend, home of Aviation Traders and the original Carvair program.

BKS competed against Aer Lingus with the Bristol Freighter before the introduction of the Carvair. Because of financial problems the Liverpool-Dublin service was withdrawn at the end of the 1961 season. Even if BKS had not withdrawn from the car-ferry market it is doubtful it could have competed against the Aer Lingus Carvairs that were introduced in early 1963. Conversely, Aer Lingus even with the Carvair was never able to overcome the foothold that BKS had in the Bloodstock Charter market. In a strange turn about in 1971, Keegan returned to the car-ferry business with the Carvair, which is the aircraft he would have competed against seven years earlier if BKS Transport had survived.

Another interesting note in the Keegan Carvair story transpired in December 1959 when he purchased the Vickers Viking fleet from the bankrupt Independant Air Travel through his company Bembridge Car Hire. The Vikings were sold or leased to other carriers. He was unable to obtain the three DC-4s owned by Independant. However, Freddie Laker purchased one of the Skymasters, G-APNH for Air Charter where it served prior to conversion to Carvair 11. It was later written off in an incident at Le Touquet just eight months before Keegan took over BAF.

Throughout the end of 1971 the future of BAF remained in limbo fueled by rumors and speculation. It was assumed that BAF would be lost in the Transmeridian operation. However, BAF was operated as a separate company. Combined advertising literature showed the Carvair but it was dominated by the ability of the Transmeridian CL-44s. British Air Ferries was slowly merged into TMAC operations but remained under BAF colors.

Transmeridian advertised in Flight International for 10 more CL-44 crews to increase the fleet to nine aircraft. Keegan assigned Alastair Pugh to evaluate the CL-44 for car-ferry service. Pugh had been Deputy Editor of *Flight Magazine* in the 1950s prior to joining Channel Air Bridge as Product and Development manager. It was announced to the press that the five remaining active BAF Carvairs would soon be retired. The local press was advised the CL-44s would expand the Southend service to Basle and Geneva. Sadly BAFs Chief Pilot Captain Robert Langley, who had been with Carvairs since the beginning, was terminated. Keegan put pressure on Southend airport officials that unless the runways were extended, facilities improved, and landing fees reduced the operation would be moved to Stansted home of Transmeridian. Southend officials held firm knowing that Stansted did not have freight or passenger facilities to handle the volume. They did agree to reduce landing fees for BAF aircraft operating the Le Touquet and Ostend services.

Keegan backed off and announced a plan to replace the Carvair with the CL-44s on the Stansted to Basle and Ostend service. The market was already unprofitable but two CL-44s were converted. Plans were announced in February 1972 for two CL-44 aircraft to be set up in a five-car 75-seat configuration. Carvair builder Aviation Traders was contracted to fabricate a bulkhead to separate the passenger compartment from the cars. The passenger cabins were installed at Southend. The service began in March 1972 and the second aircraft came on line in May. The loads did not materialize and the aircraft was not suited for the 35-minute flight to Ostend. The CL-44 could not be unloaded and turned quickly. It was a perfect example of how a long-range aircraft is not suited for the short to intermediate range market. The service was terminated in July. Keegan had been convinced by Alastair Pugh's report that the CL-44 was suited for car-ferry service. Although it may have performed well on very long haul it proved a poor choice for short car-ferry segments.

The car-ferry market was shrinking by 1973 and BAF was the only airline still offering the service. A 1973 issue of Aviation News stated, "BAF is led by Mike Keegan one of the most dynamic men in British independent airline operations." Keegan stated that BAF was waiting government approval to operate another long haul route to Bordeaux France and possibly Germany by the end of the year. Also the short haul routes to Le Touquet, Ostend, and Rotterdam would continue and were as popular as ever.

BAF had eight line engineers at the time. They were responsible for covering all the flying and maintenance. They did not fly on the Le Touquet, Ostend, and Rotterdam service unless there was a problem. Any off route, long haul or charters required an engineer, who had to be proficient on engines, airframe and electrical. Keegan expected him to get the aircraft around the route without calling for help or using too many spare parts.

After the CL-44s were withdrawn from car-ferry routes Keegan experienced a shortage of aircraft for the service. He did not have enough serviceable Carvairs to maintain the routes. A deal was struck with French carrier TAR in 1973 to purchase the two ships that BAF had been leasing for a year. Transmeridian had a DC-7F that was surplused and had become too expensive to operated since it burned 115/145 Av-gas. The grade of fuel was becoming increasing hard to obtain and was quite expensive. Keegan made a deal to trade the DC-7 as part of the purchase for the two Carvairs. The deal almost collapsed when the DC-7 was to be delivered to Nice. The crew taxied the aircraft out to runway four at Stansted but could not keep Number Three engine running without it backfiring and dying. The Captain taxied back to the ramp where the plugs were cleaned and adjustments were made. The second time the ferry flight taxied out it began to fowl plugs again. Keegan told the crew to get it out of here the best way you can. It took off on runway four with Number Three not performing. After climb out it was shut down and flown to Nice on three engines. It was written up at Nice as running rough. The exchange for the Carvairs was completed and nothing was mentioned about the DC-7 problems. With the Keegan takeover aviation trade journals stated that BAF had an improved image with the seven Carvairs. Ships 5, 7, 9, 10, 12, 13, 17 were being refurbished with new interiors, piped music, better sound proofing, and new color scheme. The seating capacity was reduced from the current 21 (originally 23) to 17. Optional configurations of 17, 40, and 65 were available depending on bookings. BAF owned seven Carvairs and two DC-4s at this time. The Carvairs were AREK, ASDC, ASHZ, ASKG, AOFW, ASKN, and AXAI. All eventually received the new livery except–ASDC, which was the first of two ships that were stripped of all paint for cargo service.

Twice in the 1973 press releases and news articles stated that an additional DC-4 was being presently converted to Carvair standards for BAF. The press was also told it was on the conversion line at Southend and would fly shortly.[2] This is quite puzzling and not possible since the Carvair production line at Stansted ended with Carvair 20 in 1965. The last one built, Carvair 21, was completed three years later in 1968 at Southend. Carvair 17, G-AXAI, pictured in the article in the new livery was actually the last one physically completed. It was built in 1964 then stored less engines and instruments until 1969 when it was completed. This was prior to Keegan gaining control of BAF. Obviously there was no Carvair being converted. The press statements in reference to the refurbishing were either wishful thinking or misstated by BAF officials.

It was no secret that the Carvair was outdated and a more modern replacement was needed if car ferry service was to survive. BAF management left open the possibility that the Britten-Norman Mainlander would be the replacement. Britten-Norman displayed a model of the aircraft in BAF colors at the 1972 Paris Air Show.

During the Carvair era and until 1973 it was always accepted that Aviation Traders did all overhauls and refurbishing since it was an Air Holdings company. Keegan did not purchased Aviation Traders along with BAF. He announced that he would take over the two old Channel Airways hangars at Southend. Channel Airways was originally founded at Southend in 1946 as East Anglian Flying Services Ltd. He intended to establish a new aircraft maintenance company and completely refurbished the hangars and hired more than 100 experienced aircraft engineers. Keegan was well experienced in production and factory operations and previously set up a factory to produce farm equipment, wheelbarrows and pig troughs.

The new organization was named Hawke Aircraft Parts. The company not only manufactured and repaired aircraft parts but also established a fiberglass production facility. Keegan's youngest son Rupert became involved in car racing and within a short time the company began producing fiberglass shells and racecar components. He purchased Hawk Motor Racing and delegated a Carvair to transport the company sponsored formula racecar driven by his son Rupert. The aircraft "Big John" even had a special modified color scheme similar to the BAF livery of the time. The upper fuselage was painted white with a light gray bottom and the stripe the length of the fuselage was eliminated. The BAF letters were on the forward fuselage and tail and British Air Ferries titles above the windows.

The demand for car-ferry service had been declining for some time because of the cross-channel roll-on roll-off sea ferries yet BAF continued to operate from Lydd concentrating on bulk freight. Company officials reasoned that if it relinquished the routes to Rotterdam, Ostend and Le Touquet a competitor could petition for the routes and BAF would not be able to get them back. Although Bristols and Carvairs operated from Lydd it is not geographically suitable to compete since it is not on the coast and does not have a rail link for freight transfer. Eventually economics won out and after 17 years the final Carvair service from Lydd operated to Ostend on 29 January 1971. The closing of the base at Lydd meant all the Carvairs would operate from Southend.

In 1972 several attempts were made to resurrect the Carvair by inaugurating Coventry—Jersey and Bournemouth—Channel Island service. Basle service was also re-instated. The attempts were not successful and the Coventry, Jersey, Bournemouth, Channel Island service was soon dropped.

BAF car ferry business declined to less than 5 percent of total traffic by 1975. Keegan announced that he was keeping only four Carvairs. BAF officials were aware of a decline for some time and concentrated on the cargo business and on-call charter service. The Carvair got a reprieve in the summer of 1975 when there was a sudden increase in demand for car-ferry service to Basle Switzerland. Although BUAF dropped this service in 1967 because of lack of traffic, BAF attempted to resurrect it in 1972. In an effort to capitalize on an unexpected and sudden demand Keegan's BAF re-introduced the service again for 1975. The traffic was nothing close to that of 1967. However, Carvairs operated three times per week in the 55-passenger two-car configuration. BAF had six Carvairs at this time with only four in the car-ferry configuration. Carvair five was on wet lease to Paulings International in Oman and seven was converted to an all cargo configuration, leaving only ships nine, 12, 13, and 17 to operate the service.

It was obvious at Transmeridian-BAF that the Carvair had seen its day. Several ships had already been offered for sale prior to the 1975 and as the season ended BAF began placing individual Carvairs up for sale or lease. During the 1975 season Caroline Frost flew as F/O on many of the cross channel flights. In July of 1976 at the age of 26, Frost was promoted to Captain and commanded the first all female crew on cross channel service. Once again the Carvair and BAF was credited with an aviation first. It is most unfortunate that this landmark event came just before the end of the BAF car-ferry era.

The End of the British Carvair Era

Even though the Carvair was given a reprieve for one more season, Keegan confirmed to the press on 30 January 1975 that the Carvair was being phased out. As he stepped off a Handley Page Herald from Gander he stated, "When the Carvair goes the motorist will have to rely on the seabound ferries."[3] The Carvair played a unique roll with Channel Air Bridge, British United, and British Air Ferries for nearly 17 years but its time was over.

BAF continued on under the direction of Keegan and the domination of Transmeridian. The combined group financial report for 1975 revealed a £1.25 million profit ($3,501,250.) with the BAF portion of only £250,000 ($700,250). The airline had only four Carvairs in service at this time. It was decided to leave one in the car-ferry role and operate the other three in a 100 percent freight role. Carvair seven was stripped of all paint and the cabin was removed becoming the first BAF all cargo Carvair.

Only five percent of BAF revenue was generated by the car-ferry in 1976. A final moment in commercial aviation history passed without notice, fanfare or ceremony, when the era ended with the final Carvair car-ferry flight on 01 January 1977. Twenty-eight years earlier in July 1948 the first revenue automobile was uplifted across the channel when Commodore Taffy Powell transported Sammy Norman's Bentley for a fare of £22 ($61.60) with great fanfare. After the car-ferry operation was phased out at the end of 1976 a second aircraft was converted to all cargo. Both were repainted in the yellow and black sash livery. From that time on the Carvair was only used in cargo and charter service and never transported another passenger with BAF.

British Air Ferries became a separate airline again in June 1977 when Keegan sold Transmeridian for £3.37 million ($8,357,600) to the Cunard Shipping Company. The Carvairs continued in various cargo roles for sometime until all were sold off. BAF operated only two Carvairs in 1978, G-ASDC and G-ASHZ, in all cargo service. In 1979 these last two ships, seven and nine, were withdrawn from service and sold. BAF continued to operate various other aircraft in passenger charter service with several changes in livery.

Keegan sold BAF in 1982 to Jadepoint for £2 Million. The company is a London property development group associated with Grants Department Store in Croydon. By 1987 BAF became the world's largest Viscount operator. The BAF outlook was good as it continued to expand and show a profit until 1988. Because of a series of market incidents it went under the British equivalent of bankruptcy and continued to operate until acquired by Mostjet Limited in 1989. Despite many problems BAF continued as one of the few profitable European airlines. The British Air Ferries name was dropped on 06 April 1993 in favor of a more up to date title of British World Airlines. The car-ferry image was finally put to rest 15 years after the last car was transported. By the turn of the century 2000 British World Airlines was in decline and eventually bankrupt.

Keegan retired after selling British Air Ferries in 1982. He moved to Spain and continued to dabble in aviation from a distance. He remained active until his death on 07 March 2003.

3

Aviation Traders
Engineering Limited

History and Diversity

Freddie Laker registered Aviation Traders on 17 October 1947 with £100 of capital. In the early days the main activity was the salvage and scrapping of surplus aircraft. Laker was given a great opportunity to expand ATL with the blockade of Berlin and the Airlift. On 08 March 1949 he registered an engineering sub-company as Aviation Traders Engineering Limited (ATEL). Over time the company developed into a first-class maintenance, overhaul and manufacturing facility. By 1952 ATEL obtained a major contract fabricating center wing sections for the Bristol Aircraft company.

Laker sold controlling interest of Aviation Traders to the Air Holdings group in 1958. The quality of work remained high because of the dedicated craftsman. He then became Executive Director of British United Airways, which was in the Air Holdings group. When he conceived the idea of the Carvair to replace the Bristol the contract went to his old company Aviation Traders. The ATEL staff was very fluid and able to move on a day to day basis from one project to another. Laker set this policy in the early days after the company was formed. By the 1960s if there was insufficient work at Southend the craftsman were bused to Stansted to work on the ATL-98s. The north side hangar at Stansted was responsible for maintenance on seven British United Britannias. In the event that none were in for work the staff went over and helped on the Carvair line. The group was known for working like a family. One day they were overhauling an aircraft and the next day they were chopping up Tudors or military equipment for the melting pot.

Laker often made surprise visits to ATEL at Southend with his dog. Actually it was not a surprise since he drove a bright yellow Aston Martin DB, which stood out from the other mundane cars in the parking lot. His rather portly Labrador always entered the building first and ambled along about twenty yards ahead of him.[1]

John Breeze, who became ATEL marketing director in 1963, transferred from Air Holdings associate company BUA and carried on Laker's policies. More than 1000 workers were employed by ATL at Southend. Half of those were involved with Carvair production as well as another 280 craftsman at Stansted that attached the nose sections and built the aircraft.

Aviation Traders was still considered a leading engineering company when G.H.C. Fisher became managing director in the 1970s. He reinforced the theory that you have to get eight hours of work into an eight-hour day. Annual revenue by 1970 was £5.5 million ($15,405,000) employing 1500 engineers, craftsman and technicians at Southend. The Southend operation claimed the largest portion of contract maintenance in the British market. Through the practice of constant shuffling of manpower the ATEL staff became remarkably diverse. Not only was ATEL a leader in aircraft maintenance but highly regarded in fabrication, conversion and manufacturing of ground support equipment. A lot of the skill can be attributed to the Carvair production, which prompted the development of cargo loaders that could lift automobiles. To handle the demand for these different services ATL was separated into four divisions.

The aircraft maintenance division was geographically split between Southend and Stansted. About 200 technicians conducted jet maintenance at Stansted because of runway limitations at Southend. The servicing of all other transports like the Carvairs for BAF were handled at Southend. The maintenance division serviced many types including the HS 748, DC-6, DC-7, CL-44 and BAC One-Eleven. All Check 3s and 4s were conducted at Southend with 1s and 2s at Stansted. During the early 1970s ATEL secured the contract to convert Vickers Vanguard aircraft to freighters. The installation of freight doors and floor systems was performed at Southend. As late as 1974 ATEL provided maintenance for the CL-44s and Britannia's of Young Air Cargo.

There were two manufacturing divisions. The first specialized in aircraft components for airframe conversions. The most notable is the Carvair nose unit. Equally complex is the cargo door assemblies developed for the Britannia and the Vanguard-Merchantman cargo door conversion. The other manufacturing division designed and built ground support equipment. One of the earlier examples is the "Hylo" loader developed for the Carvair. In addition cargo loaders, pallet dollies, and belt loaders were built for the 707 and 747. Major customers included Aer Lingus, VARIG, and BOAC that signed a £1 million contract in 1970. In addition CL-44 operator Seaboard World employed ATEL cargo loaders for the loading of automobile prototypes and racing cars. In 1971 ATEL had more than 80 cargo loaders on order from major airlines with BEA alone placing orders for fifty proving that the company continued to grow long after Carvair production ended.

The forth division is comprised of component overhaul. In 1969 plans were implemented for an electronics test shop developed from within. This division supported the aircraft maintenance activity eliminating the need to send out sensitive electronic units for service.

The sale of BAF to T. D. (Mike) Keegan in 1971 affected the Carvair maintenance program that was assumed would always be with Aviation Traders. Keegan was unable to purchase Aviation Traders since it was showing a considerable profit and Air Holdings was not interested in divesting itself of a company with growth potential. Keegan then formed Hawke Aircraft, at Southend to handle the Carvair maintenance and other work.

During the period Keegan controlled both British Air Ferries and Transmeridian he attempted to use the CL-44s to revive the car-ferry market. The Aviation Ministry required the installation of a structural bulkhead to prevent cars from accidentally crashing through the passenger compartment. Carvair builder ATEL was called in to build the bulkhead, which did not sit well with Keegan. The ATEL designed unit was heavy and difficult to install adding a weight penalty and displacing payload. Aviation Traders Maintenance Division had gained experience working on CL-44s of other carriers and put it to good use when contracted to perform schedule work on the CL-44s owned by Blue Bell to transport Wrangler denim jean products.

Aviation Traders (ATEL) remained under Airwork in the Air Holdings group until 1976. At that time the operation was moved to Stansted under Britavia, which had been the holding company for Silver City in 1947. In 1990 Britavia and Airwork were combined and the next year moved to Bournemouth. In 1993 Airwork was acquired by Short Brothers, which was acquired by Bombardier. The Aviation Traders (ATL) name was revived in 1996 when it became a separate company based at Bournemouth Airport. It exists today as an aircraft engineering and maintenance company. In a strange turn of events one of the corporate officers is John Thomas, the son of Gordon Thomas, who worked with Laker at Short Brothers in the 1930s. Gordon Thomas became the Chief Engineer for Decca overseeing the installation of the Navigator System on the Carvair. It has been falsely reported and commonly believed that FLS Aerospace absorbed Aviation Traders and the name retired.

Carvair/Design/Development

Aviation Traders (ATEL) as an associate company under Air Holdings was the only choice to build the next generation car-ferry because of considerable experience in aircraft modification of existing types and design/development experience of one new aircraft known as the Accountant ATL-90. Aviation Traders was firmly established in the skills of aircraft overhaul, repair and conversion at Southend since 1948. The more notable ATEL projects by design numbers are[2]:

ATL-8X: Manufacture of 30 center-wing sections for Bristol Freighters.

ATL-8X: Modification of two Bristol MK.31s to MK.32s in the fall of 1957.

ATL 90: The prototype Accountant, The only complete aircraft built from scratch by Aviation Traders. The ATL-90 is A Freddie Laker concept to capture the replacement market for the DC-3 plus an executive transport design. The Accountant first flew 9 July 1957 and made a final flight 10 January 1958. No orders were ever received forcing the project to be cancelled. The only Accountant completed was placed in storage at Southend and eventually broken up 02 February 1960.

ATL-91: Auditor, a two-seater tricycle undercarriage trainer.

ATL-92 & 93: Accountant Military (OR23) low wing design with various application projects and design changes.

ATL-94: Not Allocated

ATL-95: Accountant II with thinner wing and a 10-foot longer fuselage to accommodate 42 passengers. Number reassigned to a double-deck transport.

ATL-96: Freighter version and stretched ATL-90 with swing nose for car transport. Design stage only.

ATL-97: Not Allocated

ATL-98: Carvair C54/DC-4 conversion. Early design had triple tail layout.

ATL-99: Britannia C.P.F. Conversion. Freighter conversion alternate versions to the ATL-95 project consisted of a double deck airliner and a car-ferry. Also included in the design was an engine transporter for Rolls Royce. It would carry a complete RB211, which loaded through a roof door.

The above numbering system was discontinued at ATL-99

The entire concept of a new car-ferry aircraft is the result of a single airline searching for an inexpensive replacement for a specialized market. It was based around one major factor, nose door loading. It is of interest to note the parallels of Carvair and the 747, which first flew in 1968. The 747 is similar in profile for the same reason as the Carvair, straight in loading. Boeing gambled its future on the development of the 747 and believed that it would be relegated to an all cargo role after the SST (Super Sonic Transport) became operational assuming a decline in passenger traffic. The 747 was designed as a freighter with elevated cockpit and nose door to allow straight in loading imitating the Carvair. Even more intriguing is the development of a special transporter truck to move the wing spars some 45 miles to the Boeing plant similar to the ATEL special transporter to move the nose sections from Southend to Stansted. The 747 nose is built in two sections, an upper and lower half, which is split horizontally at the cockpit floor similar to the ATEL two piece engineering technique to manufacture the Carvair nose section.

There is no evidence that Boeing formally studied the Carvair but the ATEL design is easily

Aviation Traders developed a special transporter for delivery of the completed nose units from Southend to Stansted. Transporting of airframe assemblies is now common with manufactures like Boeing and Airbus (courtesy Guy Craven).

seen in the profile of the 747 and there are curious similarities in concept and design. Another design that shares some ATL-98 Carvair engineering concepts is the proposed Convair C-99 commercial production version, which is similar to the much smaller Carvair. The elevated cockpit sits above nose loading doors. Because of size the doors are clamshell type more like the military C-124 Globemaster II. The nose gear design is very close to the Carvair. It retracts to the horizontal line and requires blister type doors to cover the portion outside the fuselage. The commercial version was never built and only one XC-99 was completed for the USAF. The front loading clamshell doors and nose gear blister was deleted for a different design on the single aircraft built.[3]

Evolution of Car-Ferry

The Carvair evolved from a need for a more dependable aircraft for a specialized market. The car-ferry began with the upgrading of the Bristol 170 (Mk.21) which was developed in 1945 for the Royal Air Force for operations into short unimproved airfields. Later it was adopted by smaller British carriers for cross channel work. The original design had a solid nose and did not allow for loading of oversized cargo. The Bristol Aeroplane Company re-designed the nose adding clamshell doors with a new designation of "Mk.31." The "Mk.32" version came next with extended nose doors creating a cargo hold of 42 feet to allow for an additional automobile and had a small cabin for 15 passengers.

In the fall of 1957 Air Charter transferred two short-nose Bristol Mk.31 aircraft to ATEL for modification. It was reasoned that it would be more cost effective to convert them to Mk.32 long-nose Superfreighters rather than order new aircraft. This gave Air Charter a fleet of eight Superfreighters. Aviation Traders gained valuable experience in modifying airframes and adding extended nose sections. Two years later it proved beneficial in developing the Carvair nose conversion. More experience was gained when the Bristol Aeroplane Company contracted ATEL to produce and install wing modification sets for the 170 Freighter. This was in addition to the more than 30 center-wing assemblies manufactured for the Bristol Freighter.

The Bristol was never really suited for car-ferry work. It had dependability problems and stress produced fatigue failure of the wing structure causing wing spar limitations. These design shortcomings made it imperative that a replacement be found. The primary routes operated were Southend/Lydd to Calais, Ostend, Rotterdam and Le Touquet. All were short routes between 70 to 165 miles. The Bristol utilization was around 1000 hours per year.

The requirement for a second-generation car-ferry was twofold. It had to be a high utilization larger capacity aircraft with greater range in order to reduce cost. It was determined that flying 2000 vehicles from Southend to Geneva or 10,000 vehicles from Southend to Calais can produce the same amount of revenue. The Bristols were not capable of the range or utilization. A new aircraft either reciprocating or turbine would cost between £500,000 ($1,400,500) and £700,000 ($1,960,700). The cost amortized over seven years at £50 ($140) per hour was too prohibitive.

The acceptable cost could not exceed £200,000 ($560,200) amortized over the utilization rate. After studies were completed as suggested by Freddie Laker it was concluded that converting an existing aircraft was the only alternative and the C-54 was the best candidate. The final projected price worked out to £120,000 ($336.000) not considering zero time engines, periodic tank seal, and radios. The total cost when adding engines and tank seal using the existing DC-4/C-54 aircraft was estimated at a total cost of £160,000 ($448,160) per aircraft. Since a total of 1163 C-54/DC-4s were built, many surplus DC-4s were available for around £40,000 ($112,040). Comparatively a factory new DC-4-1009 from Douglas Aircraft cost $385,000 after World War Two.

When the final numbers were presented the four zero time Pratt & Whitney R-2000 engines and Hamilton Standard propellers cost £12,000 ($33,612). Also to be considered is the cost of tank seal, which is required every 16,000 hours on the DC-4. Even with the variables being considered the cost came in under the £200,000 ($560,200) initial cost requirement.

Freddie Laker was quite familiar with the Bristol design of cockpit above the cargo compartment and nose loading doors. The profile of the forward fuselage of both the Bristol and the Armstrong Whitworth Argosy with raised cockpit are similar. Comparing these it is fairly easy to vision his idea to change the DC-4 nose. The larger aircraft would be less noisy, more comfortable, and have longer range. It has been reported that Freddie Laker conceived the idea of converting the DC-4 while in the shower. It is clear that he had been toying with the idea of a Bristol replacement for sometime. Drawings and sketches from 1957 show the ATL-90 Accountant with a swing nose for car loading. The Accountant was deliberately designed with the nose gear set noticeably reward and a humped fuselage to accommodate large cargo items. The proposed stretched swing nose car-ferry Accountant was issued design number ATL-96. This early version was most likely aimed at Air-Charter/Silver City since it would only transport two cars. The Bristol Mk.32 already could transport three cars.

ATEL chief engineer Arthur Leftley had the liberty to recommend any other aircraft evaluated. After study he came to the same conclusion as Laker. The DC-4 was the only possible cost-effective candidate for conversion by re-positioning the cockpit 6 feet 10 inches above the original position.

The conversion did not have a name at the time and was assigned the designation ATL-98

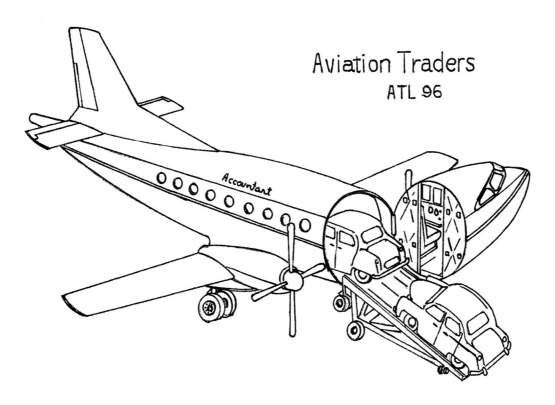

Aviation Traders
ATL 96

Accountant

Laker toyed with the idea of a Bristol replacement based on the ATL-90 Accountant. This early drawing of the proposed ATL-96 with a swing nose for car loading indicates he considered every possibility. The Accountant had the nose gear set noticeably reward and fuselage humped to accommodate large
cargo items. The proposed stretched version could only accommodate two cars (courtesy Aviation
Traders Ltd.).

or AT(E)L-98 since the company had both engineering and sales division. Originally the design
team expected to use only surplus C-54 military aircraft for conversions. The true DC-4 aircraft were still in commercial service and generally more expensive. Additionally the postwar
DC-4-1009 is 2600 pounds heavier than the C-54, which reduces payload. It would have been
easier if all conversion had been done using only C-54 airframes, which the conversion engineering was based on. Later on in the program two DC-4-1009s were used and additional problems were encountered.

The aerodynamic design was completed with wind tunnel test carried out at The College
of Aeronautics at Cranfield. The data was compiled and analyzed by Aviation Traders Chief Aerodynamicist Brain Kerry. Also tested in the wind tunnel at that time was a proposed Dart powered Carvair. The data was reviewed, but only used for the test data, promotion and a future
production possibility. The Dart powered Carvair was reviewed as the next stage DC-6 or DC-
7 conversions. The Carvair tail fin area established in the wind tunnel is almost identical to the
DC-7. Tail plane fins were first considered as an option before the development of the DC-7
shaped tail. Since slipstream effects were unknown, the vortices created on the large nose tend
to grow in intensity during yaw adding pressure to the tail plane fins. Further testing determined them to be unnecessary and proved the increased rudder and tail height to be more efficient. The rudder travel was reduced 20 percent to 16 degrees from the original 20 degrees of
travel.

Initial wind tunnel model testing was conducted at the College of Aeronautics at Cranfield in 1959. This early model has the standard DC-4 vertical fin, as it had not been established to add endplates to the horizontal tail plane or increase the height of the fin (courtesy Brian Kerry).

As early as 1959 Channel Air Bridge released to the press the positive aspects of the ATL-98. It was stated that the C-54 conversion had an excellent fatigue life and no crossbeam floor strengthening would be required. It was determined that some of the aircraft did not have the factory reinforced floor panels and the old corrugated C-54 floor was not suitable for vehicle ferry application. A composite three-quarter inch flooring was designed using a plywood face and aluminum alloy base. It was filled with a plastic material to complete the laminate. The panels ranged in size from five to eight feet in length with an all-new cargo tie down system.

The design standards required a cargo compartment capable of handling five vehicles and a rear passenger compartment for their owners and passengers. With 68 feet of the cargo compartment assigned to vehicles, 12 feet remained for passenger accommodations. The original design called for a 23-seat cabin with 18 seats facing forward and five facing rear. In production it was reduced to 22 and in time some cabins were fitted with as few as 17 seats. Some carriers and later BAF specified additional forward cabins with total aircraft capacity of 34, 40, 55 and 65 seat configurations. In the all cargo configuration the aircraft cargo hold is 80 feet long with a cargo

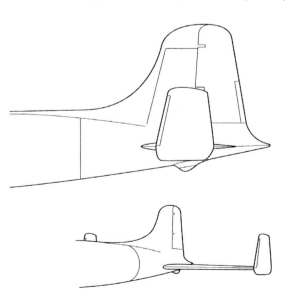

The original DC-4 fin and rudder were found to be inadequate. Studies were done to add vertical stabilizers to the ends of the horizontal tail plane. Further tests proved the same could be achieved by increasing the vertical fin by 34 inches. Pictured is the original DC-4 fin with the proposed stabilizers (courtesy Michael Zoeller).

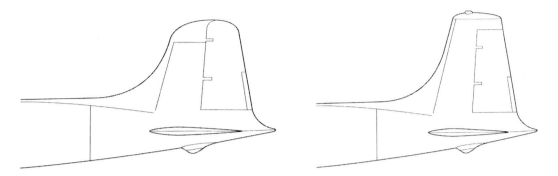

A comparison of the DC-4 (left) empennage and the production ATL-98 Carvair (right) clearly show the increased fin height, which required considerable structural reinforcement to support the weight and stress (courtesy Michael Zoeller).

capacity of 4,350 cu/ft or 4630 cu/ft with raised ceiling. The car-ferry 22-seat version is 68 feet with 3,428 cubic feet.

Aviation Traders was required to obtain permission from Douglas aircraft at Santa Monica California to modify and sell the new design. After review Douglas officials were enthusiastic with the concept. It was to their benefit that a new design could be given to an existing type that was nearing the end of its operational life. The possibility of conversion of DC-6 and 7s could also benefit Douglas if the industry found favor with the Carvair. Fleets of DC-6s and 7s would eventually be traded for newer jets. A market for otherwise obsolete aircraft held promise. Douglas wanted to be released from any liability after modification and only authorized the one design. ATEL also needed to consult with Douglas on stress and design data. It was imperative that the Carvair be seen in a favorable manner, since there was the possibility of DC-6 and DC-7 conversions if the DC-4 was successful.

Arthur Leftley flew to Santa Monica California to meet with Douglas officials in March 1959. While negotiating with Douglas a cost of $10,000 was agreed upon to supply DC-4/C-54 data and drawings. Leftley agreed considering it a fair amount, however he was thinking £10,000 Sterling ($28,010). Because of this error Leftley obtained the data and drawings for £3570 Sterling ($10,000). This was a third of what Leftley thought he had offered and a windfall for Aviation Traders.[4]

Leftley immediately returned to Southend with as much data as Douglas could provide on short notice in order to begin the drawings for conversion. The Aviation Traders design concept based on the C-54 was somewhat of a problem since the aircraft was long out of production. Most of the data immediately available was for the postwar DC-4-1009 leaving Douglas to retrieve the C-54 data from archives and send at a later date. This created a problem since ATEL needed all the data to develop the conversion for both C-54 and DC-4 depending on the surplus aircraft available. It also hampered the wind tunnel test being done by Brian Kerry at the College of Aeronautics at Cranfield. Ultimately ATEL conducted 76 hours and 33 minutes of its own testing to established lift of the new forward fuselage. The test resulted in the CG being moved 4 percent forward of the original C-54 airframe.

Flight test data was needed on the C-54 to use as a comparison on to the Carvair modified airframe. Because of the C-54 being out of production Douglas was delayed in providing the needed data and when it did finally arrive it was determined that it could not be used for comparison but only as reference. The data was from a pre-war factory fresh C-54, which was not the same as the aging airframes used for Carvair conversion. Leftley and his team decided to conduct test on one of the aircraft scheduled for conversion to obtain the needed data. G-ANYB became the obvious choice and an accelerated flight program began.

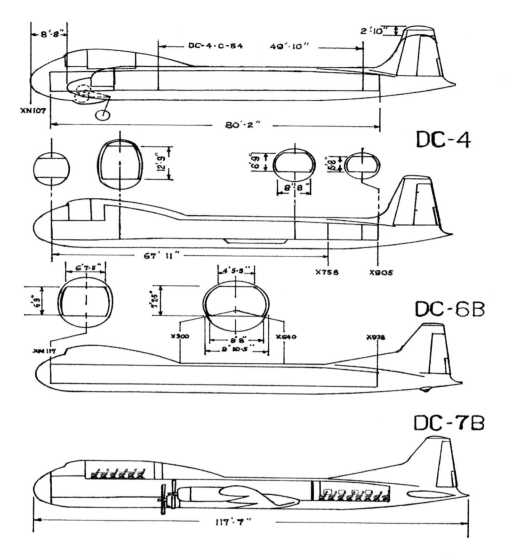

Comparison profiles represent the extent of ATEL research into Carvair possibilities and demonstrate the improved straight in loading over the DC-4. The second shows the increased capacity of the raised ceiling in the hump. The proposed DC-6B conversion had a larger fin. The DC-7 is designed with 20 seats behind the cockpit resembling the 747 upper deck plus 36 seats in the rear cabin (courtesy Aviation Traders Ltd.).

Near the end of 1959 a photographic unit was formed to document the design and test work. Guy Cravens, who came to ATEL in 1959, was set up as the photographer of the unit. Most of the major structural design work was completed and full-scale drawing frame members were produced and transferred to chip board, which was cut out and assembled to form the mock-up.

The photo section was not well equipped with only one MPP 5x4 stand camera. It was large and hard to use in confined spaces. Cravens loaned ATEL two of his cameras, enlarger, and processing equipment. Late in 1960 he was called into a meeting with the Chief Draughtsman and test engineers. There was concern on how to record the gauge readings and test data from an

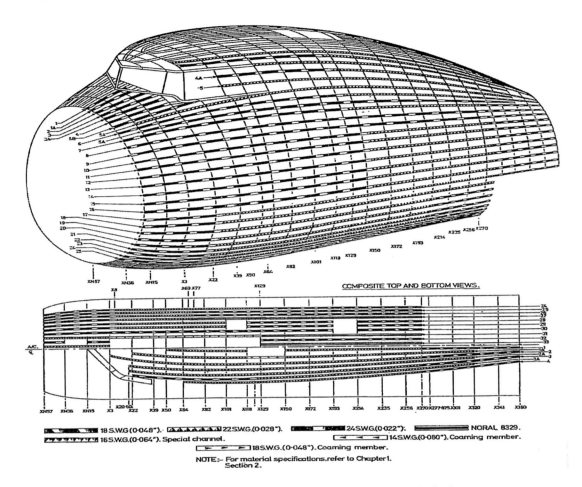

COMPOSITE TOP AND BOTTOM VIEWS.

NOTE:- For material specifications, refer to Chapter 1.
Section 2.

The design of the nose and engineering drawings showing each stringer and rib placement were essentially complete in 1959. A.C. Leftley visited Douglas Aircraft in California to obtain DC-4 stress data and permission to modify the existing airframe by adding the new nose (courtesy Aviation Traders Ltd.).

auxiliary panel mounted in the area behind the flight deck. Someone remembered that ATEL had stored a collection of ex–RAF camera equipment at Stansted that was acquired by Freddie Laker during his days in the salvage business. Cravens and the Chief Draftsman drove to Stansted and found a treasure chest of photo equipment, returning to Southend with a full van.[5]

After a number of modifications and fabrications for lens mountings a system was devised to photograph the test flights and record the data on film. In addition three Hussenot Recorders were mounted in the space behind the crew seats to record control column and rudder pedal forces, elevator and rudder angles, and trim tab positions. After the initial test program photography was completed the Photographic Unit became the Reprographic Section providing print service to Channel Air Bridge, BUA at Gatwick and photographing prominent people flying through Southend.[6]

When the flight program began Aviation Traders test pilot D. B. Cartlidge, who was a graduate of Empire Test Pilots School at Farnborough, had not been officially appointed. He left his position as Lord Derbys personal pilot in order to be considered for the position. Because of the urgency to obtain the flight test data, he was very much needed for the test program beginning on 21 September 1960. The following day Cartlidge came on board at Aviation Traders.

After the major structural design work was completed in 1959, full-scale drawings of frame members were produced and transferred to chipboard. They were then cut out to form the mock-up. The DC-4 tail cone of PH-DBZ that was purchased from KLM is visible in the background between the two units (courtesy Guy Craven).

Captain Cartlidge was teamed with Captain Robert Langley, who was chief pilot and flight manager for Channel Air Bridge. Prior to coming to Channel Air Bridge Langley served with British Overseas Airways (BOAC). While at BOAC Langley landed a Handley Page Hermes, G-ALDN, in the Sahara 110 kilometers southeast of Atar Mauritania when it ran out of fuel. It was damaged beyond repair. The flight on 26 May 1952 was en route from Tripoli to Kano Nigeria. Bob Langley resigned from BOAC as an honorable response to the incident. He then went to Air Charter where he received command of a York.

The C-54/DC-4 test program was conducted as follows:

On 21 September Captain Robert Langley, Chief pilot and Flight Manager for Channel Air Bridge took command of flight trials. He was assisted by F/O P. Drescher, A. M. Wallis, and Flight Test Engineer Ken Smith. The flight trial consisted of engine out performance at 15 degrees flaps with gear in retracted and extended position.

On 22 September Captains Robert Langley and Don Cartlidge, F/O P. Drescher, and Flight Test Engineer Ken Smith evaluated stalls for three hours and five minutes. Later in the day another test flight was conducted in various configurations of extended and retracted gear, power and flap settings and turns.

On 24 September Captains Robert Langley and Don Cartlidge, F/O R. Salvage, and Flight Test Engineer Ken Smith evaluated side-slips, yaw, and landings on a test flight of four hours and 40 minutes.

On 29 September Captain Robert Langley, F/O A.M. Wallis and P. Dresher, and Flight Test Engineer Ken Smith evaluated performance climbs from varying altitudes. The tests were conducted in numerous configurations for four hours and 50 minutes.

On 29 September (2nd flight) Captains Don Cartlidge and Robert Langley, and Flight Test Engineer Ken Smith evaluated change of stick force with varying flaps and power, stalls, sideslips, stability and landings. The test flight lasted two hours 15 minutes. Also on board were ARB Captains G. Howitt and R. Bryder.

On 03 October Captains Robert Langley and Don Cartlidge, F/O Ken Smith, and ATEL Aerodynamicist Brian Kerry evaluated stalls, landings, and trim speeds on the final two hour test flight.

After the data from the flight performance checks and handling characteristics was analyzed, the team determined that the C-54/DC-4 operated within range of the manufacture's tables. The design team now had standards of an aging C-54/DC-4 for comparison to an ATL-98 test flight.

G-ANYB was grounded at Southend and disassembly began. Except for sawing off the nose, the airframe was virtually disassembled outside before the bare fuselage was moved into the hangar. Although ATEL was starting with an existing airframe the finished product would technically be a new aircraft. All parts required inspection and if found defective or unserviceable repaired or replaced. The spars had to be x-rayed and the floor removed. A new laminated floor was designed for vehicle ferry use with 96 tie down stations utilizing a number of the Douglas pre-drilled C-54 floor beam positions.

Aviation Traders purchased the fuselage of retired KLM DC-4 PH-DBZ on 20 April 1959. It arrived at Southend via Tilbury docks in December and a full-scale mock-up was built around it. The rear fuselage of PH-DBZ was stored for possible use if the wind tunnel test proved a tri-fin empennage would be required. Tests later proved that the tri-fin design was not necessary. The rear fuselage tail cone of PH-DBZ was later required to repair G-ANYB after a forklift struck the horizontal stabilizer in August 1961.

In February 1960 Aviation Traders issued a progress report stating they were hopeful the ATL-98 would fly by October. The wooden jig for the nose had been completed at Southend. The first production nose was about to begin fabrication. All control runs were originally proposed to run along the upper ceiling surface of the fuselage. This became a major engineering problem and all except the elevator control cables were changed.[7] A series of pulleys directed the control cables from under the new cockpit floor to the sidewall re-connecting to the existing DC-4 system of pulleys under the cargo hold floor of the fuselage. The report that it would fly was obviously overly optimistic since disassembly of G-ANYB did not begin until October after the flight data trials were completed and the control run problems found to exist.

Initially only one complete ATL-98 was to be built at Southend although all conversion parts were being produced there. Since ATEL did not possess the facilities for volume conversions, a kit was planned for field assembly and offered to DC-4 operators with suitable overhaul facilities. More than 60 overseas operators expressed interest in conversions completed by Aviation Traders or at the facilities of individual carriers.

Potential Buyers

It is quite easy to understand why ATEL believed early on that the Carvair would be produced with success in large numbers. Many carriers and companies expressed interest in the conversion and the proposed variants. Interest ranged from casual inquiries to serious proposals and contract negotiations. The exact number of aircraft is not known, however the list is quite impressive with more than 60 interested carriers. It was projected that as many as 100 Carvairs would be produced.[8] The list included:

Continentale Deutsche Luftreederei expressed interest in January 1961 to consider conversion of many of its DC-4s for work in the Congo.

The Aircraft Leasing Company of George Street London inquired in July 1961 for one conversion.

Iberia Airlines made inquires in July 1961. In 1964 the carrier ordered two with an option for five more for subsidiary AVIACO.

The Royal Canadian Air Force officials inspected G-ANYB, in 1962 and reviewed the facilities and production line. The possibility of converting the C-54GM North Star was considered.

Portuguese and Spanish Air Forces in 1962 considered converting its entire C-54 military fleets.

Carlos Tejera of Caracas Venezuela expressed interest in purchasing one Carvair in 1962 for work in the developing oil industry.

The State of Hawaii recommended and approved the Carvair for inter-island ferry service in early 1962. Aloha and Hawaiian Airlines representatives visited Aviation Traders and were considered very interested. The state board of transportation favored sea-going surface transport and no commitment was made. The Carvair eventually flew with Pacific Air Express in 1983 and Hawaii Pacific Air in 1990, which were based at Honolulu. Both carriers competed against Aloha and Hawaiian Airlines cargo operations.

Sociedale Comercial Romar Ida of Lisbon Portugal made inquires in January 1962 for two C-54 conversions.

Lebanese based Middle East Airlines (MEA) expressed interest in January 1962. The carrier was expanding and had become the first pure jet operator in the region with Comet 4s in 1960. MEA was owned by Sheik Najib Alamuddin. Yousif Bedas purchased control in 1961 in an effort to merge it with Air Liban. Financial problems emerged and after the initial inquiry there was no follow up with Carvair orders.

Trans Mediterranean Airlines (TMA) of Beirut inquired in January 1962 and made several additional inquiries, the last in 1966. There are some interesting aspects to the TMA inquiries. In 1963 Aviation Traders purchased a TMA DC-4 that was converted to Carvair 15. The carrier continued to consider Carvair conversion for its DC-4s and was hopeful the DC-6 conversion was forthcoming. After consideration and the realization that a fleet upgrade was needed TMA began leasing CL-44s in 1967.

East Anglian Flying Service (traded as Channel Airways) placed a tentative order for four Carvairs in February 1962. The first two ships had a delivery date set for May 1962. Channel Airways intended to compete with Aer Lingus in the car-ferry market between Bristol/Liverpool and Cork/Dublin. ATLB licenses were approved for the service and fares/schedules were given to travel agents. Three daily round-trips were planned with an order placed for two ships valued at £360,000 ($1,008,360) to be delivered in May. In anticipation of service, DC-4 N33679 was purchased and British United Captain Bob Langley conducted crew training. The carrier applied for registration G-ARSF for conversion three, but it was eventually used by Channel Air Bridge.

Irish authorities restricted authority prompting Channel Airways to put car-ferry service on hold. Supplemental passengers could be transported on Bristol-Dublin but not Bristol-Cork. Channel Airways did not finalize the order and never operated Carvairs.

Air Fret and Air Corse in March 1962 considered joint ownership of one Carvair. At the same time Air Fret was also reviewing a possible Caravelle cargoliner model. In 1972 Air Fret actually purchased Carvair 15 which was in non-flyable condition at Nimes-Garons France. Later that year it was broken up by BAF for spares.

Airnautic requested quotes in March 1962 for converting two of the carriers C-54s.

Fairways-Rotterdam held discussions and showed a firm interest in April 1962.

Great Circle Air Charters of New York made inquires in April 1962. No further talks were held after the initial inquiry.

Aero Consult of Buenos Aires considered in April 1962 operating Carvairs from Buenos Aires to River Plate resorts.

An Indian newspaper, *The Hindu*, published in Delhi, expressed interest in July 1962 in purchasing one Carvair for delivery of newspapers throughout India. Indian Airlines aircraft were being chartered at the time.

Politeonia Aerea SA, Madrid negotiated to purchase one Carvair in August 1962. The contract was contingent on the carrier receiving government approval of Barcelona — Palma service. Approval was not granted and the contract expired.

Linjeflyg inquired in September 1962 and expressed interest in initiating a Carvair operation.

Pacific Western Airlines of Canada had VP of Operations Duncan McLaren review the Carvair for Canadian and Arctic operations in 1962. The carrier opted for the standard DC-4, which was available at a lower cost.

Autair of Luton inquired in October 1962 into the possibility of a Carvair built as a helicopter transporter. The company was an overseas helicopter operator established in 1952. It began flying Dakotas in 1960 and eventually became a successful scheduled passenger carrier with a network of inclusive tour services. The Carvair later proved to be suited for helicopter transport as well as other types of aircraft.

SAGAT, Torino discussed purchasing one or two Carvairs in October 1962. Market potential was evaluated before dropping plans.

L. Asthner of Malmo late in 1962 proposed to operate Malmo — Hamburg car-ferry service. The Plans were to operate three Carvairs as Scandinavian Air Ferries, initially leasing one from BUAF beginning in May 1963. Financial backing came from a wood-pulp enterprise. The new carrier was never established.

Ansor Corporation of London (incorporated in the U.S.) discussed ordering three Carvairs for delivery in June, July, and September 1963. Two C-54B aircraft were to be traded in after delivery as partial payment.

Tunis-Air filed a letter of intent to purchase one Carvair late in 1964. The carrier received approval and planned car-ferry service on 01 June 1965 between Tunis and Marseilles. Aviation Traders reviewed the proposal to convert a C-54A to five-car 22-seats with a convertible feature to 84-passnger configuration. Tunis-Air also reviewed an all cargo Caravelle at the same time but no orders were placed for either aircraft.

Austrian Airlines submitted plans late in 1964 to purchase and operate a long-range Carvair (possibly G-APNH). The proposed route was Dublin — Manchester — Southend — Vienna.

Luft Lloyd reviewed the purchase of one aircraft in 1964.

Faucett Airlines inquired in December 1967 into the possibility of leasing one Carvair for an oil field support contract with Mobil Oil in Peru.

Aerovias Nacionales Transcontinental Ecuador (ANTENA) in May 1968 negotiated to advance stages a contract to purchase at least one Carvair for all cargo service between Ecuador and the United States (Miami). The contract was never signed. Prior to ATEL negotiations, ANTENA negotiated to purchase Carvair eight, EI-AMR, from Aer Lingus in September 1967. Company officials determined a new aircraft could be purchased at only a slightly higher cost. The transaction with Aer Lingus was never completed.

Air Congo of Leopoldville (Kinshasa) reviewed the potential of the Carvair. The DC-4 operator was formed in 1961 to assume the operations of SABENA-Afrique. It needed an aircraft capable of transporting lengthy cargo and cars, but considered the Carvair too expensive. During the same period the swing-tail CL-44 was reviewed. While searching for a less costly alternative to the Carvair and CL-44, Air Congo officials conceived the idea of a swing-tail modification for its DC-4. The idea was presented to SABENA engineering and a kit was developed. The less expensive swing-tail modification was potential competition for the Carvair. Air Congo and SABENA relations deteriorated over time and were terminated in 1965 when Congo political leadership changed. In spite of this the plans for a swing-tail DC-4 conversion progressed.

SABENA built the first swing-tail DC-4 for Air Congo in 1966 at Brussels. Like the Carvair the straight in loading could accommodate five pallets or 4–5 cars. Only a small number of conversions were built, like the Carvair it was an idea too late. Most carriers were phasing out the Douglas piston series aircraft for more modern equipment. In the 1980s Aero Services of the Congo owned two Carvairs and the first Swing-tail DC-4 at the same time.

With so much interest in the Carvair conversion it is easy to understand how ATEL officials were overly optimistic in their projection of sales. Prior to the merger with Silver City, Channel Air Bridge initially ordered 10 with an option for eight more. Channel Air Bridge took delivery of the first three Carvairs built and placed a firm order for seven more. Initial enthusiasm was strong with many interested carriers. There was also high interest in special models and variants such as DC-6 and DC-7 conversions.

French carrier Compagnie Air Transport (CAT; known as Cie Air Transport) purchased Bristols from Silver City in 1961 to operate car-ferry service. The market exchange agreement gave CAT a 25 percent stake in the English — French car-ferry market. After the Carvair was introduced CAT held talks with ATEL stating it would order DC-6 conversions for contract work and car-ferry service to Corsica. The order was never acted on and no DC-6 conversions were produced. CAT eventually purchased second hand DC-4 Carvairs.

Although there was great enthusiasm and optimism at ATL the merger of associate company Channel Air Bridge with Silver City Airways possibly contributed to the demise of the Carvair. The short haul car-ferry traffic peaked in 1962. The decline was because of multiple reasons. The road systems in England and Europe were greatly improved and the drive on/off sea ferries grabbed market share. There were many competing freight companies lowering the fares. This hurt the short-haul car-ferry, which was already being sustained by freight. Only the long haul could show a profit on vehicles alone. Despite these facts no one seemed to notice and the potential for sales was believed to be good.

Linea Expresa Bolívar CA of Caracas requested a quote to convert one or more of its DC-4M aircraft to Carvairs. The conversion included re-engine with Wright R-2600 and re-certification. Aviation Traders gave a quote of £400,000 ($1,120,400) each but the carrier never responded.

Aviation Traders investigated the possibility of converting the DC-4M North Star, which was powered by liquid, cooled Merlin engines. The project was dropped because of their being only 23 un-pressurized version built for the Royal Canadian Air Force (RCAF). Because of the low number of types available ATEL would have needed to acquire the entire fleet. Twenty commercial pressurized DC-4M versions were built but not available.

Despite a tremendous amount of interest from many carriers generating great optimism, and aviation trade publications printing many stories and progress reports, ATEL did not receive any major orders.

Additional Carriers and Corporations expressed interest in converting their fleets to Carvair standards under license. Aviation Traders would produce the nose and conversion kits, which would be shipped to the particulars for conversion. The list is as varied as those interested in the complete converted aircraft.

Avianca of Columbia operated a large fleet of DC-4 types and seriously considered the option of purchasing kits.

As early as May 1961 Fairey Aviation of Halifax Canada expressed interest in becoming the conversion licensee for the American continent. Lockheed Aircraft Services also expressed the same interest in June 1961. Pacaero Engineering of Santa Monica California, the city where many DC-4s were manufactured, also reviewed the possibility of becoming the licensed conversion base in America.

In 1965 after the success of the Carvair with several carriers and its performance in special situations Air France expressed interest in the conversion. The plan was to purchase the kits from ATEL and convert the aircraft at Toulouse. Several factors were considered in putting the project together. Air France was soliciting orders from associate companies in North Africa and other areas, which would actually cut in to the ATEL market of potential buyers. However, if the orders materialized 24 DC-4/C-54s of the Air France fleet would also be converted resulting in additional sales.

The project advanced to the serious planning stage leaving ATEL officials quite optimistic. Unfortunately the orders did not materialize and the plan was terminated by the French when lack of orders on unrelated projects at the Toulouse plant caused massive cut backs in manpower. The situation became so serious French officials considered closing the plant.

Car-ferry service proved to be seasonal on all but a few routes. The Royal Aeronautical Society conducted seminars to try and interest any carriers in the Southern Hemisphere to consider the Carvair. The idea was that even if Philippine and Indonesian carriers were not interested in purchasing Carvairs, they could be leased during the English winter off-season.

The order of the first Carvair for Ansett in 1965 prompted ATEL to draw up contingency plans for contract production by Hong Kong Aircraft Engineering (HAECO). It was anticipated that Ansett would require eight conversions. It was further speculated that the Australian Aviation Ministry would stipulate that Trans Australian Airways (TAA) would require eight as well.

Aviation Traders previously conducted design test and feasibility studies on DC-6 and DC-7 Carvair conversions with Rolls-Royce Dart engines as an upgrade to the less reliable radial engines. It was proposed to re-engine an existing C-54 Carvair with G-APNH as the obvious choice. Subsequent studies in 1964 revealed that without pressurization and with limited fuselage strength the C-54 airframe was not suitable for turbine power. The project was cancelled in the early design stages and G-APNH retained the Pratt & Whitney R-2000-7M2 engines.

Engineers continued studies to improve the design of the Carvair and increase payload and range. Aviation Traders received inquires as early as 1961 for possible conversion of the DC-7. United States carrier Eastern Airlines operated DC-7s and had a large number of surplus aircraft they were unable to sell in the secondary market. The carrier seriously considered conversions to extend the utilization and operational life of the fleet in a cargo role. Alaska airlines, under Charles Willis Jr. expressed interest in the early Carvair but reviewed the possibility of converting its DC-6 and -7s with an extended top with room for 20 passengers.

On 03 December 1964 Freddie Laker announced that ATEL was proceeding with a £500,000 development program to re-engine the Carvair with 1.740 e.s.h.p Rolls-Royce Dart 510 engines. It was estimated that operating cost would be reduced by 25 percent. Studies for a DC-7 Carvair had been done in early development but abandoned. Now ATEL considered the DC-7 necessary to increase sales since many operators were reluctant to purchase piston powered aircraft in the turbine age. A number of operators had inquired and expressed strong interest in turbine models as well as DC-6 and -7 conversions. It was reasoned that surplus Dart engines were available for as little as £5,000. The Super Carvair would be more comfortable and quiet when the 13-foot propellers were replaced with 10-foot units. The performance, payload and range would be increased and the Dart engines would move the c.g. forward easing balancing problems with the current model.

In reality a Dart powered DC-4 version was never offered. The inquiries came to maturity in 1965 when ATEL design and engineering offered three models of advanced Carvair conversion. The study stated that the new type could be built, test flown and certified within 18 months of firm orders. The three models were:

1. DC-6B un-pressurized with Carvair nose.
2. DC-6B pressurized with Carvair nose and Rolls Royce Dart 8 engines.
3. DC-7B un-pressurized with Carvair nose retaining the Wright R-3350 power plant. Fuselage length would increase from 108 feet 11 inches to 117 feet 6 inches.

In 1964 Freddie Laker announced that ATEL would go forward with a Dart powered DC-7 Carvair. Wind tunnel test had been conducted at Cranfield in 1959. A model of the Carvair 7 was displayed at Farnborough in 1966 (courtesy Guy Craven).

Aviation Traders displayed a model of a DC-7 version at Farnborough in 1966 with Rolls Royce 3,180 hp Dart Rda.14s. The increased capacity would hold five cars and 58 to 82 passengers in rear cabin and upper deck. A swing-tail version of the DC-6 and -7 was also reviewed. The operational cost of the DC-6 was lower than the DC-4 however the conversion cost of the non pressurized DC-7 with piston engines was only 50 percent of the DC-6 prompting several operators to express interest. The primary reason DC-6 and -7 Carvairs were never built was cost. The second hand market price for a DC-7 was between $100,000 and $150,000 compared to $225,000 to $330,000 for the older DC-6. Add to that the conversion cost for a Dart powered DC-6B estimated at £250,000 ($700,250). Compared to £120,000 ($336,120) for the standard version it did not make economic sense in the emerging turbine age.

Aviation Traders had publicly announced as early as 1960 that Ansett-ANA was interested in obtaining conversion kits for all cargo configurations. Ansett-ANA then reviewed and considered ordering one DC-6 conversion. The DC-7 was not an option since they were never certified in Australia. Fred Olsen Air of Norway an operator of DC-6As expressed interest and was expected to place an order for the DC-7B conversion. No firm orders were ever placed and the project of a Super Carvair was eventually dropped.

Three versions of the original C-54 Carvair conversions were planned. The car-ferry, the combination mixed passenger/cargo freighter, and a long-range freighter. A moveable bulkhead allowed for changing roles of the aircraft between five car/20 passengers (actually 22) or 32 passenger/cargo combi configuration. Once they were produced the seating configurations were altered to meet the needs of the carriers up to 65 passengers.

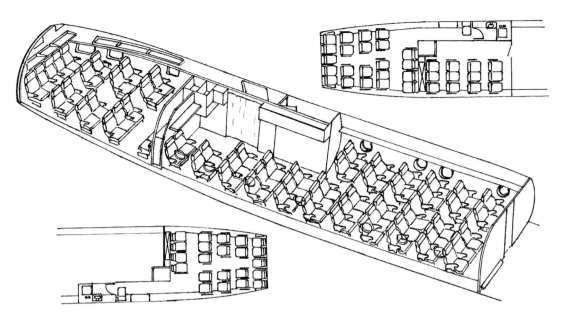

The Carvair was introduced in a 22-seat five-car configuration (lower left). During the period that BAF and Transmeridian were owned by Mike Keegan, at least one aircraft was fitted in a 65-seat combi configuration, which left 34.5 feet for bulk cargo but no cars. By comparison, Aer Lingus operated in a two cabin 34-seat four-car set up (upper right) (courtesy Aviation Traders Ltd.).

The full freighter version has the passenger cabin and lavatory removed leaving a cargo hold length of 80 feet with no bulkheads. It was believed that this version would appeal to the air ministry to transport rockets and military cargo. In reality only the three ships built for Ansett were pure freighters and they did not have long range fuel tanks. The only aircraft with long-range additional fuel tanks is Carvair 11 and it was not a pure freighter.

The freighter concept was not well accepted and a total of only five ships were ever built as cargoliners. Two of them, ships four and five, were converted back to car-ferry standards in 1963 after returning from United Nations work in the Congo. The few ships that survived into the late 1970s and 80s became freighters by removing the passenger cabins. All of the later cargo conversion retained some of the rear cabin windows and some retained the last row of cabin seats.

The three Ansett ships never had the rear cabin windows and were built with four roller tracks from the nose to station X798 just behind the side-loading door leaving a manual cargo area at the extreme aft. The rollers were mounted on a metal/balsa sandwich floor. This option allowed the loading of seven 88 in. × 108 in. pallets. Two of the ex–Ansett ships later had rollers mounted in the extreme aft cargo position from station X798–X891.

The Carvair pure freighter version had the following Payload/Range and Endurance:

Payload (lbs.)	Range (NM)	Endurance (Hrs)
18,000	1300	8.7
17,000	1500	9.5
16,000	1700	10.4
15,000	1800	11.2
14,000	2000	12.0
13,000	2200	12.9
12,000	2300	13.7
11,000	2400	14.4

Each version had the following performance ratings.

	Pure Freighter	*Standard Carvair*	*Long Range Freighter*
OEW	42,554	41,824	41,365
ZFW	60,700	59,000	n/a
Max Landing	63,500	64,169	64,170
Max Takeoff	71,000	73,800	n/a
Max Payload	18,146	17,176	17,635
Cruise Speed	185 Mph	184 Mph	185 Mph

The Carvair in a freighter role presented somewhat of a problem for the loader operator. The pallet loader was lined up at the same level of the floor. As the pallet entered the aircraft the nose would drop several inches grounding the pallet. The loader operator had to anticipate the drop depending on the weight of the pallet and adjust accordingly. The problem was often compounded when the pallet was pushed in past the aircraft CG. The nose would then rise and the operator had to use care not to send the tail down hard on the stand. The basic rule was never push the first pallet in past the second position until the next pallet was loaded. The reverse problem existed when unloading. As soon as the first pallet was on the loader the nose would rise on the tail-heavy aircraft. The cargo handlers would now have to push the pallets uphill. When the handlers got to the last pallet they were pushing uphill until they passed the CG. As soon as the pallet passed the CG the nose would drop and the loader operator would have to be alert to lower the loader platform to match the floor level of the aircraft. Since this was a time before loaders were equipped with an automatic leveler, the operator had to be quite skilled and anticipate the constant need to adjust the loader.[9]

Aviation Traders became aware early on of the difficulty of loading an aircraft with the floor level nine feet above the ground. A sloping ramp similar to that used on the Bristol was not practical because of the extreme length required. Engineers developed a system using the Bristol ramp with an additional section at the higher end. It served as the prototype for developing the "on-board" ramps for the Interocean Carvairs. They proved to be heavy and awkward and were used intermittently. It soon became evident that ramps for the Carvair were not practical.

A secondary business was actually created to design and produce a suitable "lifting" type loader. Several types of loaders were evaluated including a hydraulic scissors type, a four-jack screw, and a cable type. Each new Carvair was delivered with an ATEL "Hylo" loader. The design and building of the first ATEL loader took considerably more time than first anticipated. Officials decided a second vendor should be found to produce an inexpensive unit without hydraulics in the event ATEL could not complete a unit in time. Alastair Pugh, who was the Product and Development manager at Channel Air Bridge, negotiated with Little Green Engineering named for the community where it was located. Little Green's chief engineer, Heath Robinson, designed a low cost chain drive unit that resembled a large bed. Although it appeared rather clumsy it worked quite well. Jokes were made about it since it was not little or green but bright yellow. It proved to be quite dependable and many were still in use along side the ATEL hydraulic scissors loaders at the end of the Carvair era. The third type tested was cable operated and not self-propelled, which proved difficult to operate.

The ATEL "Hylo" Mark II loader was powered by a Coventry Victor engine with a top speed of 10 mph and weighed 3.5 tons with a lifting capacity of five tons. The platform was six feet nine inches wide by 16 feet eight inches long and raised to a maximum height of 11 feet six inches and lowered to 18 inches off the ground. It required additional ramps to allow the car to be driven on the platform. With the ramps attached the overall length of the loader was 28 feet seven inches. The Carvair indirectly demonstrated the versatility of ATEL generating contracts for other loaders, 747 baggage pallets and the production of Jet Bridges under contract with Aviabridges.

4

Engineering

Fabrication

With the success of the first Carvair and potential orders the program was underway with great optimism. The large stock of spares acquired from Resort Airlines along with three C-54's contributed to keeping cost in line. The total number of projected aircraft conversions was not known at the outset restricting expenditures for metal forming production tools. Each skin panel was hand formed on a wheeling horse by very skilled craftsman. This is no easy task considering the complex angles and curves of the Carvair nose. Because of the eventual low sales volume of only 21 aircraft, all skin panels were hand formed using the wheeling horse, validating the statement that the Carvair was hand built.

ATEL tried whenever possible to use "Off-the-shelf" parts. This is evident from the latch on the cockpit entry trap door and the roof emergency exit borrowed from the Bristol.[1] The brakes and main gear oleo struts were taken from the DC-6 plus the windscreens, yokes, and instruments robbed from discarded DC-4 nose sections.

The nose production was set up at Southend in a workshop adjacent to the rear of the main hangar. The completed units were moved by road to Stansted by a special transporter on Sunday mornings to avoid any traffic problems. An engineer rode in the cockpit to warn the transporter driver of any low objects. Technicians from the electricity board and telephone linesmen would accompany each nose delivery raising or disconnecting overhead lines to allow clearance.

The nose section took nearly five weeks to produce. They are built in two sections split longitudinally along the cockpit floor. The upper and lower main halves are built up of stringers and half frames on a jig. The top half is completed first with the cockpit floor then mated to the bottom. Both the upper and lower sections were not skinned down to the point where they are joined. After joining the two halves, the wide area at the mid height level is filled with skin panels, which were hand wheeled by metal craftsmen to the correct contour.

The nose is skinned back to station X270. Stringers extend at random 18 to 24 inch lengths past the rib for attachment to the DC-4 fuselage. The upper half stringers are extended an additional seven and a half feet to form the rear hump ending at station X360. After attachment to the fuselage, this area is skinned with hand-formed panels.

The production of the nose was set up on 01 October 1960. After the first conversion of G-ANYB, all nose units were built in groups of four at Southend. At least that was the plan, however only 23 were built (Three groups of four plus one before the change to raised control runs. Two groups of four plus two with elevated controls before production closed). The last two noses were never fitted to an aircraft, although parts were used from one of them for repairs. Adding to the confusion is the raised ceiling option. The first 12 ships were built with standard control runs (ships six and eight were later retrofitted with elevated control runs). All conversion after 12 were to have elevated control runs and high ceilings. The exception is 15, and 17, which were built on speculation using the last two earlier produced noses from stock.

Top: The upper and lower main halves are built up of stringers and half frames on a jig. The top half is completed first with the cockpit floor, then mated to the bottom. A completed upper half of the nose sits in the jigs at Southend awaiting the installation of reconditioned windscreens (courtesy Guy Craven). *Bottom*: The lower half of the nose as viewed from the rear. Originally each unit was planned to be built inverted, then righted, but as work progressed they were completed upright, then mated and skinned (courtesy Guy Craven).

The stringers on the lower portion of Carvair 19 are attached to the DC-4 fuselage. The upper half stringers will be extended an additional seven and a half feet to station X360 to form the rear hump; then the area can be skinned with hand-formed panels (courtesy Lindsay Wise).

The last three units were transported to Stansted and stored awaiting orders. After the line at Stansted was shut down a nose was needed back at Southend for number 21. The extra noses, 22 and 23, were transported back to Southend and stored in the M.F.O. hangar where they remained until at least 1971. Aviation Traders engineers believed that if a Carvair was ever damaged in England that it would be possible to rebuild it using one of the spares. This plan was seriously considered in March of 1971 when G-APNH was damaged at Le Touquet France when the nose wheel collapsed on landing. After review it was concluded that cost would be too prohibitive and -APNH was broken up. Subsequently with the Carvair long out of production the two noses and the transporter were declared surplus and moved outside with other airframes waiting their day with the scrap man.

Fortunately the two nose sections were not immediately reduced to scrap and were still on hand at Southend in May of 1972 when Eastern Provincial Carvair eight, CF-EPV, was severely damaged on landing at Gander. Parts were salvaged from one of the noses and shipped to Eastern Provincial Airlines in Canada to complete the repairs.

The random purchases of C-54 airframes for conversion resulted in aircraft arriving in extremely poor condition requiring additional work. Aviation Traders intentionally acquired very tired airframes as they were going to be totally stripped out. However, some were in considerably worse condition than others. Only two Carvairs were converted from postwar DC-4-1009 airframes. All the others were originally built as C-54s for the military and later upgraded to DC-4 standards when surplused after the war. The surplus C-54 airframes were lighter than the postwar DC-4s by 2600 pounds. The C-54s came in five variations, with four and six wing fuel tanks. Some had been fitted in the past with fuselage tanks. There was no standard for fuel line plumbing because of operators modifying routings and fitting A, B, models with E model wings.

The airframe overhauls during Carvair conversion by ATEL proved to be quite a process.

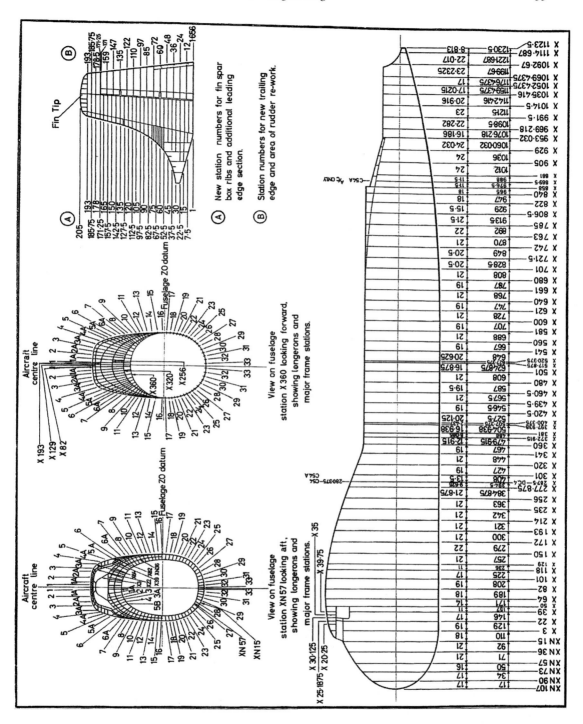

Aviation Traders maintained a policy of converting only C-54 airframes because DC-4-1009s were 2600 pounds heavier and some stringers and ribs did not match as noted on station number drawings at positions X287-5 and X869-5 (courtesy Aviation Traders Ltd.).

Sometimes information was established about the aircraft that had not been previously recorded. One aircraft was discovered to have suffered a wheels-up landing incident that was never recorded resulting in considerable repair work before it could be converted. Aviation Traders experienced multiple problems early on in the program finding suitable aircraft to convert.

It is noteworthy that at one point when reasonably priced C-54 type DC-4 airframes could not be found, ATEL did not consider converting DC-4-1009 G-AOXK, which it owned. It was purchased in July 1963 with the notion it would be used for conversion. After engineers pointed out the problem of using postwar DC-4s, it was sold to Williamson Diamond Mines in the same month. In addition to being 2600 pounds heavier reducing payload the DC-4-1009 has some stringers and ribs that do not match requiring additional engineering in those areas as noted on the station numbers drawing. Aviation Traders maintained a policy of only converting C-54 airframes until confronted with the Australian Directorate of Civil Aviation (DCA) ruling involving Ansett. The ruling forced ATEL to accept two Ansett DC-4-1009s for conversions 19 and 20.

All airframes were stripped of paint during conversion and most were repainted at Stansted after Conversion. Aviation Traders contracted to paint specialist Reg Taylor to apply all lettering and artwork at both Stansted and Southend. Taylor was a gifted craftsman, who designed many individual logos and recommended scheme layouts. Drawings were created covering paint schemes and titles but were subject to the painter's creative interpretation. The Channel Air Bridge ships were fairly uniform but subtle differences began to appear with first British United colors. By the time the two-tone band up the tail was adopted there were noticeable differences in each aircraft. The design, which was an interpretation of sweeping jet fins, was awkward on the Carvair.[2]

All of the Carvairs known as "Bridge" aircraft and originally delivered to Channel Air Bridge, British United, and British Air Ferries were fitted with the Decca navigation log system. The Chief Engineer for Decca was Gordon Thomas whom Laker had worked with at Short Brothers as a young man. The Decca unit was installed on the right center console behind the throttle levers. As the Carvair aged and was sold off by the original carriers to second level operators it was updated and required engineering changes and modifications.

The three Aer Lingus aircraft were modified to begin cold weather operations with Eastern Provincial. Twin heaters were installed, the single intake duct was removed from the nose door and insulation added (courtesy Tony Rogers).

The aircraft were continually being upgraded as demonstrated with the three Aer Lingus ships. The first two were built without the "Rolamat" floor system and raised ceiling then retrofitted for horse charters and palletized cargo with the delivery of the third aircraft. These three ships were not designed for cold weather. After purchase by Eastern Provincial in 1968 they were returned to ATEL at Southend for cold weather modifications, which included doubling the heater capacity and nose door insulation to reduce heat loss in flight.

The three EPA aircraft were fitted with a unique feature of a hatch in the cargo bay floor forward of the nose gear. It was installed for mail and parcel drops at small settlements and construction sites where there is no adequate runway. However crew and maintenance personnel soon began using it for easy entry into the aircraft. It is reported that EPA aircraft were configured for up to 70 passenger by adding a 48-seats ahead of the existing 22-seat cabin. In reality EPA Carvairs had 46 seats as verified with ATEL and in EPA advertising materials.

Disassembly and Conversion

Only three conversions were done at Southend. Once the program got underway 18 conversions were built at Stansted. High interest and many inquiries received by Aviation Traders led to the belief there would be a large number of conversions. The ATEL facilities at Stansted were not sufficient for a large conversion line. The old RAF T2 hangar would only accommodate three aircraft with the outer wings removed while the dismantling work took place outside on the north side of the ATEL hangar.

After arrival the C-54s were de-fueled; engines, outer wings, and control surfaces were removed. Next the tail-planes and fin were removed and a trestling device with wheels similar in size to the main gear was attached through the tail spar. Once in place and secured the DC-4 nose could be removed. It was first removed in a clean single station cut at station X270. Then a cut was made at station X270 mid-height of the fuselage horizontally above stringer 18 back to

Carvairs two, three, and four under construction on 26 January 1962. They were the first produced at Stansted with number three, G-ARSF, in the foreground. Only six months after completion it was lost at Rotterdam in a tragic accident (courtesy Guy Craven).

Carvair seven on 16 November 1962 with the nose, which arrived five days earlier, aligned for grafting to the fuselage. Conversion number six for Aer Lingus is in the background (courtesy Guy Craven).

station X360. At that point the vertical cut resumed around the top of the fuselage. The skin was removed around the bottom between stations X270 and X301 leaving the stringers exposed. A temporary frame unit was installed across the cut up to floor level at station X270 to hold the stringers in place and prevent distortion. After the new nose was mounted half frames were mounted above stringer 18 to allow for skinning to fill the area between the hump and DC-4 fuselage.

The bare fuselage including center wing was mounted on support stands in the hangar after removing the DC-4 nose. Care had to be taken to prevent any distortion in the fuselage with the nose removed. The old DC-4 cockpit was robbed of control columns, pedestal, pedals, windscreens and frames, seats, instruments etc. As part of the cost saving plan all were inspected and overhauled for re-installation in the elevated Carvair cockpit.

Windscreens

The windshield frames were quite difficult and costly to produce. To reduce cost ATEL technicians recovered and re-condition the DC-4 frames and reinstalled them in the new Carvair nose. After being cut away, the DC-4 nose sections were transported to Southend to be stripped. A team of three mechanics led by a technician identified as "Ned" very carefully drilled out the rivets to recycle the windscreens. After the noses were declared scrap some were actually transported back to Stansted where a melting pot was set up.

All of the DC-4 aircraft control and support systems (heater, fuselage oil tank, nose landing gear etc.) that were mounted forward of station X270 (lower half of fuselage) and X360 (upper half of fuselage) had to be redesigned or deleted.

Oil Tank

One example of a deleted item for the Carvair is the C-54/DC-4 fuselage oil tank. The DC-4 has a 50-gallon (41.3 Imp) fuselage oil storage tank located in the upper cargo compartment with a four-position manual selector valve and lines below the cabin floor. This unit allows the crew to transfer oil from the central tank to each 20-gallon (16.3 Imp) nacelle tank if supply drops below one half. The transfer lines and selector are permanently factory installed on all C-54/DC-4. The fuselage oil tank components were shipped loose with the aircraft giving the owner the option to install the system. The tank is deleted from all ships converted to Carvair standards and factory plumbing and selector capped and left in place. As far as can be determined no Carvair was ever permanently retrofitted for the original owner. The exception is Carvairs five and nine, which were modified many years after production by Bob McSwiggan for Custom Air Service. McSwiggan utilized the transfer lines on N83FA by mounting a 40-gallon tank in the rear lower cargo compartment adjacent to the Anti-Icing Alcohol tank and routing a line to the selector valve. On N89FA two tanks were welded together to create an 80-gallon oil tank. These are the only two examples to ever utilize the option of a central oil tank.[3]

Cockpit Design

The cockpit was re-designed and modernized for a two-man crew with a jump seat installed behind the pilots. An engineer was not required but often assigned to the aircraft to reduce the workload for the pilots since the Carvair definitely required both pilots.

A navigators station, two aft-facing sleeperette seats or extra crew seats were available for use on long range charters and special missions. A small galley and lavatory was also designed for the cockpit of the long-range version. Configuration drawings were done but no records exist of a cockpit lavatory option being installed on any aircraft. Any crew lavatory is most likely a small portable toilet mounted in the nose door similar to Interocean aircraft four and five or seven and nine as cargoliners. Carvair 11, as the only long range Carvair, was fitted with a cockpit galley. Ships four and five were fitted with a cockpit galley position for a coffee pot and hot plate when built for Interocean as freighters.

Nose Door

There are three different nose door bulkhead designs. The original or standard is found on ships one, two, three and six through 18. These bulkheads are uniform around the circumference with alignment pins near the six, seven, 10 and 11 o'clock positions. The second design was fitted to ships four and five. The bulkhead is notched and squared at the two and 10 o'clock position to allow for loading of military trucks. The alignment pins are located near the six, seven, nine and 11 o'clock positions. The third design is fitted to ships 19, 20, and 21, which are pure cargoliners. The bulkhead is squared at the bottom tapering to floor level to allow for loading of the 88 × 108 (P9) cargo pallets. The alignment pins are located near the six, eight, 10, and 12 o'clock positions.

The one-piece clam shell door is hinged on the left side and will open through 180 degrees. It is operated by two cylinders connected to a hydraulic hand pump located on the right side of the fuselage behind the doorsill. The main air intake for heating and ventilation is located in the center of the door with a duct that runs upward to engage the other half when the door is closed. A shelf is fitted in the door at cargo deck level. This shelf was used to mount the chemical toilet for the crew on ships so equipped. Early nose doors were all metal skinned. A di-electric patch was later fitted to the lower half.

Nose Gear

The nose gear required complete relocation at a position forward of the DC-4 mounting with the range of up travel and stowage re-engineered. The DC-4 nose gear retracted to a position 19 degrees above the horizontal line of the fuselage. The nose gear bay would no longer accommodate the assembly because of the level of the cargo bay floor. As a cost saving measure the actual DC-4 nose gear was reused with changes in geometry. The nose gear retraction travel was reduced from 94 degrees to 75 by lowering the yoke mounting to the fuselage by four inches. A steel tube was installed in the piston of the actuating strut to limit travel. The re-design worked well placing the retracted strut parallel to the fuselage horizontal datum (ZO) line and below cargo floor level. However, the lower half of the strut and nose wheel protruded outside the fuselage. Blister type nose gear doors were designed to cover the strut assembly when retracted.

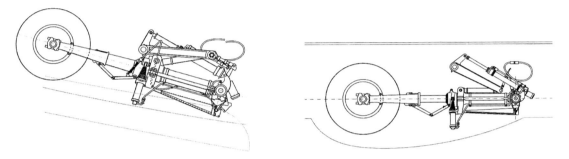

The DC-4 (left) nose gear retracted up 19 degrees above centerline and stored in the fuselage. The ATL-98 (right) retracted parallel to the horizontal datum (ZO) line with the lower half outside the fuselage. The retraction travel was reduced from 94 degrees to 75 by lowering the yoke mounting to the fuselage by four inches. A steel tube was installed in the piston of the actuating strut to limit travel. Blister type nose gear doors were designed to cover the Carvair strut assembly when retracted (courtesy Michael Zoeller).

Heating System

The DC-4 heater system was re-designed to supply air to both cockpit and rear cabin. The twin Janitrol system was removed, a new air intake system designed into the nose door and a 100,000 BTU Dragonair combustion heater installed to supply warm air to both the flight deck and rear passenger cabin. The engineering of the intake and heater ducting is similar to the original DC-4 with the intake in the center of the nose. It is also similar to the ducting in the nose door of the Bristol MK.32. The duct is routed along the left side of the cockpit floor, between the electrical panel and wall to station X320. It then makes a 45 degree turn to the center top of the fuselage and connects to the existing DC-4 duct along the cargo bay ceiling to the rear cabin and down the bulkhead. The warm air flows through floor level grilles at the front and rear of the cabin. Cool air is supplied through louvers at each row of seats. On ships equipped with a forward passenger cabin (Aer Lingus) an additional duct extends forward supplying warm air on the starboard side.

Aviation Traders re-engineered the heater system for the three Aer Lingus aircraft when they were purchased by Eastern Provincial. The heater capacity was doubled for Arctic operations by installing a twin unit in an area behind the flight deck. The EPA Carvairs six, eight, and 14 have a black exhaust track at the rear of the flight deck hump on the port side. A small stack extends through the aircraft skin at the forward end of the track. The exhaust was previously vented through a vent below the port side cockpit window. The nose door is also insulated to reduce heat loss in flight and the air intake and duct is skinned over. Twin intakes are position on each side of the fuselage behind the flight deck to supply the new high capacity heaters.

Cabin Seating, Lighting, Galley, Lavatory

Seating is quite cozy on the Carvair with only 50 square feet for 22 passengers. The rear DC-4 bulkhead is moved back 48 inches to accommodate the rear cabin and maintain five car positions. The cabin seats were manufactured by Flying Services Engineering and Equipment Limited in Chesham Buckinghamshire. The seats are non-reclining with a 30-inch pitch, 15½ inch width and stressed for 9g loading. They are built in two (39.25 lb), three (57.75 lb), four (95.8 lb) and five (90 lb) seat units that attached to track floor fittings. One armrest is fitted to each seat with the cabin sidewall armrest mounted on the wall. All aircraft that were later fitted with additional forward cabins used a different type of seat mounted on rails with dual armrest and ashtrays.

A number of passenger, car, and cargo configurations were offered and operated until the end of the car-ferry era. The original configuration ATEL offered was 23-pax (actual 22) 5-car, which was installed on the Channel Air Bridge and BUA aircraft. In 1964 BUA also used a 55-pax 3-car and proposed a 32-pax cargo-combi. BAF operated at least two ships in 1969 with 34-pax (2 cabin) 4-car set up. Under Keegan, BAF operated a 17-pax 5-car/40-pax (2 cabin) 3-

Seating is quite cozy in the Carvair cabin with only 50 square feet for 22 passengers. The rear DC-4 bulkhead was moved back 48 inches to accommodate the cabin and allow for the loading of five automobiles (courtesy Guy Craven).

This 55-seat three car or cargo combi was created by adding a 43-seat cabin ahead of the 22-seat rear cabin. The lavatory and entry door is beside the two flight attendants. The lighted ceiling and cabin walls are permanently attached even though the forward bulkhead is moveable and seats can be removed for standard car-ferry (courtesy BAF).

After returning from the Congo, Interocean opted in 1963 for a 17-seat cabin when its two Carvairs were upgraded from freighters to car-ferry configuration (courtesy Interocean/Leif Hellström collection).

car/and 65 pax two cabin bulk cargo (22 rear cabin 43 forward and later 17 rear 48 forward). Interocean opted for 17-seats when the passenger cabin was installed after returning from Congo service. Aer Lingus operated a 34-pax 4-car (2 cabin with 22 rear — 12 forward) prompting the addition of an additional emergency exit in the forward cabin. Eastern Provincial operated and advertised 46 passengers. A 70-seat configuration cargo combi (2 cabin with 22 rear — 48 forward) was considered but no record exist of it been installed. This arrangement is the "Keegan" BAF 17 rear 48 forward cabin with the rear increased to 22.

In May 1964 BUA began experimenting with a 55-seat three-car version for the long-haul deep continent routes. The passenger cabin paneling and lighted ceiling were permanently attached and extended into the cargo area with a removable forward bulkhead. As bookings dictated the aircraft could be set up for the 55-seat 3-car Ostend service or 22-seat 5-car standard car-ferry. A padded snap-in fabric liner was fabricated to cover and protect the walls when the moveable bulkhead and seats were removed for transporting five cars or cargo. The lighted ceiling panels were permanently installed and remained exposed in either configuration. Cargo handlers were instructed to use care and not allow cargo to contact the ceiling panels when loading.

Cabin lighting consist of three concealed overhead units on each side of the cabin. There are no individual lights for each seat although the overhead lights can be dimmed.

A small galley is fitted in the forward passenger cabin bulkhead opening into the passageway across from the entry door. It has storage for bottles, glasses and box lunches and a

The Carvair pioneered modern "Quick Change" interiors. This 55-seat interior has been converted to 22 seats by removing the forward bulkhead and seats. Padded liners are snapped in to protect the cabin walls from opening automobile doors. The permanent passenger cabin lighted ceiling is visible at the top (courtesy Terry Leighton/Aviation Traders).

The lavatory protrudes into the cargo compartment with the gravity flow wash basin water supply tank mounted seven feet up between stations X581 and X600. The channel trays for the elevator control cables run along each side of the fresh air and heater duct in the overhead. The floor level fresh air ducts for the passenger cabin are visible on the rear bulkhead (courtesy Interocean/Leif Hellström collection).

two-gallon coffee dispenser. The Aer Lingus ships have stainless-steel work surfaces, Steibel hot containers for tea and coffee, and a waste disposal unit. Those ships are also fitted with Stewardess-call chimes.

The lavatory is forward of the entry door in a small corridor that extends into the cargo compartment. The door opens inward to the right with the washbasin on the left. The toilet is a Mark III re-circulating unit mounted on the forward wall, which is supplied by C.F. Taylor Limited of Wokingham, Berkshire, England. It does not require external water supply and dispenses disinfectant each time the lid is raised. The unit is provided with carrying handles for removal and emptying. The wash basin is fitted with hot and cold taps and is supplied by a 12 gallon (10 IMP.) tank. The gravity feed tank is mounted seven feet up between stations X581 and X600 in the cargo compartment. The hot water is obtained by passing the cold water through a ½ kilowatt G.E.C. storage heater located under the wash basin. Waste water from the basin passes to an overboard outlet.

The flight attendant is seated forwarded of the cabin entry door adjacent to the lavatory. Across from the flight attendant seat is a small sliding door accessing the cargo compartment or curtain on aircraft with a forward cabin. After takeoff the flight attendant is required to enter the cargo compartment to inspect that the vehicles are secured and there are no fuel leaks. After inspection an "All Clear" call to the Captain is required.

Electrical System

The Air Registration Board required a re-design of the DC-4 electrical system. The basic generating complex and components was allowed to remain in the section of the fuselage behind the cut. The distribution system remains a 24 to 28 volt direct current single-wire system in which the aircraft structure forms the negative. It was upgraded to main and essential bus-bars with isolating controls. The system is segregated and protected in the cockpit junction box by means of a current limiting fuse and bus-bar relay. Each engine driven generator feeds the main bus-bar and has its own current limiting fuse. The essential bus-bar is connected to the batteries by a battery relay. This allows the essential equipment to be fed from either the battery or generators.

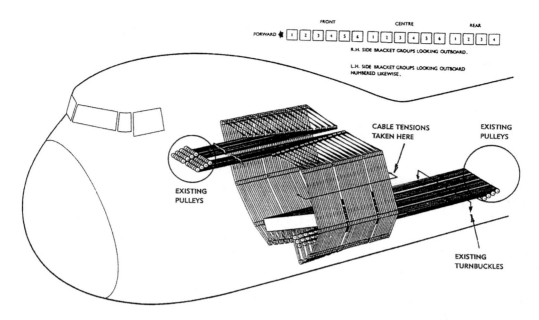

The control cables turned 90 degrees over to the sidewall before running down the wall and back under the cargo hold floor, changing directions six times before following the original routings (courtesy Aviation Traders Ltd.).

Control Cables

The original cockpit equipment easily re-installed in the new nose because the design is unchanged from the original DC-4. The exception is the 150 control runs, which required re-routing under the raised cockpit floor to connect to the original control cables.

Since the aircraft is un-pressurized, the structural design is straightforward following the basic standards of the C-54. The size and height of the new nose created mechanical design problems that were not anticipated. The controls of the C-54 are mechanically operated running under the floor leaving the cargo compartment unobstructed. The same design had to be achieved with the Carvair to enable vehicles to load through the front under the flight deck. This is achieved by running the controls cables under the flight deck floor to a pulley harp. The cables turn 90 degrees over to the sidewall then run down the wall to pulleys back and under the cargo floor changing directions again running toward the rear of the aircraft. All of this is achieved ahead of the wing with the cable changing directions six times before following the original routings through the wings.

The control cable re-routing requires hundreds of brackets and six hundred additional pulleys. To accommodate all of the pulleys and brackets the fuselage height was increased an additional 18 inches between the flight deck floor and cargo bay. The average increase in each cable tension is 1.5 pounds per cable with no perceptible loss of motion.

The relocating of the rear passenger cabin bulkhead required changes in pulley mounting stations in the tail for rudder and elevator control. The DC-4 pulley mounting located in the floor at station X905 was strengthened and relocated to the backside of the rib at station X953. This was done in order for the cables to clear the passenger cabin bulkhead, which was moved 48 inches rearward.

The exception to the under cargo bay floor re-routing is the six cables that run to the rear of the aircraft for elevator and tab control. These cables run directly back from under the new

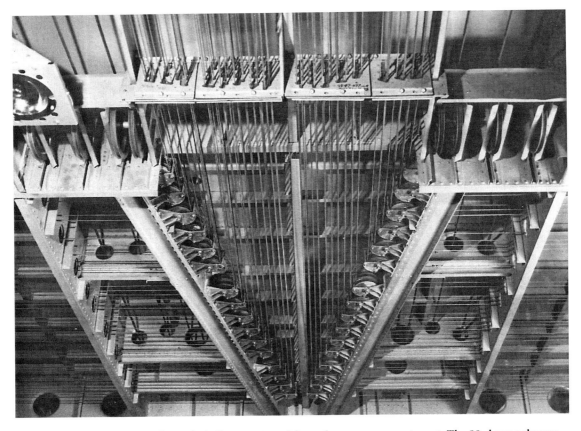

The pulley harp under the cockpit floor as viewed from the cargo compartment. The 90-degree change in direction routes the control cables over to the fuselage wall where they are routed down and back under the cargo floor (courtesy Guy Craven).

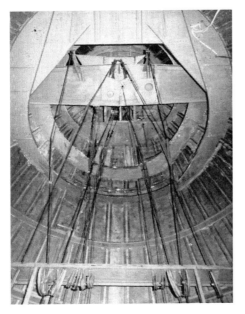

cockpit floor through channel beam trays and along the roof of the cargo compartment to the rear of the aircraft. After passing through the relocated bulkhead at the rear of the passenger cabin at station X905 the cables are routed through pulleys in the empennage ceiling down to the floor. The elevator cables pass through the re-positioned pulleys at station X953 to a mounting bar that was fabricated to attach to the rib in the floor. The cables are routed in this manner to reduce the number of pulleys required and duplicate the same amount of pressure required for operation on the DC-4. The rudder trim cables remained in the DC-4 routing under the floor. They were routed from the cockpit mount under the cockpit floor through the pulley harp and down the walls to under the cargo bay floor.

This original ATEL design of running the elevator

A composite photo of the rudder and elevator pulleys of the DC-4, which are located on a mounting bar in the floor at station X905 (author's photograph).

Top: To clear the Carvair rear cabin bulkhead that is relocated 48 inches rearward, a strengthened mounting bar is relocated on the backside of the rib in the floor at station X953. Unlike the DC-4 the elevator cables come down from the ceiling (author's photograph). *Bottom*: The Carvair elevator cables run directly back from under the cockpit floor under the hump through channel beam trays and along the roof of the cargo compartment to the rear of the aircraft. The fresh air heater duct runs between the cable trays to the rear of the aircraft (author's photograph).

cables along the ceiling worked quite well. However it left unused space in the rear third of the hump behind the cockpit. At the request of Aer Lingus, engineers designed a new routing for the cables to utilize this space by raising the ceiling in the cargo compartment. This allowed for the standing up of cargo or a higher ceiling for horses. Nine aircraft were fitted with this feature, seven on the line and two retro-fitted. Conversion 13, G-ASKN, is the only British United

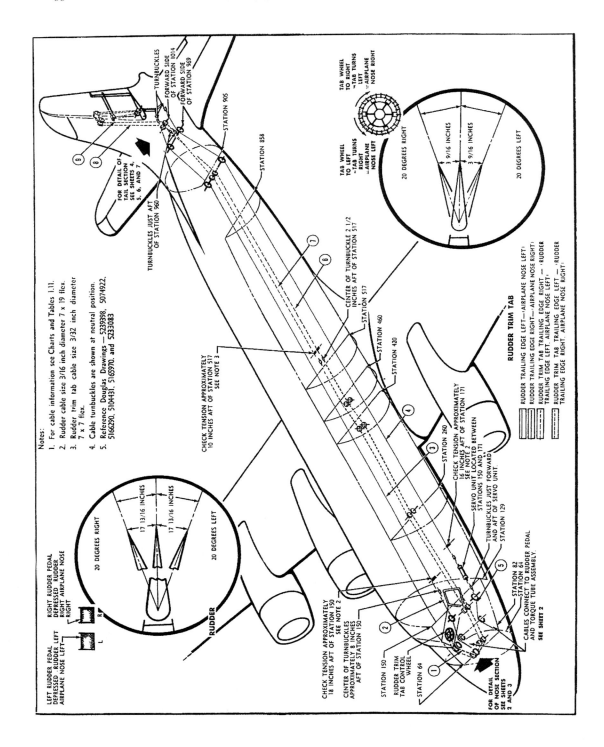

The standard C-54 control cables are mechanically operated, running under the floor, leaving the cargo compartment unobstructed as shown in the rudder cable routing (courtesy Aviation Traders Ltd./Douglas).

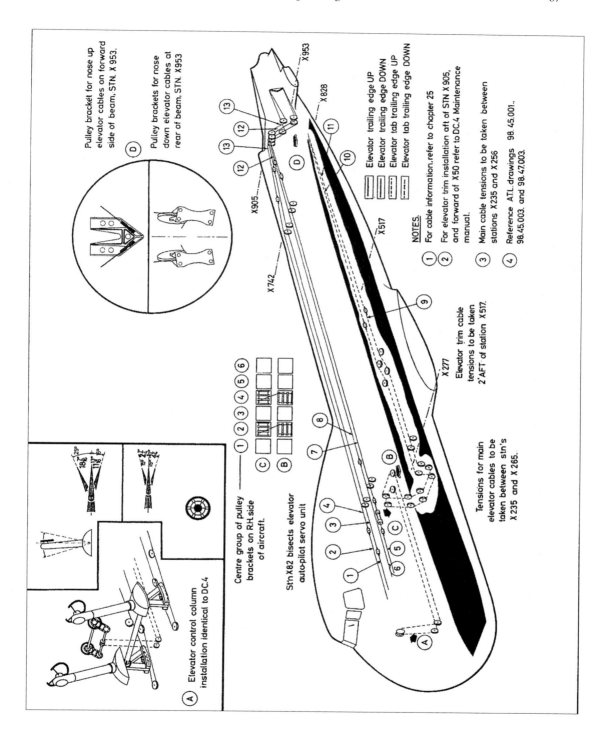

Only the elevator control cables are routed straight back from the cockpit through the hump to pulleys at station X905 behind the relocated rear cabin wall. By contrast the rudder cables are routed through the pulley harp and under the cargo compartment floor (courtesy Aviation Traders Ltd.).

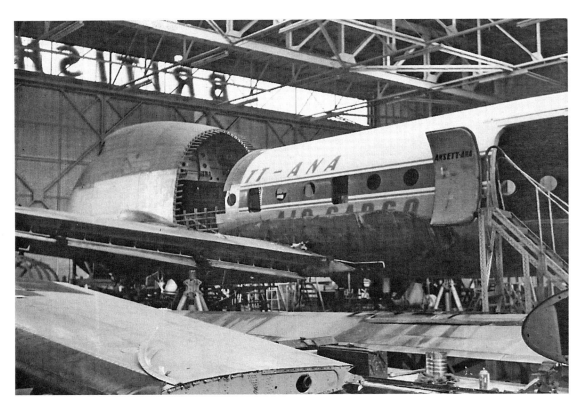

To utilize the space in the hump to transport horses, Aer Lingus engineers and later Ansett requested a re-designed raised ceiling requiring the re-routing of the control cables. The pulley-mounting device is visible on the rear upper deck wall, routing the cables up to follow the contour of the ceiling. Nine aircraft were fitted with this feature, seven on the line and two retrofitted (courtesy Ken S. Elliott).

aircraft built with the raised ceiling. Aer Lingus Carvair 14 was built with raised ceiling while aircraft six and eight retrofitted. The two AVIACO aircraft 16 and 18 plus the three for Ansett rounded out the group. Conversion 17 was unsold when built and completed as cheaply as possible without the raised ceiling.

All the airframes with the raised ceiling in the cargo bay behind the cockpit are obviously different when viewed inside the cargo compartment. The cables are routed up at station X235 as they exit the raised cockpit floor. They follow the contour of the raised ceiling in a single tray down the hump to station X360 where they follow the ceiling of the cargo bay to station X785. At that point they pass through the overhead crossbeam and split into two trays the same as all other conversions.

Vertical Fin and Rudder

The DC-4 fin proved to be inadequate to maintain proper yaw control. The original design called for endplates on the tail planes for directional stability. After wind tunnel test and study it was determined that excessive pressure would be put on the end plates requiring additional stress engineering. The units would not be needed if the vertical fin were extended to the same profile as the DC-7 unit. This proved to be more cost effective and required considerably less engineering. Because of the additional weight and stress exerted by the larger fin, reinforce-

Chief Aerodynamicist Brian Kerry demonstrates the characteristics of the redesigned vertical fin with increased height in 1959. The profile resembles the DC-7 fin (courtesy Brian Kerry).

ment was required in certain areas of the empennage. The extra strength is provided with additional or heavier stringers and double skin.

Fuel Systems

All conversion airframes and power plants received a major overhaul and were zero timed. Aviation Traders policy was to purchase the lowest cost units for conversion. Some of the DC-4s were purchased for the airframes only. The engines were removed and returned to the previous owners. Many of the aircraft purchased for conversion arrived at Aviation Traders with different series Pratt & Whitney R-2000 engines. The Carvair standards specify the R-2000-7M2 series. This engine differed only slightly from the standard R-2000. Primarily the -7M2 has a two-speed supercharger that is locked in the low speed condition. Particular attention was given to the fuel line plumbing system in relation to the choice of series of R-2000 engine specified. There were some isolated incidents where DC-4s had experienced fuel starvation problems when older aircraft were equipped with later model engines or retrofitted with wings from different series altering the fuel line plumbing.

The aircraft used for Carvair conversion were selected at random creating additional problems. The fuel systems on the C-54s were upgraded and changed on the different models as they were converted to civilian DC-4s. Consequently the C-54 "A," "B," "E" and DC-4-1009 models were not the same presenting many problems. The C-54A fuel line plumbing was originally mounted on the front of the wing spar. As running changes were made the lines and valves were located in many different locations on the spar. This was due in part because of the six and eight tank configurations and some models with fuselage tanks. The Air Force converted some C-54s to eight tank systems with rear spar mountings of valves and lines. Some operators also made modifications, which resulted in mixed systems with valves, lines, and fittings mounted on both front and rear spar.

The original C-54A system had provision for four auxiliary fuselage mounted fuel tanks. The C-54B had two of the fuselage tanks removed and integral tanks in the outer wing panel. The C-54E had collapsible tanks in the inner wings and no fuselage tanks. The Navy version C-54 designated as R5D-1 from which Carvair five was converted had another system. That system was similar to the C-54B with certain modifications to meet navy standards. Ships 19 and 20 were DC-4-1009s, which have an even different system with the least fuel capacity of only 2,868 U.S. gallons. The C-54A capacity is 3,620 U.S. gallons, C-54B at 3,740 U.S. gallons, and the C-54E at 3,520 U.S. gallons. Aviation Traders had no choice but to standardized the fuel tank plumbing systems for the Carvair with specifications of a six-tank model with capacity of 2,390 Imp. (2,870 U.S.) or eight-tank long range model of 2,993 Imp. (3,595 U.S.).

Since ATEL engineers were confronted with an extremely time consuming problem they adopted the C-54A front spar supply system for all Carvair conversion. As if the numerous series of C-54s used for conversion with different fuel tank configurations was not enough, over the years operators had re-fitted A, B, and E wings to the different series of airframes. Some of the DC-4s used for conversion had also been fitted with wing hard points for attachment of external cargo or JATO bottles. This modification also had to be considered when developing a common wing configuration for the Carvair.

When engineers decided that all the fuel line plumbing, fixtures, and valves had to be standardized it was concluded that the possibility of fuel starvation could be eliminated by a uniform system. The chance of failure would be highly unlikely even if the new owner substituted different series engines. The DC-4 fuel starvation problem is reviewed further in the history of Carvair five.

Carburetor Air Intake

The carburetor intake on the early DC-4 has several different profiles. The majority of C-54s built and all DC-4-1009s have the non-ram air type scoop. The system is identified by a long fairing that extends from the carburetor air intake above the cowl flaps back along the length of the nacelle and is the most common. It was originally designed for hot and cold air plus a filter feature controlled by a lever on the lower left of the console in the cockpit and is presumably for the military to operate in high dust conditions. Installing of the filter, cockpit lever, and rigging added weight and also required an engineer to check and change filters at prescribed intervals. The filter system was removed from aircraft so fitted when they were converted from military C-54 to DC-4 standards. Some C-54 aircraft were built with the ram air short scoops. After the filter system was removed from aircraft so fitted, the revised non-ram air system became basically the same as the ram air. Both systems have hot and cold air controlled in the cockpit by four separate levers with blue knobs between the fuel selectors on the forward pedestal. They are rather awkwardly placed low and forward of the pitch levers.

The Carvair was built with the idea of keeping it simple. Some DC-4s arrived for conversion with the short ram-air carburetor intakes but only five retained them for life of the

Opposite—top: After the filter system was removed and blocked off, the revised long fairing non-ram air intakes were essentially the same as the ram air system (courtesy Aviation Traders Ltd.). *Bottom:* The short fairing ram air intake only appeared on six Carvairs, three Aer Lingus, two AVIACOs and one Channel Air Bridge (G-ARSD). After 18 months of service G-ARSD was converted to the long revised non-ram air type fairing (courtesy Aviation Traders Ltd.).

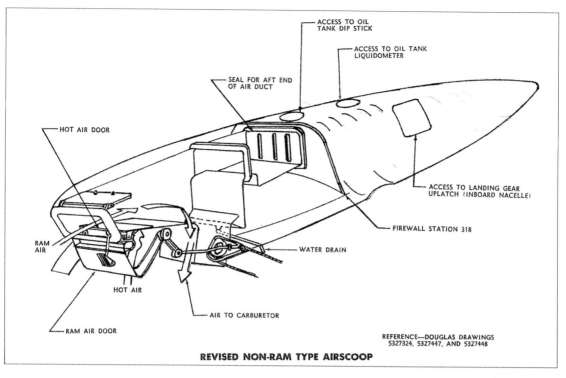

ACCESS TO OIL
TANK DIP STICK

ACCESS TO OIL TANK
LIQUIDOMETER

SEAL FOR AFT END
OF AIR DUCT

HOT AIR DOOR

ACCESS TO LANDING GEAR
UPLATCH (INBOARD NACELLE)

FIREWALL STATION 318

RAM
AIR

WATER DRAIN

HOT AIR

AIR TO CARBURETOR

RAM AIR DOOR

REFERENCE—DOUGLAS DRAWINGS
5327324, 5327447, AND 5327448

REVISED NON-RAM TYPE AIRSCOOP

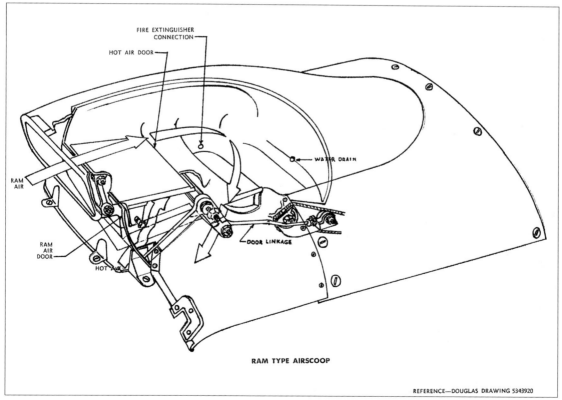

FIRE EXTINGUISHER
CONNECTION

HOT AIR DOOR

RAM
AIR

WATER DRAIN

RAM
AIR
DOOR

HOT AIR

DOOR LINKAGE

RAM TYPE AIRSCOOP

REFERENCE—DOUGLAS DRAWING 5343920

aircraft after Carvair conversion. Most of the C-54s were purchased as airframes only not including engines leaving it to the purchaser or ATEL to supply engines. Conversion eight for Aer Lingus is the first of only two Carvairs converted from the long fairings to short ram-air systems. A total of six Carvairs had this system with short carburetor scoops after conversion, three Aer Lingus, two AVIACO and one Channel Air Bridge. The AVIACO ships arrived with short ram-air scoops and maintained them after conversion. Channel Air Bridge's G-ARSD is the only DC-4 that retained the short carburetor scoops after conversion only to have them changed 18 months later to the long fairings when in for overhaul. The one odd exception to all of the Carvairs is G-ASKN with a hybrid intake with a long forward portion and short taper over the nacelle.

Wing Bob Weights

Another unique engineering feature is ATL MOD 98/5150 "Wing Bob-Weights," which were installed on the British Carvairs used for short range cross channel service. The new nose and vertical stabilizer modifications added 2300 pounds to the DC-4 aircraft dry weight. The wing bob-weights brought the total increase to 2800 pounds. This resulted in the ATOG being reduced from 73,800 to 72,900 for the United States FAA certification issued in September 1962.

All aircraft fitted with the weights are designated as ATL-98-A models. These "Air Bridge Aircraft" used on the cross channel routes were modified for the purpose of increasing the maximum zero fuel weight (MZFW). This would allow the loading of up to 1000 pounds of additional revenue cargo. Lead weights (ATL MOD 98/5150) totaling 250 pounds are placed in the outboard wings in a canister type device, which is accessible through a panel under the wing. On occasion when the Carvair mission was changed, such as long range routes deep into the continent and Mediterranean, the weights were deleted. This is a simple procedure by removing the weights, re-weighing the aircraft and revising the manuals back to standard maximum ZFW.

The weights allowed the airframe to transport more payload while maintaining the same stress loading of the wing. More simply put, when the aircraft lifts off, the wings deflect as they pick up the body of the fuselage. The tendency to stress the wing roots must remain within limits. By placing the 250 pounds of weight far from the centerline of the airframe at the extreme end of the wing, the wing loading remains basically the same. The extra payload is distributed along the length of the wing reducing wing deflection and the root is not stressed beyond limits. The only penalty is fuel since the maximum takeoff gross (ATOG) remains the same. Fuel is not an issue on the cross channel flights because of short duration.

Flooring and Cargo Hold Tie-Downs

A new floor system was designed because of the poor condition of the original corrugated DC-4 floors. The system has 96 D-ring lashing points on all but six of the 46 laminated floor panels. A number of them matched up with the pre-drilled floor beam attachment points of the DC-4.

Aircraft Specifications and General Data

Wing Span 117 feet 6 in (35.82 m). All metal three-spar, 7 degree dihedral.
Overall Length 102 feet 7 in (31.27 m) all metal.

Overall Height 29 feet 10 in (9.09 m).

Area including dorsal fin 202.9 sq ft (18.85 sq m).

Wing area 1,462 sq ft with all metal flaps.

Tail Unit — Metal with fabric covered rudder and elevators.

Tailplane span 39 ft 6 in area 324.9 sq ft (30.18 sq m).

Belly Cargo Compartment (Rear) 151 cu ft 2,520 lbs.

Main Deck Cargo Capacity: 3428 sq ft Car-Ferry, 4350 sq ft freighter, (4630 sq ft w/raised ceiling).

Main Deck floor area 665 sq ft (including psgr cabin).

Main Cargo Compartment length 68 ft (80 ft 1 in freighter).

Rear passenger cabin length 13 ft 2 in (17, 22, 23 psgr). *Optional twin cabin capactiy (34, 40, 55, 65, 70).

Maximum width 9 ft 8 in.

Maximum interior height 6 ft 9 in (except with raised ceiling option).

Landing Gear — Tricycle type. Cleveland Pneumatic Oleo-pneumatic shock absorbers struts. Goodyear 44 in (112 cm) nose wheel. Goodyear or Firestone twin 15.50 × 20 main wheels. Wheel track 24 ft 8 in (7.52 m) Wheelbase 27 ft 4⅝ in (8.35 m).

Power Plants Pratt & Whitney 1,450 hp R-2000-7M2 twin row 14 cylinder. Hamilton Standard Hydromatic three-blade metal 13 ft propellers.

The Carvair cruising speed was reduced by 4 knots from that of the DC-4. The three-engine climb remains the same.

Weights

Max takeoff 73,800 (33,475 kg).

Max landing 64,170 (29,100 kg) (with 9,000 pounds of fuel remaining).

Max zero fuel 59,000 (26,760 kg).

or 60,700 (27,530 kg).

Empty wt 41,365 (18,762 kg).

Performance

Max speed (Vn^e) 250 mph (402 kmh) (Never to be exceeded).

Max permissible diving speed 275 mph (442 kmh).

Max normal operating speed or cruising speed (Vn^o) 213 mph (342 kmh).

Max speed w/landing lights extended 145 mph (233 kmh).

Max speed w/landing gear extended 180 mph (289 kmh).

Autopilot speed 209 mph (335 kmh) (When Auto pilot engaged do not exceed).

Econo cruise 207 mph (334 kmh).

Stall speed (Vs) 119 mph (191 kmh) (Wheels and flaps up).

Flaps max speed 20° 202 mph (324 kmh).

30° 158 mph (254 kmh).

40° and over 154 mph (247 kmh).

Service Ceiling at 73,000 lbs. 18,700 ft (5,700 m).

Takeoff run 3400 ft (1035 m).

All engines operative ISA 73,800 lbs.—4200 ft (1280 m) + 10 percent.

ISA + 15° C 73,800 lbs.—5000 ft (1524 m) + 10 percent.

One engine out ISA 72,900 lbs.—5080 ft (1548) + 10 percent.

Landing distance from 50 ft 3,120 ft (950 m).

Landing run 2,130 ft (650 m).

Range with max fuel, 10,000 lb. payload at 10,000 ft 3,455 miles (5,560 km).

Range with max payload at 10,000 ft 2,300 miles (3,700 km).

5

Carvair Operators

Original Five Carriers After British United

Channel Air Bridge/British United Air Ferries

The Carvair was developed primarily for the Channel Air Bridge division of British United, which is covered in detail throughout the individual aircraft history. Although optimism was quite high because of the many inquiries by DC-4 operators, there were only five carriers that purchased the Carvair. A sixth carrier, Alisud, leased Carvair 10 the day after it was delivered to BUAF. Each of these original carriers had specific needs that prompted orders. Channel Air Bridge (CAB) division of British United took delivery on the first three ships. They were transferred to BUAF with the merger of Silver City Airways. British United Air Ferries companies (CAB BUAF BUA BAF) took delivery of nine new Carvairs and five second-hand. One of those five was never flown and salvaged for spares. Fourteen of the 21 airframes built were owned by British United and operated under CAB, BUAF, BUA and BAF making it the largest operator of the Carvair.

Interocean Airways — Intercontinental/Winston Factors

Interocean, a DC-4 operator, became the second of the five carriers to order the Carvair by way of associate company Intercontinental. Both Carvairs were for Interocean but one was ordered by Intercontinental and the other by Winston Factors. The U.S. carrier Intercontinental, Luxembourg based Interocean and financing group Winston Factors all shared corporate officers and owners. They were also closely associated with U.S. Transport Corporation, which was the registered owner of another Intercontinental DC-4. The two carriers were short lived yet played a major role in the Carvair story.

Intercontinental operated as a passenger and cargo charter airline depending on mass movements as the primary source of revenue. Interocean concentrated on the annual pilgrimage to Mecca and United Nations relief missions. It also pursued minor charter contracts such as transporting gold for the British Monetary Authority.

Interocean was registered in Luxembourg in March 1960 naming John Chapman as President. Service was launched in June with two Intercontinental DC-4s. Intercontinental began operations in 1959 under the same owners and management. Financial backing came from three New York attorneys, Benjamin B. Peck, Bernard Nathanial Goldberg, and Lawrence M. Kesselman, who were also officers of Winston Factors. Interocean management was shuffled late in 1960 when John O'Connell became president and John Chapman moved to Director of Operations. Captain Gerry Morris, who was formerly with TWA became chief pilot. Both airlines operated the readily available DC-4s, which were inexpensive to operate and easy to repair with an abundant supply of spares on the market. Interocean rapidly became a growing charter carrier in Europe and secured United Nations relief work contracts in Africa transporting equipment and passengers.

Seven Seas Airlines, also based at Findel Airport, was a major bidder on United Nations contracts. It ceased operations in 1961 leaving a clear opportunity. With less competition Interocean purchased four more DC-4s in order to fill the void left by Seven Seas for the lucrative contracts. Interocean needed aircraft with more cargo capacity to insure growth and bid on specialized cargo charters. Intercontinental could no longer rely on charters to Mecca for its primary income. The management team was aware the relief work in Africa would eventually end and considered equipment for future cargo charters in Europe. In June 1962 Intercontinental placed an order with Aviation Traders for two Carvairs with an option for a third. Preliminary talks were held with automobile manufactures to transport assembly parts after the Congo work was completed.

Engineers at ATEL were experimenting with an onboard ramp system that got the attention of Peck and Goldberg. The prototype system employed the old Bristol loading ramps with a new fabricated top extension. Both aircraft were ordered with special features for use at unimproved fields, which included an improved version of the self-contained ramp system. The Interocean ships are the only two Carvairs fitted with these units.

Interocean expressed an urgent need prompting the acceptance of two Carvairs in progress off the line rather than wait to have Interocean DC-4s converted. After securing the United Nations ONUC contract Interocean had acquired three more DC-4s, LX-BBP, -BNG, and -LMK, for African work.

When Interocean placed an order for two Carvairs in June 1962, ATEL was developing an on board ramp system. On 14 February 1962 this experimental ramp was tested. The upper section is longer but closely resembles the production model developed for Interocean. The lower section is the Bristol loading ramps (courtesy Guy Craven)

The final design of the ramp system was tested on N9758F prior to delivery. The two Interocean ships were equipped with the system, which proved to be heavy and difficult to set up. The door mounted toilet is in place but the privacy curtain has not been installed (courtesy Terry Leighton/Aviation Traders Ltd.).

Flight crews were dispatched to Southend for Carvair training and certification. On 04 October 1962 Interocean Congo Division General Manager Walter B McCarthy requested UN identification for flight crews. The list included the first three Interocean pilots to qualify on the Carvair, Gerry R. Morris, Arvid P. Fairchild, and Robert Dedman.[1] The trio arrived from Luxembourg on 12 September 1962 on board Benjamin Peck's Aero Commander N9368R. Crew training was conducted at Southend, Stansted, and Manston. After training Fairchild was made Director of Congo Operations division for a short time. He was replaced by the well respected Captain Morris, who was a very capable operations manager. Dedman took all his training at Stansted between 16 and 20 September before going to the Congo to train other pilots and later became the sole pilot for Benjaman Peck's Aero Commander. He eventually left Interocean and flew for TWA for 28 years.[2] Fairchild was involved with at least three other Carvair operators later in his career. Morris became the first American to hold an FAA certification for the type.

Interocean traded in to ATEL two standard DC-4s, LX-BNG (BNG the initials of Bernard N. Goldberg) in August and LX-BBP (BBP the initials for Benjamin B. Peck) on 11 November 1962. LX-BBP was actually a substitution for LX-LMK (LMK the initials of Lawrence M. Kesselman).

Intercontinental purchased Carvair four on 20 September 1962. It received United States registration N9758F. The second Carvair, conversion five, was purchased 20 November 1962 and registered as N9757F. Interocean listed seven DC-4s with Luxembourg registrations and the two Carvairs with U.S. registrations allocated to the Congo contract.[3] Two months later, December 1962, N9758F was registered to Interocean as LX-IOG. In January 1963 it was transferred to Interocean and registered as LX-IOH. The registrations were reversed on the paperwork and it was easier to change the letters on the aircraft rather than try and sort out the applications. The LX-IOG registration was used for the second time and given to N9757F, which was purchased by Interocean on 17 December 1962.

The dual registration of the Carvairs was only for a short time as verified in a letter from Bernard Goldberg to the FAA dated 25 February 1963. Long after the demise of Interocean the registration is further noted on a title abstract dated 29 March 1985 when the aircraft returned to the U.S. after being sold to Ruth May for Falcon Airways. The title had not been cleared and the FAA was not notified for more than 20 years.[4] In February 1963 Intercontinental also transferred to Interocean DC-4 N30042 prior to sale to Aviation Traders.

Intercontinental and Interocean were so closely associated that administrative changes affected both companies. Benjamin Peck, an original financial backer of Intercontinental, became president succeeding John O'Connell. Henry Pransky assumed the Managing Director position at Interocean. The management team was planning expansion with additional Carvairs. Air Pictorial Magazine reported in May 1963 that Carvair nine would probably be for Interocean assuming the carrier would exercise the option for a third conversion. Aviation Traders did not publicly mention it again and Carvair nine went to BUAF.

Interocean was awarded a new United Nations contract effective 01 May 1963 while Captain Morris was Director of Congo Operations. He informed the UN three DC-4s, LX-IOA — LX-IOE — LX-IOF and the two Carvairs, LX-IOG — LX-IOH would comprise the Interocean Congo fleet reducing the complement by four DC-4s from the original commitment.[5]

Intercontinental, the parent company, was under scrutiny in the U.S. by late 1963 for alleged violations of its passenger transport licensing authority. Efforts were made to circumvent the problem by forming a third airline company in early 1964 named Tran-State. Captain Robert Dedman was called back from the Congo to assist in setting up the U.S. airline.[6] The plan was to assume all of the Intercontinental contracts. The Intercontinental certificate expired in March 1964 and the CAB refused to renew it. With the U.S. operating certificate denied the Intercontinental operations ceased on 17 March 1964 leaving Interocean to survive on its own. The airline continued service mostly in Britain and the Continent actively pursuing both military and civilian contracts.

During the summer of 1963 Alisud, a small Italian carrier based at Naples, signed a contract with Interocean. Alisud operated in the Naples to Palermo market with one leased Carvair from BUAF. The carrier planned to expand service with three Carvairs. Because of licensing problems the expansion never materialized and Alisud terminated the service after only five months. The two Interocean Carvairs were once again out of work. Interocean appeared to be holding on, but its fate had already been cast.

In addition to owning the two Carvairs Interocean leased a Constellation, DC-6, and DC-7, which were painted in very smart new liveries. It was only a matter of time however before the Carvairs as well as the airline became history. The Carvairs had originally been delivered to Interocean via Intercontinental in very subtle white upper fuselage. The idea was to be inconspicuous because of the politically sensitive contracts the aircraft were flying. The Congo livery for the Carvairs consisted of white upper half, thin blue stripe and bare metal bottom with no logo or lettering. During United Nations operations in Africa, ONUC was added in large white letters on the forward lower bare metal fuselage. After the Congo the ONUC was removed and Interocean Airways was added to the upper sides of the nose along with the blue and white stripe arrowhead logo and red lion. The exact date of this change is not clear but the new livery was applied to the aircraft by mid 1964.

Magazine advertisements of 1963 showed one aircraft in the new livery and the other with Interocean Airways in large dark block letters with a slim V shaped arrow in a circle on the forward fuselage. This livery and logo was a re-touched photo and never applied to the aircraft. Examining the "W" in "Airways" further supports this argument. The "W" was always made by laying two "V"'s one over the other to form a "W." Another print advertisement photo from 1963 has the correct printing and logo. However, the lettering is much smaller and high above the cheat line. Both aircraft returned from Congo United Nations work in Plain white. In a rush to generate business with print advertisement it was easier to add a logo to an existing photo than wait until the aircraft was painted. A limited number of charters were operated before repainting and the installation of the rear passenger cabin and windows.

Interocean operated other aircraft but was primarily a DC-4 operator. The Carvair being a derivative of the DC-4 easily fit its maintenance program and shared standard DC-4 parts. The Carvairs were not generating sufficient revenue by 1965. Adding to the problems a leased Interocean Constellation LX-IOK, inbound from Dublin, crashed on landing at Addis Ababa Ethiopia on 02 October 1964 with one crew fatality. With this loss Interocean was again a DC-4/Carvair operator.

During the summer of 1963 both Carvairs were back at Aviation Traders for maintenance and a retro fit of car-ferry standards with the addition of the rear passenger cabin and galley. Both were originally delivered as pure freighters with self contained ramp systems. The print advertisement of August 1963 emphasized their versatility with the portable ramp systems. Even with the passenger cabins Interocean was unable to generate enough business to support the Carvairs and they were put up for sale. Both ships ferried to Stansted on 28 April 1965 where they underwent maintenance checks prior to the purchase by French operator Cie Air Transport (CAT). LX-IOH, was delivered to CAT on 31 May and LX-IOG followed on 19 June 1965.

After the sale of the Carvairs, Interocean continued to slip downward, operating only three DC-4s, one leased DC-6, and one leased DC-7. By September 1965 the Interocean fleet was two DC-4s and a DC-6 insuring the end was near. Interocean ceased operations in March 1966 after six years, marking the end of one of only five original Carvair operators.

It is noteworthy that three Interocean DC-4s were converted to other Carvairs. Both LX-BNG and LX-BBP, which were added to its fleet in 1961 were traded in on Carvairs four and five and became conversions seven and ten. The third DC-4, LX-IOF, was purchased by Interocean in 1962 and operated in the Congo beside the Carvairs before becoming conversion seventeen in 1964. Ironically LX-BBP, which became Carvair 10, did United Nations work in Rhodesia as a DC-4 during the UDI blockade. Overall Interocean operated two Carvairs and 13 DC-4s making it a major player in the history of five Carvairs. Another DC-4 LX-LMK was sold to ATEL in 1964 but instead of being converted it was re-sold to Trans-Australian Airlines.

Interocean DC-4 LX-BBP, Carvair ten, returned to Africa in 1975 and passed through several owners in Zaire and the Congo. It was fully restored in 1993 and operated by EclAir on tobacco distribution contracts in Zaire. In 1995 after only two years it was withdrawn from service. It was destroyed in Kinshasa (formerly Leopoldville) Zaire later that year.

Interocean DC-4 LX-IOH, converted to Carvair 17 passed through many owners until purchase by Bob McSwiggan for Custom Air Service. It was broken up at Naples Florida in 1993 for spares to support Custom Air Service Carvair five, which was built originally for Intercontinental/Interocean as LX-IOG. Spares from Carvair 17 were also used to support Carvair seven which was converted from Interocean DC-4 LX-BNG.

Interocean ceased operations in 1966 after only six years, its aircraft were 25 percent of the Carvair fleet continuing to be a part of the Carvair story for another 30 years.

Aer Lingus

Aer Lingus reluctantly became the first flag carrier outside Britain to order the Carvair. Aviation Traders was very pleased with the order and considered it a positive sign for potential sales. The decision to purchase was made under the direction of the Irish government and the influence of the Irish tourist authority, Bord Failte.

During the summer of 1960–61 BKS Air Transport operated car-ferry service between Liverpool and Dublin using the Bristol Freighter. BKS was co-founded by T. D. "Mike" Keegan, who became the owner of British Air Ferries in 1971. Operating up to ten flights per day in each direction BKS transported 1628 automobiles during the summer of 1961. There were no drive-on sea ferries operating across the Irish Sea during this period. This was slightly more than six and a half percent of the traffic in the market prompting the opinion the car-ferry operation had great potential for growth.

Aer Lingus in an effort to expand into new markets reviewed four potential aircraft for car-ferry work and for possible replacement of the DC-3 cargo aircraft. The choices were the Bristol Freighter, which was already in car-ferry service, the Armstrong Whitworth AW650 Argosy, the planned AW670, and the Carvair. It was concluded that the Bristol was not suited because of high maintenance and low capacity. The AW650 was too expensive and the AW670 project was eventually cancelled. The carrier was quite interested in the AW670 and worked closely with the manufacturer on reviewing its potential. It soon became obvious that it was not cost effective for car-ferry work and the Carvair was the only suitable aircraft. Aer Lingus officials concluded that even the Carvair would not be profitable in dedicated car-ferry service and considered a mixed role of cars, horses and cargo. It was decided that the economical DC-3s, already in service, were quite sufficient in the cargo role until at least 1966. After further review it was clear that the Carvair would not be profitable and would require government subsidy.

BKS Air Transport withdrew the Liverpool-Dublin service at the end of the 1961 season because of financial problems. Southend based Channel Airways already an operator of car-ferry Bristol Freighters applied to take over the Liverpool-Dublin service. With this change in circumstances Bord Failte became concerned with the lack of car-ferry service and the prospect of Channel Airways expanding into the Irish market. The Irish government concluded that Bord Failte was correct and directed Aer Lingus to provide car-ferry service even if not profitable.

On 31 August 1962 Aer Lingus placed the first order for two Carvairs and support equipment (including eight "Hylo" Mark II loaders) at a cost of £700,000 ($1,960,700). The carrier also took an option on a third Carvair to be exercised at an unspecified date. They were scheduled to begin service in the spring of 1963 to capitalize on the tourist season. The purchase of all Carvairs included one "Hylo" loader. Aer Lingus positioned the pair acquired with the first two aircraft at Dublin and ordered five additional loaders to be positioned at Cork, Liverpool, Manchester, Bristol, and Cherbourg. The option for a third Carvair added the ninth loader, which was positioned at Bristol's alternate city of Cardiff.

Irish specifications required the Carvair fuel systems to be uniform on all ships. Aviation Traders was acutely aware of fuel system problems since they were not the same on the any of the DC-4 conversion candidates to date. Engineers at ATEL were already involved in standardizing the Carvair fuel system but in order to meet the Aer Lingus requirement the fuel selectors were permanently set on tank to engine. Aer Lingus also specified the Carvairs be equipped with Collins Flight Director 101, which was used on the Viscount fleet. The FD 101 ran off the battery through a separate inverter.

To capitalize on all markets and to attempt to make the Carvair profitable Aer Lingus became the first carrier to order the "Rolamat" floor system. The system was installed along with the raised ceiling option during the winter of 1963–64. This system allowed the aircraft to

In addition to the two loaders received with the first two Carvairs, Aer Lingus ordered five more "Hylo" loaders to be positioned at Cork, Liverpool, Manchester, Bristol and Cherbourg (courtesy Aviation Traders Ltd./Guy Craven).

be quickly loaded with palletized cargo, horses, or automobiles. The "Rolamat" system was designed in such a way that the auto tires fit between the roller tracks when in car-ferry service.

It is also the first carrier to order the four-car 34-passenger cabin to be used on the Dublin — Cherbourg route or special bookings. The 34-seat cabin was also used on a passenger —freighter service to Liverpool (pallet cargo only). Normal car-ferry service was operated in 22-passenger, five-car or 34-passenger four-car configuration. The passenger cabin decor was very similar to BUAF and consisted of deep-blue carpets with light blue with gold Lurex seats. The lower cabin wall was coral Vynide with white on the upper walls and ceiling. The window curtains were light beige color. An extra emergency exit was installed on the right side of the fuselage in the forward cabin.

The cargo bay sliding door, which is directly across from the passenger entry door, was secured in the open position for the 34-seat four-car configuration. A sliding curtain was installed over the opening and a small door was installed on the forward bulkhead to allow access to the cargo bay. After takeoff the flight attendant was required to enter the cargo bay and inspect the cars for fuel leaks. Upon completion the flight attendant was required to call the captain and state, "Freight hold free from fumes." At that time the "No Smoking" signs were turned off. When horses were transported a groom also rode in the cargo bay forward of the cabin. The groom carried a stun gun in the event a horse became unmanageable and placed the aircraft in danger. The captain had the authority to order the groom to destroy the horse in such an occurrence.

Aer Lingus officials considered every option for maximum utilization of the Carvair even convincing Silver City Managing Director W.C. (Bill) Franklin to take a position with the Irish carrier when the Carvair was delivered. He remained with Aer Lingus as the Cargo and Ferry Manager for many years.

The first Carvair was delivered 14 March 1963 followed by the second on 29 April. Both aircraft were immediately put into crew training service under the direction of Captain J.J.

Sullivan, who became and held the position of Training Captain during the entire Carvair tenure. He later became the Training Captain on the Aer Lingus 747 fleet often pointing out the ATL-98 cockpit similarity.

Aer Lingus crews soon learned that flying the Carvair was most definitely a two handed affair. The non-flying pilot was asked to set the power, especially during approach. In cruise the captain usually flew because the "Autopilot" control unit was on his side. It was little more than a stabilization device run by vacuum gyroscopes and hydraulic servos under the cockpit. In cruise he would set 650 BHP and accept the airspeed, which was dictated by the current weight of the aircraft.[7]

Use of carburetor heat automatically reduced the boost because of lower air density. The pilot had to add more boost to compensate, which changed the inlet temperature again. These delicate adjustments were made until the temperature stabilized at 20°C. If you popped out of a cloud into warmer air you had to instantly reduce the carburetor heat to avoid detonation and reduce boost at the same time to avoid over-boosting the engines (eight levers to play with).[8]

As the aircraft was rounded out for landing the Captain would call for a "Slow Bleed" to reduce power. If it was done to quickly you arrived with a bump. It could be counter acted with a heave on the elevator to stop the sink, which was difficult because of the weight of the controls and reduced airflow over the elevators. Too slow and you flew to the ground very gently. Too fast landed you long requiring hard braking with no anti-skid increasing stopping distance on a rapidly shortening runway.[9]

The biggest problem not foreseen by Aer Lingus officials, was reliability. The flight crews although well qualified were never really comfortable with the Carvair. The braking system was very effective but required careful handling to avoid tire damage since the brakes were not fitted with anti-skid. The crews had considerably more experience with the eight Viscounts and seven F-27 aircraft operated by Aer Lingus and preferred the Rolls-Royce Dart turboprop, which was more forgiving and easier to manage. It was acknowledged that the Carvair takeoff and climb performance was very good but experienced considerable engine failures because of what was perceived as the more temperamental R-2000 radials. The responsibility for the problems was generally laid upon the crews. The argument has been made that the American made Pratt & Whitney engines were considerably more complex.

The practice of cutting power on approach and letting the prop turn the engine causes lubrication problems and valve backlash resulting in a high rate of engine failure. The crews were instructed to insure the manifold boost pressure in inches was not to drop below the engine rpm (i.e. if engine rpm is 2000 the minimum boost is 20 inches). It was also decided to reduce T/O Boost from 50 inches to 48, which lessened stresses on the engines producing a noticeable result. It was noted that at the end of the maximum allowable time for T/O boost all engine parameters, cylinder head and oil temperature, were at maximum allowable. Overheating caused the oil pressure to drop making damage inevitable. After a period of learning adjustment engine failures were somewhat reduced but still remained at an unacceptable level.

The Carvair is a larger aircraft than the crews were generally experienced in operating and it was thought to require a softer touch. It was theorized that these factors combined with critical airflow considerations because of the large nose reduced reliability. In reality the engines could not be stressed as much as the Dart engines on other aircraft. In addition the aircraft were operated on short segments not intended for the long range aircraft and fields with marginal takeoff with a full load. Bristol was one such field that tested the resolve of the Carvair crew. Captain J.J. Sullivan noted such an incident at Bristol when the aircraft approached V1; the F/O called engine fire. Captain Murphy responded "Give me more boost" and continued a takeoff and climb on three engines immediately returning to the field, having completed a fire drill and feathering procedure. The incident ended without damage or injury but demonstrated the point that there was no margin for error.[10]

Aer Lingus pilots were not comfortable with the Carvair in these marginal situations and company officials were never really pleased with the Carvair service. The company always viewed the operation as a marginal government stop-gap measure to help Irish tourism until the arrival of Irish Sea car-ferries.

Early on some of the mechanical problems were traced to the magnetos because of the Carvair being operated in the high moisture atmosphere of Ireland. The moisture caused arching and misfiring damaging the engine cylinders. The most effective approach to resolve the problem was the use of a red sealer on the magneto covers to prevent the moisture from entering.[11] Aer Lingus continued to deal with the engine failures while the flight crews

A fresh overhauled Pratt & Whitney R-2000 awaiting installation of the QEC kit. At a cost of $20,000 each in 1963 the high rate of failures was considered unacceptable (author's photograph).

became very apprehensive about the problem and preferred the Viscount. Engine failures became so common that one Captain experienced 12 failures in 12 months. Captain Joe Dible wrote in the front of his log. "A Gentleman's aeroplane, which does not require one to worry unduly unless one has a thing about engines!"[12]

Eventually it was learned that faulty engine overhaul practices by a contractor was the culprit. During the majority of the five years the Carvair was with Aer Lingus, replacement engines were supplied by Scottish Aviation. In mid 1965, near the end of the Irish Carvair operation, engine replacements were supplied by Air France overhaul. At this point the problems were greatly reduced and reliability improved to the point that engine failures were rare. The crews were exonerated and the culprit was identified as unreliable practices by contract maintenance.

Reliability improved so much that EI-AMP operated for an entire month without incident. The performance became so good that the flight crews became concerned and some feared a failure may be on the next flight. Crews often stated, "It has to break sometime."

Aer Lingus took over the BKS car ferry terminal at Dublin. The two aircraft were scheduled initially for 7 flights per week beginning 08 May, increasing to 13 flights per week during the peak period form Dublin-Liverpool. Three services per week were scheduled Dublin-Bristol beginning 11 May increasing to six during peak months. Cork-Bristol began 12 May with two per week later increasing to four. The forth car ferry route from Dublin to Cherbourg was originally announced as once weekly on Thursday from 20 June to 12 September. Advanced bookings were so strong it was increased to twice per week. The Dublin-Bristol-Cork was set up as circle trip utilizing the aircraft for more than eight hours per day. Several combinations were used to attempt to maximize utilization and yield. The cross channel flights transported only passengers with cars in the five-car 22-passenger set up. The Dublin-Cherbourg flights operated in the three-car 34-passenger set up accepting passenger without cars.

The rates were set up with varying rates for size of car and day of the week travel. Accompanying passengers were charged the normal airfare.

Dublin — Liverpool	Cars up to 11 ft. in length	£10.0s
	(Mid Week)	£9.0s
	Cars up to 15 ft. in length	£22.0s
	(Mid Week)	£19.10s
Dublin — Bristol	Cars up to 11 ft. in length	£11.10s
	(Mid Week)	£10.10s
	Cars up to 15 ft. in length	£28.0s
	(Mid Week)	£26.0s
Cork — Bristol	Cars up to 11 ft. in length	£12.10s
	(Mid Week)	£10.10s
	Cars up to 15 ft. in length	£29.0s
	(Mid Week)	£27.0s
Dublin — Cherbourg	Cars up to 11 ft. in length	£17.10s
	Cars up to 13 ft. in length	£26.0s
	Cars over 13 ft./under 15.5	£42.0s

Initial pre-season bookings for 2,160 cars were met with optimism. This was almost half of the capacity on the cross channel routes. In addition slightly more than half of the reservations of 4000 cars Aer Lingus estimated would be transported during the first season were already sold. Officials were quite pleased that by the end of the season 4,324 cars and 12,532 passengers were transported. This was better than planned traffic and the not yet corrected reliability problems prompted the airline to exercise the third Carvair option in November 1963. Consideration of a fourth Carvair was discussed with ATEL. It was believed a third aircraft would better the Aer Lingus position for the 1964 season. It would also improve schedule reliability in the event the mechanical failures could not be reduced.

Aer Lingus Carvairs were delivered in a 34-passenger four-car configuration with the standard ceiling in the cargo hold. The first two ships began service in an all cargo role supplementing the DC-3s. They operated on overnight mail service from Manchester to Dublin and cargo flights to France, which was more profitable because of the longer flight segment. The program was successful enough that two of the DC-3s were surplused and sold. The two Carvairs were returned to Southend during the winter of 1963–64 for maintenance and modifications. In a further attempt to transport more cargo and increase market share the "Rolamat" floor system was installed and the ceiling was raised in the cargo compartment behind the cockpit hump. The modification gave a cathedral effect to the forward section of the cargo hold. This was done in an effort to obtain horse charters, which BKS Air Transport had dominated. Aer Lingus officials believed they could expand into a new market and increase revenue.

After the delivery of the third Carvair the remaining DC-3 cargo aircraft were sold and the car-ferry frequency was increased. The three ships were employed on car-ferry service between Dublin, Liverpool and Bristol and many bloodstock charters to Cambridge. During the 1964 season 5101 cars were transported, however aircraft reliability still remained a problem. The three aircraft were used for all cargo service during the winter of 1964–65.

As the cheaper drive on Irish sea-ferry operators increased market share, Aer Lingus was unable to compete and began to scale back the Carvair car-ferry service. The fleet was upgraded with Viscounts, which were much more reliable, and a freighter version was planned to begin service in 1966. The Carvairs were scheduled to operate through the 1966 season with the cancellation of the Dublin-Liverpool service. The traffic was further reduced with the increased

sea ferries. The Irish government agreed to allow Aer Lingus to discontinue the car-ferry service in September at the end of the 1966 season.

The Carvairs operated in all cargo work through October 1966 then the cargo operation was contracted to Aer Turas DC-4s and Bristols until the Viscount cargo aircraft could be delivered. Aer Lingus never considered the Carvair successful even though it was instrumental in developing the Irish tour industry and palletized cargo transport.

Top: The low ceiling is evident on this cargo version without the "Rolamat" floor system. The twin overhead channels for the elevator and rudder control cables are visible as they run to the rear of the aircraft (courtesy Ruth May). *Bottom*: The raised ceiling option originally ordered by Aer Lingus for transporting horses and later seen on this Ansett Carvair is obvious. The rudder and elevator cable runs are in a single channel following the contour of the hump and cargo hold. The "Rolamat" floor expedites the loading of palletized cargo (author's photograph).

The three aircraft transported more than 16,000 cars, 41,000 passengers, 700 horses, and 21,500 tons of cargo. The transport of 700 horses was not considered a success because BKS Air Transport dominated the market. It does demonstrate the efforts of Aer Lingus to maximize utilization in order to show a profit with the Carvair. The carrier originally planned to purchase a fourth Carvair but because of perceived dependability problems, changing market conditions, and fleet upgrades it was never ordered.

The three Aer Lingus Carvairs were placed up for sale and stored at Dublin. One of the ships was used for recovery work and returned to storage. The three remained idle for nearly two years at Dublin until a tentative purchase by Eastern Provincial was announced in January 1968 for £70,000 ($196,070) each. Eastern Provincial negotiated with ATEL for new Carvairs in June 1967 but no contract was signed because the purchase price was thought to be cost prohibitive at £120,000 ($336,120) each. Aer Lingus brought one aircraft into service in 1967 pending sale to ANTENA of Ecuador. The sale never materialized and the aircraft was returned to storage.

The three aircraft were scheduled to be delivered to EPA beginning with EI-AMP, Carvair six. Because of an engine malfunction and fire during trials it became the last aircraft delivered. Carvair eight, EI-AMR, was delivered first followed by Carvair 14, EI-ANJ. The Aer Lingus Carvair era ended 05 July 1968 when CF-EPX (EI-AMP) departed Southend en route to Gander and Eastern Provincial Airways.

Aer Lingus scheduled freight services continued with Viscounts that were converted to freighters by Scottish Aviation. The bloodstock charters were completely abandoned and left to Aer Turas, BUAF, and BKS Transport.

AVIACO (Iberia)

Aviación y Comercia SA (AVIACO) is a subsidiary of Iberia formed in 1948 as a private non-scheduled carrier that began scheduled passenger service in 1950. AVIACO was established with Spanish domestic routes that fanned out from Madrid. Over time it operated a mixed fleet of aircraft from Madrid on a considerable network throughout the western Mediterranean serving Barcelona, Balearic Islands, Canary Islands, and several destinations in North and West Africa. The carrier also served Algiers and Oran Algeria from Palma and Alicante on the East Coast of Spain plus Palma — Nice, Seville, Lisbon and Marseilles. Iberia maintained control holding two thirds ownership with The Instituto Nacional de Industria as the other major shareholder.

In February 1964 AVIACO became the third non–British and the fourth of the original five Carvair operators. The order was placed for two ships with an option for an additional five. Four Bristol Freighters were traded in as part of the contract. Aviaco signed an agreement with ATEL to convert two Iberia ships since it had only five DC-4s. Removing two DC-4s from service for conversion would severely impact schedule reliability. Realizing the conversion would not be completed to begin the season, AVIACO opted to lease a Carvair while waiting for two conversions. On 18 April 1964 Carvair 12 was transferred on a lease program and Interocean was contracted to operate limited service for AVIACO with ships four and five. The two Interocean Carvairs had just been converted from pure freighters to car-ferry standards after the return from United Nations Congo work.

Only the single leased Carvair was operated in AVIACO colors. After the second ship arrived the service was expanded to serve Palma de Mallorca Barcelona, Valencia and Nimes. Because of problems with route authority and slumps in traffic Iberia deliberated whether to convert the second DC-4. Delivery to ATEL was delayed while other possibilities were considered. The leased Carvair (12) was returned to Aviation Traders where it remained stored. The Interocean Carvairs were utilized on an "as needed basis." The option for five additional conversions never materialized leaving the carrier to operate only two Carvairs.

Spanish Carrier AVIACO entered into exchange agreements with C[ie] Air Transport (CAT), another Carvair operator, in order to maximize market coverage in the Mediterranean area. The two Luxembourg based Interocean Carvairs were purchased by CAT for use in the agreement. AVIACO Carvair operations only lasted about four years. The two ships were withdrawn from service and leased to Dominicana in 1968 in some type of aid to third world Caribbean countries scheme. Dominicana exercised the option to purchase in 1969. The Carvairs operated for AVIACO but the registrations remained under Iberia. AVIACO operated as a subsidiary of Iberia for many years after the Carvair era until eventually absorbed in 1999.

Ansett-ANA

Ansett is the last of the original five to select the Carvair in an effort to increase its cargo operation at a minimal cost. The expense of purchasing newer aircraft was considered prohibitive therefore after lengthy discussions on specifications Ansett ordered two aircraft. The contract was signed 26 April 1965 for conversion of two DC-4-1009s. The Australian Directorate of Civil Aviation (DCA) would only approve postwar built DC-4s for conversion. This stipulation prevented ATEL from converting less costly surplus C-54s available on the world market. The only alternative was to convert the two heavier postwar DC-4-1009s despite the fact that Ansett was operating pre-war C-54s as freighters.

Engineers at ATEL were aware that extra work was required to convert the -1009s versus the C-54 since all design and fabrication was based on the C-54s that were upgraded to DC-4 standards. The problem of some of the stringers on the DC-4-1009 airframe not lining up would require additional engineering. The basic difference in airframe structure accounts partially for the true DC-4 being 2600 pounds heavier than the C-54 reducing the payload of the Ansett Carvairs by that amount. Ansett holds the distinction of having the only two conversions, 19 and 20, from postwar DC-4 airframes. All other Carvairs including the last one ordered by Ansett were converted from DC-4 upgraded C-54 airframes.

The January 1968 issue of the Australian Air Log quoted the conversion cost of AU$ 160,000.00 per aircraft. The Directorate of Civil Aviation required the maximum takeoff gross to be reduced an additional 1000 pounds below other Carvairs. Ansett also specified an enlarged forward doorsill similar to Carvairs four and five built for Interocean. The requirement was needed to accommodate the 88" × 108" International Standard cargo pallet carried by QANTAS. This would allow the transfer of cargo on Carvair overnight routes that link Melbourne, Sydney, Adelaide, Hobart and Launceston. The installation of the "Rolamat" floor system for exclusive cargo operations reduced turn-around time of 17,000 pounds of cargo from two hours to thirty minutes.

Ansett exercised an option for a third aircraft becoming the last original Carvair operator. Ships 19, 20, and 21 were registered VH-INJ, VH-INK, and VH-INM respectively. Some registrations were skipped in the sequence but Ansett did reserve VH-INN, VH-INO, VH-INP and VH-INQ for possible conversions in 1971/72.[13] Conditions changed and no other Carvairs were ever built. The postwar DC-4-1009 registered VH-INL (N32AC) was passed over for unknown reasons. It was sold and eventually lost at Phnom Penh with Carvair 19, VH-INJ (N33AC).

It is reported that the numbers were reserved for DC-7 conversions, which is pure speculation. The DC-7 was never certified in Australia and could not have been considered for Ansett Carvair conversions. The unused registrations were intended for the DC-4 Carvairs. There is a bit of Australian aviation humor that states, Aircraft are not certified until the weight of the paperwork exceeds the MTOG, eliminating any chance of the DC-7 ever being certified.

Under the Australian two-carrier system Ansett-ANA competed with Trans-Australian Air (TAA). Prior to Ansett ordering Carvair conversions ATEL designed a 82" × 118" side door to replace the standard 67" × 95" opening on C-54s. The first door conversion was for Ansett competitor TAA. It was performed on former Interocean C-54 LX-LMK that had been acquired from Aviation Traders. It arrived at Southend on 09 April 1964 for the modification and was flight-tested using class B markings. It departed Southend for Australia on 18 December 1964 with Australian registration VH-TAF. The TAA side loading C-54s effectively competed with the Ansett front loading Carvairs for many years.

Studies were done in Australia by ATEL that concluded Ansett would require eight conversions. Since the Australian Aircraft Ministry would stipulate a requirement of eight conversions for Trans-Australian Airlines, ATEL planned for additional engineering capacity for at least 16 conversions. Hong Kong Aircraft Engineering (HAECO) was licensed for kit conversion if ATEL facilities could not handle the additional work. Ansett ordered only three aircraft so no additional manufacturing capacity was needed. HAECO was utilized to do the Carvair heavy maintenance.

Ansett has a rich history dating back to 1931 when Sir Reginald Ansett began a passenger car service. Ansett operated in Victoria between Hamilton and Ballarat. Sir Reginald had a vision of expansion and Ansett Airways was born in 1936 with a Fokker Universal operating out of Hamilton. Ansett acquired Australian National Airways in 1957, which was another pioneer airline having been founded in 1930 by Kingsford Smith. That same year after acquiring Australian National Ansett-ANA acquired Queensland Air Lines and Butler Air Transport.

The next eight years the freight business continued to grow and after considerable study the need for all cargo aircraft was apparent prompting the Carvair order in 1965. When Ansett transferred the two DC-4s to Aviation Traders in April and May, Lindsay Wise, a maintenance engineer who joined Ansett in 1941 and later became an Australian Airworthiness Surveyor, was dispatched to the U.K. to review the conversions. The carrier wanted a first hand account of every step of the conversion. Wise spent a month evaluating the maintenance systems to establish training for any procedures that were different from the DC-4. He later expressed his

The noses of Ansett DC-4s VH-INJ and -INK on the scrap heap at Southend in 1965. Australian Aircraft Surveyor Lindsay Wise expressed his concern about cutting the noses off perfectly good aircraft with a power saw (courtesy Ian Callier).

apprehension as he watched ATEL craftsman cut the front off a perfectly good DC-4 with a metal power saw.[14]

Although Ansett placed an order for two conversions on 26 April 1965 and took delivery in September, officials reviewed the specially equipped long-range Carvair in October 1965. Aviation Traders conducted trials with G-APNH on the world demonstration tour and visited Australia and New Zealand. Ansett was afforded the opportunity to evaluate all the Carvair options in an Australian operational environment.

The Ansett Carvairs were pure freighters with "Rolamat" floor system and enlarged door-sills for 88-inch wide standard pallets. The rear bulkhead was moved back 48 inches like car-ferry ships, but the windows and passenger cabin were not installed. It was completed with STC SA2IN conversion certificate, which approved combi operation with a passenger cabin. The Ansett ships had the re-routed control runs and raised ceiling behind the cockpit that was first fitted to Aer Lingus ships. Even with these extra options and raised ceilings the Ansett conversions only took four months to complete.

After nearly three years of Carvair service Ansett-ANA realized the need for a third Carvair. Because of political regulations a postwar Ansett-ANA DC-4 was expected to be converted. However, the requirement was not enforced and a C-54 upgraded to DC-4 was used. The aircraft departed Australia in February and was delivered to Southend on 01 March 1968 to begin conversion. It was the last Carvair built with line number 21 but not the last Carvair completed. The three Ansett Carvairs appeared the same but the "Rolamat" floor system on the last aircraft was different. Possibly because the first two were postwar DC-4s.

Australian National Airways (ANA) ceased to operate 21 October 1957 but Ansett Transport Industries (ATI) kept the name alive for legal and tax purposes. Ansett-ANA became Ansett Airlines of Australia 01 December 1968 and continued to operate the Carvair fleet until February 1972 when VH-INJ was withdrawn from service. The Carvairs served well but a fleet upgrade was needed. The following year ships 20 and 21 were withdrawn in February and March respectfully. The Lockheed Electra was used to replace the three Carvairs ending seven and a half years of Carvair cargo service.

All three aircraft were withdrawn from service and stored at Melbourne Tullamarine Airport in 1973. They were place up for sale with -INJ being the first to go to a new owner. It became involved in world politics and was eventually destroyed in war torn Cambodia. The other two passed through a series of owners remaining idle for long periods. Both were intermittently active until May 2007 when Carvair 20 was lost in Alaska leaving only two surviving examples.

Alisud

Alisud-Compagnia Aerea Meridionale (Alisud) technically does not qualify as one of the original five carriers although it did lease Carvair ten for a short time. The Italian carrier leased the BUAF Carvair for operations between Naples and Palermo beginning in August 1963. The idea of Alisud successfully operating a Carvair is incredible. It appears officials of the independent carrier were either short sighted or had great delusions. It was only operating one Percival Proctor and one Piaggio 136 at the time. To take on the challenge of operating a large new type four-engine transport is viewed by aviation experts as short-sighted. The proof is the fact that no independent carrier has ever survived in Italy.

Conversion ten was completed for British United and christened "Channel Bridge." The day after it was delivered the British United titles were painted out and Alisud titles added. The service was not successful and the aircraft was returned within six months. While it is true the aircraft did not operate a revenue flight for British United prior to the lease, technically Alisud was the second operator. The five original carriers, Channel Air Bridge, Interocean, Aer Lingus, AVIACO, and Ansett-ANA actually purchased Carvairs.

Second Line Carriers

After the Carvair ended service with the five original airlines, they were sold off to many second line carriers. The 21 ships have passed through at least 75 different owners in the 40 years after they were built. Most operators leased or purchased a single aircraft. There are other second line carriers but only four purchased or operated three or more ships. They include French carriers Cie Air Transport (CAT) with four and Trans Aeriens Reunis (TAR) with three; Canadian carrier Eastern Provincial (EPA) with three; U.S. carrier Falcon Airways with three, and the Hawaiian Carvairs. Although not a single carrier, the same individuals owned or operated four Carvairs in Hawaii under three separate airlines PAL, PAE, HPA and Turner Aviation.

CAT was formed in 1946 as Society Air Transport and became Compagnie (or, Cie) Air Transportation in 1947. Silver City Airways negotiated car-ferry marketing agreements with CAT, which eventually acquired 25 percent interest in the carrier. After Silver City was absorbed into British United, CAT continued to maintain a close association and is woven into the car-ferry era through a number of marketing and operational agreements. Evidence of the relationship is seen with the same logo appearing on Silver City Bristols and CAT Carvairs. The logo remained on the CAT Carvairs long after Silver City Airways was merged into the British United group.

CAT transported automobiles between Le Touquet and Calais to Lydd (Ferryfield) with additional service between Deauville and Southampton. The Bristol Mk.32 service was upgraded with the purchase of the Interocean Carvairs in 1965 with service beginning on 15 June. Three Hy-Lo loaders were acquired from ATEL at the same time. CAT operated limited service in 1964 between Nice and Corsica with the Bristol MK.32. In April 1965 it was granted a two-year limited authority for service between Nice-Ajaccio and Bastia Corsica. In addition Nimes-Corsica and Nimes-Palma (in association with AVIACO) was operated with the Carvairs. The operating certificate expired in 1970 and was not renewed. The carrier postponed the order of two additional ships in 1965 awaiting the results of a market study for the season.

Another French carrier, Transport Aeriens Reunis (TAR; formerly Air Dauphine), purchased the three remaining CAT ships in 1970. Two of them were used primarily for lease agreements with British Air Ferries. The third aircraft was parked at Nimes and eventually stripped for spares to support the other aircraft. TARs operating certificate expired on 30 June 1975.

Eastern Provincial of Canada acquired three Carvairs to service several development projects. The carrier had an established reputation of cold weather operations in the remote areas of Eastern Canada. In 1957 it secured a U.S. Air Force contract to supply DEW Line sights in Labrador and Newfoundland with DHC-3 Otters. This made EPA the natural choice to support other projects in the frozen wilderness. The three Aer Lingus Carvairs were purchased for transport work to support major hydroelectric projects and mining in Eastern Canada. One aircraft was lost in Labrador. The other two found their way back to Europe after the projects were nearing completion and Carvairs were no longer needed.

Falcon Airways based in Texas is the forth second level carrier to acquire three aircraft. It was established in 1971 by Lewis May with one Beech 18 and grew to a company with 20 pilots and 55 ground employees. It is significant in being the first to base Carvairs in the U.S. for all cargo service. All the previous second line carriers used them in mixed car-ferry or cargo combi service. One of the two surviving Carvairs in 2007 once belonged to Falcon. The two other ex–Falcon Carvairs were lost while flying cargo in the United States. Falcon fell on hard times in 1980 due to changing market and high fuel cost. Two of the Carvairs were repossessed and the third reverted back to Ms. May and Falcon was eventually liquidated. Lewis May passed away 12 October 1993.

Société Anonyme de Construction known as SOACO, although not an air carrier, is also of note. The French construction company secured major projects in Gabon and West Africa, which required large quantities of materials. The company purchased a number of aircraft including two Carvairs to transport building materials from Marseilles to Libreville Gabon. During a peak period a third Carvair was wet leased for three months from British Air Ferries.

The Hawaiian Carvairs are included in this group of second line carriers although there are three different companies, Pacific Aerolift (PAL), Pacific Air Express (PAE) and Hawaii Pacific Air (HPA). All three of the companies were connected by specific individuals and financial interest. The connection and complexity is generally described since it is not possible to cover all details. The primary personalities involved in at least two of these operators are James Turner, A.P. Fairchild, Louis Khem, Harry T. Snow, and Charles Willis Jr. who are noted in the individual ship histories 5, 7, 17, 20, and 21.

James Turner, president of Turner Aviation and Love's Bakery Honolulu, participated in compiling a feasibility study in 1983 to operate Carvairs in Hawaii. The evaluation obtained market and financial details of the Hawaii cargo market. James Turner enlisted local aviation experts as well as officials of First Hawaiian Bank to compile the report. The report was dated one year after PAE and PAL received Carvairs.

Turner as well as some of the other individuals listed on the report had interest in several different Carvair operations. Turner owned the two PAL ships plus a financial interest in PAE aircraft. He eventually sold the two PAL Carvairs to HPA. Harry T. Snow is listed in the report as President of Snow Aviation. He was also president of Air Cargo Enterprises later changed to Air Cargo Hawaii, which had financial interest in Carvair operator Pacific Air Express (PAE). Later Air Cargo Hawaii was listed as DBA Hawaii Pacific Air (HPA) with partners Harry Snow and George Crabbe. To add to the confusion A.P. Fairchild was Vice President of Pacific Air Express (PAE) and later Director of Operations at Hawaii Pacific Air (HPA). He also held a position at Pacific Aerolift (PAL) and made claims that he had started three Carvair operations in Hawaii. Charles Willis Jr. was President of PAL, leased Carvair 17 for a short period and also attempted to purchase Carvair five.

Charles Willis Jr. is a bit of an enigma with ties to A.P Fairchild since 1940. The two airmen joined the Navy in 1940 and became decorated combat veterans. After the war Willis started Willis Flying Service, a worldwide freight airline, and in 1957 acquired controlling interest of Alaska Airlines. He demonstrated an interest in the Carvair when it was introduced and discussed with ATEL converting some of Alaska Airlines DC-6 and DC-7 aircraft at a later date.[15] Fairchild went on to Seaboard and Western, Interocean, United, and Japan Airlines.

Willis became the first to fly charter flights from Alaska to Hawaii. He was known as "Whiskey" Willis due to his drinking. He had a reputation of being financially reckless taking extreme risk with the airline's assets frequently over committing to economically impossible promises. On 12 May 1972 a board of director's coup at Alaska forced him out.[16] Willis attempted to set up a Carvair cargo operation in Panama in 1980. In 1982 he and A.P. Fairchild became involved with James Turner and attempted to get into the Hawaii cargo business forming Pacific Aerolift, which was incorporated in May 1982. There are few details on this upstart but the project was abandoned leaving James Turner in debt with two partially overhauled Carvairs.

Turner turned to Pacific Air Express in an effort to establish transport of the bread for Love's Bakery. A.P. Fairchild, who was involved with Pacific Aerolift became Vice President of Pacific Air Express (PAE), which was owned by Louis Khem. Turner Aviation apparently financed the two PAE DC-4s N300JT c/n 36072 and N301JT c/n 18375. To complicate problems PAE DC-4 N300JT crashed on 29 June 1983 on approach to Maui. The NTSB report faulted the pilot for improper settings causing carburetor ice. This created additional financial problems for Turner. Louis Khem made a request to the FAA on 31 May 1983 requesting reserved registration N302JT (James Turner) be assigned to a Beech C45H formerly N9487Z with no reason given.[17]

PART II

Individual Aircraft Histories

Each of the 21 Carvairs built have a unique and separate history. The entire fleet appears to be uniform when in fact there are many differences among them. The original airlines ordered certain options and equipment and second line carriers required additional changes. This combined with the different series C-54/DC-4 selected for conversion made each aircraft different. The C-54/DC-4s were operated by many carriers prior to becoming conversion candidates. An overall accounting of each airframe from manufacture by Douglas, conversion to Carvair, and service life throughout the world tells a unique story.

6

Carvair Zero

c/n 10480-0, C-54B-1-DC, Mock-up

The DC-4 airframe used to build the mock-up was never an operational Carvair. Its history is significant since the engineering design of the Carvair was built up around the surplus airframe. The fact that this DC-4 and the first Carvair were both C-54B-1-DC airframes made it easier to transfer parts. After an incident severely damaging Carvair One, a portion of the empennage was grafted to G-ANYB. In essence part of PH-DBZ continued to fly until 1967. In addition the windscreens were used on the second production ship G-ARSD.

The C-54 was delivered to the USAAF on 21 December 1944 with military serial 42-72375. After being surplused at the end of World War II it was purchased by Douglas Aircraft at Santa Monica where it became the 67th C-54 converted to DC-4 civilian standards. On 15 January 1947 it was purchased by Santa Fe Skyways, named "Sky Chief Apache" and registered in the U.S. as NC90862. After only 16 months with Santa Fe Skyways it was sold to KLM on 24 May 1948 and re-named "Zeeland." It received Netherlands registration PH-TEZ, which it carried until 29 March 1954 when it was re-registered PH-DBZ.

The DC-4 was removed from service in 1958 and broken up for spares on 09 January 1959. The surplus airframe was purchased by Aviation Traders on 01 July 1959 and shipped to England via Tilbury Docks on 02 November 1959. The fuselage was transferred by road to Southend where it was placed in stands for use as the mock-up for the Carvair.

KLM DC-4 PH-TEZ as it appeared in 1953. It became PH-DBZ in 1954, operating until 1958, when it was removed from service and broken up for spares. Aviation Traders purchased the fuselage on 01 July 1959 and transported it to Southend for engineers to evaluate and use as the mock-up. A portion of the tail was eventually grafted to G-ANYB after it was damaged in an incident during the test phase (courtesy AAHS/William Steeneck).

7

Carvair One

c/n 10528-1, C-54B-1-DC, Channel Air Bridge

The first Carvair conversion did not become a "Hangar Queen" like most prototypes but has the distinction of being the first production ship G-ANYB. It was flown in regular service for nearly nine years before being retired. Although it is the first Carvair it was not built from the oldest DC-4 airframe. Fifteen other Carvairs were converted from older C-54/DC-4s airframes.

The C-54 was built in Chicago and delivered to the Air Force 22 January 1945 as a C-54B-1-DC c/n 10528, USAAF serial 42-72423. This series was basically the same as the C-54A models except the two fuselage fuel tanks were re-located in the outer wing. The Skymaster was built with reinforced floor but it is not known if this ship was equipped with the cargo hoist. It was also fitted with hard points under the wings for attachment of heavy cargo that could not be carried in the fuselage. Douglas Aircraft re-purchased it from the Air Force in 1946 and it was converted to DC-4 civilian standards and registered as NC88723.

Braniff Airways purchased several DC-4s after World War II for use on the Dallas—Kansas City—Chicago route. During the war the carrier flew to Central America and Panama for the Air Transport Command. Braniff petitioned the CAB in 1943 for Mexico route authority but it was postponed until after the war. With the experience gained, in 1946 the CAB awarded 7700 miles of South American routes to Braniff giving them the opportunity to compete with Pan Am. The routes included Rio de Janeiro and Buenos Aires and countries in between. Braniff established a radio station in Ecuador and set up passenger facilities at Lima Peru and La Paz Bolivia. In 1948 the route was established across the 20,000 foot Andes to Rio de Janeiro and Buenos Aires. Braniff Airways purchased NC59952 from Douglas in 1947 to operate on these South American routes.

This is the third Skymaster purchased by Braniff that eventually became a Carvair. The other two DC-4s were purchased in November 1945 became Carvairs three and four. The DC-4s remained in South American service until taken over by the DC-6 equipment in 1950.

The altitude at La Paz Bolivia presented problems, which positioned this aircraft in aviation history. The airfield is at 11,800 feet, which is the highest commercial airport in the world. The DC-4 is severely weight restricted in high altitude gross-takeoff operations. The solution was to equip the Skymaster with Rocket Assisted Take Off known as RATO or JATO. Since this aircraft was manufactured with the hard point mountings under the wings it was easy to adapt the RATO system.

Experiments had been carried out with other commercial aircraft using Rocket Assist Takeoff. The RATO system had been developed by the military during World War II for short field takeoff at high gross weights. It was first successfully tested 16 August 1941 at March Field California.

The RATO system is a steel bottle with an electric ignition. When activated the rocket burns for only twelve seconds with a constant 1000 pounds of thrust. These units weigh 200 pounds and cannot be shut off once ignited. After the aircraft reaches 100 mph the bottles are ignited. The thrust boost is approximately 350-horse power per unit, which is considerable for a loaded DC-4. The net effect for Braniff was a higher payload by taking advantage of the reduce drag with lower flap settings when operating out of La Paz with an extra 700 horsepower kick during takeoff.

Braniff operated the NC59952 for nearly six years. The United States CAB dropped the "C" from the registrations on 31 December 1948. It remained N59952 until receiving British registration G-ANYB, which it carried after Carvair conversion. On 15 May 1953, just one month prior to being sold, N59952 was damaged in a landing incident at Dallas Texas. It was rebuilt and put up for sale. Edward J. Daly purchased the re-built DC-4 in his name for World Airways in June 1953.

Daly was attempting to expand World Airways, which he purchased on 29 March 1948 with $50,000 borrowed money. When Daly purchased World it was incorporated in Delaware and was actually no more than a letter of Registration with the Civil Aeronautics Board. Daly was a colorful character that led others to believe that he won World Airways in a poker game. World, under Daly, received government contracts to supply DEW line sites in the Arctic. He also used his DC-4s for Military Air Transport Service (MATS) trooping flights.[1]

The DC-4 served with World Airways less than a year before it was sold in 1954 to Air Carrier Service Corporation, a subsidiary of California Eastern Airways. California Eastern was actually formed in 1946 as a cargo Carrier. It was plagued with financial difficulties and formed a leasing company in order to supply aircraft to other carriers. N59952 is one of four C-54/DC-4s owned by California Eastern or one of its subsidiaries that later was converted to Carvair standards. (California Eastern is covered in more detail with Carvair two.)

On 24 January 1955 the DC-4 was purchased by Freddie Laker's Air Charter and ferried from New York to Stansted. It was acquired for transport work in Germany and trooping flights from Stansted to Cyprus. It was named "ATALANTA" and re-registered in England on 31 January 1955 as G-ANYB, which it carried the balance of its operational life. In 1959 Air Charter was granted authority to fly passengers on day-trips from Southend to Calais without a passport. The demand was so great that G-ANYB was moved from trooping flights to service on the route, which later became a regular Carvair destination. It was operated by Air Charter until 01 July 1960, when the company was merged into British United.

The idea of converting an existing aircraft was conceived through necessity. Channel Air Bridge officials decided in January 1959 that a replacement was needed for the Bristols. After two years of planning and selection of the DC-4, ATEL began the task of actual conversion. The first candidate for disassembly was G-ANYB, which was owned by the associate company Air Charter and was positioned in New York in September of 1960.

Aviation Traders chief pilot Captain McKenzie and F/O Mike Jennings with Freddie Laker on board ferried the aircraft from New York via Shannon to Stansted. It was withdrawn from service and fitted with test equipment to obtain needed flight data for the Carvair test program. Although a proven aircraft, datum performance had to be established for comparison to the not yet built Carvair. The DC-4 made the first pre-conversion test flight on 21 September 1960 with Captain Robert Langley in command. It made the last flight as a conventional DC-4 on 03 October 1960.

Carvair Conversion

Once the DC-4 flight test were completed, the airframe with more than 37,000 hours was set up for disassembly at Southend. It is the first of only three ships converted at Southend with all others built at Stansted. On 01 October 1960 engineers began to dismantled the DC-4

G-ANYB being stripped of outer wings, rudder and dorsal fairing as it is prepared for Carvair conversion in October 1960 (courtesy Transportraits/Richard Goring).

After removing the nose the stripped fuselage is placed in stands in the RAF T-2 hangar at Southend, 11 November 1960. The next step is the alignment of the lower half of the nose section for attachment to the fuselage (courtesy Guy Craven).

outside. The fuel tanks were drained and purged then the outer wings, empennage, and engines removed. On 07 October 1960 the stripped airframe was towed to an ex–RAF T2 hangar of World War II vintage owned by ATEL where the airframe was placed in stands. On 25 October the DC-4 nose was removed forward of the wing at station X270. Then at stringer 18 it was cut horizontally to station X360 where another cut was made around the upper half of the fuselage. All other conversions had the noses remove outside and a trestling device was attached to the stubs of the horizontal stabilizer.

Early in the design phase engineers determined that the floor although reinforced was not suited for car-ferry use. The corrugated floor was removed and new supports, laminated floor and tie down stations were fabricated and installed. The fuel lines were removed and refitted in a standardized system designed for all future conversions. Electricians removed the wiring harnesses and laid them out in orderly fashion to repair, modify, and re-route. The ARB required the electrical system to be split into Main and Essential bus-bars.

Over the next five weeks while this work was in progress the upper and lower sections of the new nose section were fabricated at Southend. The lower section of the nose was grafted to the fuselage at stringer 18 and below while the upper section was suspended in position. The ribs and stringers were added between the upper and lower halves then skinned over. The DC-4 windows, yokes, and assorted DC-4 equipment were installed after attaching of the nose.

In December 1960 after the grafting of the nose was completed, the airframe was removed from the jacks and allowed to rest on the main and re-positioned nose gear. At this point the

The lower half of the nose is aligned and grafted to the fuselage while the upper half is suspended above on 11 November 1963. The ribs and stringers are added, then the area between the two halves is skinned. All subsequent noses were completed prior to grafting to the fuselage (courtesy Guy Craven).

outer wings, engines and tail had not been installed. It was moved by hand from the T-2 hangar on 13 May 1961 without any dorsal fin or control surfaces. The vertical stabilizer and rudder were modified and by pure coincidence are the same size and profile of the DC-7 fin. Early wind tunnel test determined a change was needed in the vertical stabilizer for better lateral control. Originally a tri-tail with vertical planes planned for the ends of the horizontal stabilizer was designed. It would have required more structural changes. Further test showed the same

Top: On 13 May 1961 G-ANYB was moved from the T-2 Hangar to the flight shed, less control surfaces, outer wings and engines, using only manpower (courtesy Guy Craven). *Bottom*: Masking and painting has begun on the fuselage, an engine has been mounted at Number Four position and a wiring harness hangs from the cockpit window. The nose door and attachments are not yet installed (picture supplied by Alan Bennett).

control could be achieved by increasing the height of the original DC-4 vertical fin. The increased height was to compensate for change in airflow because of the larger nose.

The outer wings were the first components to be reinstalled after the fuel tanks had been re-sealed. The spars had been x-rayed and defective areas were reinforced with stainless steel doubler plates. The new empennage was installed followed by the engines. The R-2000 engines were overhauled to -7M2 standards and tested at Bristol by the Siddeley Engine Company. During test stand runs the Number Three engine failed and had to be replaced. The new engine was mounted at Number Four position and the Number Four engine was moved to the Number Three position. Engines one and two were re-mounted at the original positions. The final step was the painting the Channel Air Bridge titles, which was performed my master paint specialist Reginald Taylor. The name "Golden Gate Bridge" was not applied until just before the christening ceremony. The retraction checks were performed and final adjustments were made to the controls after the painting.

After eight and one half months G-ANYB was rolled out 17 June 1961. The new empty weight was recorded at 40,855 pounds. The engine run-ups were performed over the next few days. High-speed taxi test were performed at Southend on 20 June 1961. The test went so well that -ANYB inadvertently became airborne for a short distance. The next morning, despite it

Above: Final stages on 14 June 1961, only seven days before the first flight. The new empty weight was recorded at 40,855 pounds. Note the fronts of the engine cowlings are covered and sealed (courtesy Guy Craven). *Right*: Chief test pilot Don Cartlidge, left, and captain Bob Langley complete preflight checklist 21 June 1961 for the first flight. The Decca Roller Map display system is in the middle of the console (courtesy Guy Craven).

Top: The first flight of G-ANYB on 21 June 1961, as seen from Freddie Laker's Cessna 310D G-ARAC chase plane piloted by senior captain L.P. Griffith. Laker was accompanied by A.C. Leftley and J.R. Bell. The flight of two hours went without incident, landing back at Southend (courtesy Brian Kerry). *Bottom*: Chief pilot Don Cartlidge, bottom, captain Bob Langley, middle, and flight test engineer Ken Smith exit G-ANYB after a successful first flight. Cartlidge always wore a coat and tie, even on test flights, while the others opted for flight suits (courtesy Brian Kerry).

being extremely windy, at 07:26 A.M. the first Carvair lifted off Southend runway 06 commanded by Chief Test Pilot Captain D. B. (Don) Cartlidge. He was assisted by Channel Air Bridge Flight Manager Captain Bob Langley. Langley became operations manager for British Air Ferries remaining with them until 1971. The third member of the crew was Aviation Traders Flight Test Engineer Ken W. Smith.

The maiden flight of two hours went without incident landing back at Southend at 09:26. The upgraded DC-6 Dunlop disc brakes, although not equipped with anti-skid, stopped the aircraft well and were noted as an improvement over the original DC-4 brakes. The Carvair modification was declared sound and handled as well as the original DC-4. Captain Cartlidge flew the Carvair up to 6000 feet at speeds up to 210 mph. and as slow as 97 mph. Although the aircraft was flown at low power settings near stall speed, no stalls were attempted on this first flight. The aircraft was also fitted with a five-foot long dump tube between the main gear. This was only a short-term modification for jettison of ballast during the test program.

The chase plane was Freddie Laker's personal aircraft, a Cessna 310D G-ARAC, flown by senior pilot Captain L.P. Griffith. Also on board the chase plane was chief designer A.C. Leftley and ATL technical director J.R. Batt. After the flight the decision was made to begin production on the Channel Air Bridge order for ten Carvair conversions.

G-ANYB was designated and designed to be a car-ferry ship, it was built and tested without the rear cabin and before the large windows were installed in the aft section. A second two-hour flight was flown the next day and another on 23 June. It was flown to Stansted on 26 June then to Filton for flight test on the 27th returning to Southend on 30 June. The only aerodynamic adjustment made as a result of the test flights was the relocation of the port Pitot head. It was moved six inches higher to clear the wake of one of the nose door hinges.

The ARB and FAA certification program was in process when on 28 August a forklift driven by Ronald Wilson struck the aircraft while it was parked outside the hangar. Freddie Laker was livid and fired him on the spot.[2] The severe impact on the forward side of the left horizontal tail-plane nearly separated the rear of the aircraft. The fuselage was split around the full circumference of the empennage all the way up to the tail fin just forward of the tailskid. Testing was halted while the rear section of the mock-up fuselage, KLM DC-4 PH-DBZ, was grafted to the aircraft. Fortunately the mock-up had been built up from the fuselage of another C-54B-1-DC. Using the mock-up expedited the repairs by eliminating the need to search for another DC-4 of the same series. G-ANYB flew again for the first time with the new rear fuselage on 29 September 1961 resuming flight trials.

To obtain type certification the Air Registration Board (ARB) required takeoff and landing trials at different weights and configurations. These trials were conducted at Filton since it was one of the few fields in England equipped with the necessary photographic and recording equipment. It was fitted with a five-foot long tube under the center fuselage to dump water ballast and a long Pitot tube on the port wing tip. The C of A trials were conducted between 08 and 24 October 1961.

It was originally estimated that type certification would require 60 hours of flying at different altitudes and speeds to measure control effectiveness and response. The cruising speed penalty because of the large nose had been calculated at 10 knots. The test surprised engineers with a drag penalty of only 4 knots and a rate of climb slightly better than the original DC-4. The flight test program originally projected at 60 hours eventually, required 155.27 hours to complete.

The aircraft returned to Southend where work continued. Finally the Certificate of Airworthiness was issued on 30 January 1962. A few weeks before it was issued the rear compartment passenger cabin windows were cut and installed by engineers at Southend. It is fortunate that the windows had not been fitted to the original rear fuselage prior to the forklift damage. On 12 February 1962 the name "Golden Gate Bridge" was applied to the nose prior to delivery to Channel Air Bridge.

The first Carvair was formally delivered to the Channel Air Bridge division of British United Airways on Friday 16 February 1962. The christening of the aircraft "Golden Gate Bridge" by Mme. Daeniker, wife of the Swiss ambassador to London highlighted the ceremony. Also in attendance at the naming ceremony and photo session was BUA Executive Director Freddie Laker; Mr. A. Kunz, General Manager of Swiss National Tourism; Mme. Daeniker and Dr. Daeniker, the Swiss Ambassador; and A.C. Leftley, Chief Designer. Aviation Traders director J.R. Batt officially handed over the new aircraft to Douglas A. Whybrow, Director and General Manager of Channel Air Bridge and Silver City. Mr. Whybrow complimented the Swiss ambassador on the co-operation of the two countries to promote motor car holidays to Basle and Geneva.

The naming of the first Carvair "Golden Gate Bridge" began a tradition that gave seven more Carvairs landmark "Air Bridge" names. After the ceremony the first overseas demonstration flight was not flown to Switzerland but to Ostend Belgium. The following day the Carvair flew the first revenue charter from Southend to Malaga Spain with a consignment of cars for a motor show. Other proving flights followed to Basle on 22 February and inaugural service to Rotterdam on 01 March. The new Carvair operated a proving flight to Geneva on 11 March returning on the 12th. Inaugural Geneva service was flown on 02 April the same day Carvair two was delivered to Channel Air Bridge. This was followed up with G-ANYB operating a proving flight to Strasbourg France on 29 May and full service on the 31st. Once in service Captain Bob Langley always flew G-ANYB and it was known as his aircraft. The inaugural Channel Air Bridge service to Strasbourg was flown by G-ANYB on 01 June 1962. Freddie Laker and Douglas Whybrow accompanied the press on the flight and returned to Southend on 03 June.

After the delivery of the third Carvair to Channel Air Bridge the trio settled into regular service. Rotterdam was at that time the longest vehicle air-ferry route in the world at a length of 330 kilometers. The decision was made early in the conversion program that as soon as the third ship was delivered, Rotterdam would be flown exclusively with the Carvair. The log from 18 September 1962 for G-ANYB is a typical example of a daily flying.

Southend	Flight	Rotterdam	Flight	Southend
07:50	AB500	08:55–09:35	AB501	10:35
11:15	AB502	12:15–12:55	AB503	13:55
14:40	AB506	15:40–16:20	AB507	17:20
18:00	AB510	19:00–19:50	AB511	20:50

Carvair one departed London 19 October 1962 for a proving flight to Singapore. The actual mission for the Ministry of Aviation was not disclosed, however it was generally accepted as a provisioning and troop rotation flight to demonstrate the capabilities to the military. G-ANYB returned back to Southend on 27 October. The expansion of service continued with inaugural service to Manchester (EGCC) on 26 January.

The first ship was often used for sales and demonstration flights in order to entice new Carvair orders. Captain Don Cartlidge flew G-ANYB to Turin Italy on 20 January 1964 to try and impress Fiat officials of the possibilities of the Carvair. The aircraft diverted to Genoa on the inbound leg because of fog. After some tense waiting the flight was completed to Turin where Fiat was waiting with a full load of Starfighter sub-assemblies to be transported to the plants at Gosselles and Charleroi. Parts were being transported through the Alps in the Bristol Freighter. The higher capacity Carvair demonstrated it could more than handle the work and was a suitable replacement. It is noteworthy that this is the same concept Airbus utilizes today with the Super Guppy and Beluga. Fiat officials were impressed when G-ANYB proved its ability to transport considerably more cargo in one flight than the Bristol but they did not commit to a freight contract or Carvair order.

British United Air Ferries

Channel Air Bridge became British United Air Ferries (BUAF) on 01 January 1963 requiring a change in livery for the Carvairs. Prior to the actual painting of the aircraft, retouched photos were circulated to the press with very small British United titles and a British flag on the tail. When the new livery was introduced the flag was dropped. G-ANYB returned to Southend for maintenance and was repainted in the British United livery on 17 December 1962. It was the second ship to be repainted in the new livery of dark stripe and red British Untied titles. The merger of Channel Air Bridge into British United Air Ferries was not official until 01 January.

The new color scheme was accomplished by simply changing the red stripe along the windows to black and adding a black band on the vertical stabilizer with the Carvair title. The black band and line are accented with a thin yellow border. The red "Channel Air Bridge" letters were changed to "British United" on the forward fuselage and the CAB logo was deleted. This was the second livery for G-ANYB and the last color scheme for the first Carvair. It remained on the aircraft until it was scrapped. Only days after receiving the new colors -ANYB was chosen to transport a Ferranti Orion computer valued at £250,000 ($700,250) from Manchester to Gothenberg Sweden.

It continued in regular service until 05 March 1967 when it was withdrawn and placed in open storage at Lydd Kent. A reduction in traffic prompted the cancellation of the seven BUAF long-range routes. The British registration was cancelled on 15 March 1967. BUAF became British Air Ferries (BAF) on 01 October 1968 but -ANYB remained in storage in the British United livery and never returned to service or flew again. While in storage it was parked with G-ARSD beside a hangar at Lydd. During a windstorm it was blown into the side of the hangar and damaged. It was also vandalized during the 1968 Lydd Air Show when many easily removed interior items were stolen. Eventually -ANYB was moved into a field on the west side of

After only five years of service, the first Carvair G-ANYB was in its final days at Lydd Kent in May 1969. Carvair two, G-ARSD, is parked behind it with blanked out BUA titles (courtesy J.W. Bossenbroek).

runway 04 at Lydd. It remained there with Carvair two, G-ARSD, and several deteriorating Bristol Freighters for three years, finally being broken up in August 1970.

The first Carvair carried the name "Golden Gate Bridge" its entire operational life. All parts were sold for scrap or salvaged for spares. Only a few items still exist today such as the manufactures data plate, which is held by Captain Bob Langley. He was one of the Carvair program test pilots contributing much of the data used in certification and development of the type. Tony Rogers of Southend, formerly with ATEL, discovered a captain's instrument panel in 2006 at Northampton that is believed to be from G-ANYB.

8

Carvair Two

c/n 10311-2, C-54A-10-DC, Channel Air Bridge

Carvair two is the first production ship and the first of 18 produced at Stansted. The first conversion, G-ANYB doubled as the prototype and was built at Southend. The C-54A-10-DC was originally built in Chicago and delivered to the USAAF on 27 May 1944. It was fitted with double rear cargo doors, cargo hoist, reinforced floor, and powered by Pratt & Whitney R-2000-7 engines. Only two Carvair conversions were built from this series of C-54. The Skymaster was delivered to the Army Air Force with serial 42-72206 and remained with the military after the war. On 22 September 1949 it was declared surplused and purchased by Federated Airlines. It was converted to DC-4 standards and received civilian registration NC57670. The carrier operated the Skymaster for approximately two and a half years before it was sold in May 1952 to California Eastern Airways.

California Eastern/Transocean/Resort Airlines

California Eastern Airways was formed in 1946 by Jorge E. Carnicero and based in Washington DC. The carrier initially operated as Transcontinental Cargo Service transporting fresh fruit. Most of the postwar start-ups took advantage of the availability of the large number of surplus military transports. Many wartime C-54s were converted to DC-4 standards by the Air Transport Division of Matson Navigation Company, which was established in 1943. In 1944 Matson acquired contracts for maintenance and overhaul of the R5D Navy version of the C-54. When the war ended the Navy cancelled the contract leaving Matson to commercial maintenance and conversion of military C-54s to DC-4 standards for civilian carriers. Two of the many carriers Matson performed maintenance for were California Eastern and Transocean. Both of these carriers owned DC-4s that were later converted to Carvairs. In mid 1948 California Eastern declared bankruptcy and ceased operations. The aircraft were leased out until the carrier exited bankruptcy and resumed service in December 1950.

California Eastern operated Trans-Pacific service after being awarded a government contract in 1950. The service continued until October 1953 when operations were once again suspended. Transocean leased and operated N57670 for a short time on Military Air Transport flights in 1953. After the Pacific contract ended Samuel Soloman moved into the senior management position at California Eastern. Under his direction the carrier operated contract service for MATS between California and Tokyo from February to May 1954. Additional flights were also operated to the Philippines.

During the Pacific operation the carrier was struck with a major setback when one of the DC-4s caught fire on the ground because of a fuel leak. The right wing was burned off outboard of Number Three engine. California Eastern was so in need of equipment to maintain the Tokyo contract that Vice President and Director of operations Robert Caskey decided to rebuild the aircraft. In retrospect this was probably a mistake in judgment. It took more than four months and 35,000 man-hours to complete. During the overhaul the wings were changed and it was converted from C-54A to C54E standards. The shortage of aircraft combined with the financial condition of the carrier forced the remaining fleet including N54670 to maximum utilization and minimum maintenance. This is one of many examples of C-54s being refitted and upgraded by changing wings with later series models.

California Eastern also operated ad hoc contracts and domestic charters. A short-term contract was awarded to operate DEW line flights for a two-month period in 1953. The level of performance was good enough to be considered for additional trooping flights. A contract was obtained from the Navy in 1955 for trans-con troop transport. In 1954 California Eastern was granted a five-year operating certificate. Skymaster N57670 was utilized for nearly two years before being leased to Resort Airlines on 29 October 1954. Resort also operated Logair trooping flights for the military. The DC-4 remained on lease for the next two years until returned to California Eastern in 1956. As carriers attempted to modernize throughout the world DC-4s were less in demand and readily available from surplus fleets. California Eastern took advantage of the available aircraft and became more active in leasing aircraft to other carriers rather than operating its own fleet. In a joint venture Aero Leases was formed by California Eastern and North American Airlines as a way of providing aircraft for other carriers.

The leasing operation was successful enough that it marked the end of operations of any aircraft under California Eastern. Another DC-4 that became Carvair 13 passed through Aero Leases. The Skymaster that was converted to Carvair one transited still another California Eastern subsidiary known as Air Carrier Service Corp. By 1961 Skymaster N57670 was idle and placed in open storage at Oakland.

California Eastern expanded its leasing operations to other carriers in 1956 by purchasing five Lockheed L-1049 Connies. Three were painted in California Eastern colors but were soon leased to TWA. The other two were transferred to another associate carrier in Argentina known as Transcontinental SA. This is a variation of the name that California Eastern operated under after being formed in 1946.

N57670 remained in Resort Airline colors after returning to California Eastern in 1956. It was placed in open storage at Oakland awaiting the scrap dealer until rescued by Aviation Traders as the second of five Resort DC-4s that were converted to Carvairs (via Jennifer M. Gradidge).

Although N57670 was returned to California Eastern in 1956 it was not utilized and remained in Resort Airline colors in open storage at Oakland. The DC-4 was considered surplus and relegated to a date with the scrap man while continuing to deteriorate. It was rescued by Aviation Traders to be the second of five Resort DC-4s that were later converted to Carvair standards. As the Carvair program was getting started California Eastern once again was in financial trouble. It was re-organized and assets placed under the re-named Dynalectron Corp in 1961. The new company sold off aircraft and assets no longer necessary for the operation. In time the company moved into other ventures and moved out of the airline industry.

Channel Air Bridge

The Channel Air Bridge projected plan of expansion required ten Carvair conversions. With the success of G-ANYB the conversion line was set up at Stansted. Air Charter became part of British United, the parent company of Channel Air Bridge, and did not have spare DC-4s available. Aviation Traders began shopping on the open market for surplus C-54/DC-4 aircraft for conversion. Skymaster N57670 was withdrawn from service and was essentially scrap stored for years in the open at Oakland California. FAA records indicate the last owner as California Eastern, however ATEL records indicate it was purchased from Resort. The owner is not clear but ATEL negotiated the purchase believing they were obtaining a worn but suitable airframe. It was the practice of ATEL to buy airframes that were not in the best condition. It was reasoned the purchase price would be lower and condition was not an issue since the airframe would be completely overhauled. However, this DC-4 still painted in rather weary Resort colors presented considerable more problems than expected before it could be ferried to Stansted for conversion.

Carvair Conversion

The Skymaster was transferred to Aviation Traders on 18 June 1961 and registered G-ARSD before ferrying to Stansted on 20 July. Only three older C-54/DC-4 airframes were used for conversion. The R-2000 engines on this early DC-4 were equipped with the short ram-air carburetor intakes easily identified by small scoops on top of the nacelle above the cowl flaps. The system was retained for a short time after conversion while the aircraft flew in the Channel Air Bridge colors. In early 1965, after being repainted in the British United colors, the carburetor air intake fairings were retrofitted to the more streamline non-ram air system. The reason for the retrofit is a mystery since there is no functional change with the longer fairings.

G-ARSD is the only Carvair to have a window in the rear section of the double side entry cargo doors. The DC-4 arrived at Aviation Traders with this option and was never modified. Many other C-54/DC-4s had one or two windows in the rear half of the cargo doors when they arrived at ATEL but they were removed during conversion. Since this was the second Carvair built, uniform configuration standards were still not an issue demonstrating how each Carvair was hand built. Although outwardly all ships appear the same each one is unique. The three ships built for Aer Lingus are more alike than any others because of the carriers rigid standards for uniformity.

The C-54A-10-DC airframe had the reinforced cargo floor from Douglas. It was determined that the corrugated floor would have to be removed and a new laminated floor with cargo tie downs for car-ferry service installed. This series C-54 airframe is very similar to that used to produce G-ANYB, the first conversion.

Disassembly began late July 1961 and two days later ownership was transferred to Channel Air Bridge. The nose was delivered by road from Southend to Stansted on 21 October 1961. This is the first nose assembly transported to Stansted since Carvair one was built at Southend.

The windshield and frames that were installed in the new Carvair nose were originally from PH-DBZ. They were removed from the KLM DC-4 used as the mock-up and reconditioned. Because of a shortage of suitable windscreens care was taken to remove them from the DC-4 nose sections for use in production. The remaining nose section of PH-DBZ was later melted down at Southend. The windscreens removed from the DC-4 nose of N57670 were reconditioned for use in another conversion. This practice saved production time by having a set of windscreens ready for installation.

The conversion was completed at Stansted in eight months, which is two weeks less than the first ship. It was fitted with six fuel tanks although the DC-4 airframe was configured for eight. The new Carvair was fully painted at Stansted except for the name of "Chelsea Bridge" and the new weight recorded at 40,519 pounds. Officials at ATEL felt comfortable with the production time since it was shorter than ship one and held optimism that future production times could be reduced. In order to establish performance standards for all future production the Aviation Registration Board (ARB) scheduled clearance flights.

It flew for the first time as a Carvair G-ARSD from Stansted at 5:27 P.M. 25 March 1962. It did not perform well and had so many malfunctions and system failures it was diverted to Southend immediately. Every measure was taken to insure quality but some of the problems were attributed to the deplorable condition of the Resort DC-4 used for conversion. The incident stressed officials at Aviation Traders, who were over confidant after the success of the first ship. Production could not continue without ARB approval. The aircraft underwent three days of intense maintenance to bring it up to standards and insure all systems were functioning correctly.

On 28 March 1962 G-ARSD began a four-day ARB certification program, which was flown by Captain Don Cartlidge with two ARB inspectors on board. After very intense and stressful

Three days after the first flight, G-ARSD begins ARB certification at Southend on 28 March 1962. Note the short ram-air carburetor intake scoops, which were later changed to long non–ram air type. Only three Carvairs wore this livery (courtesy Guy Craven).

trials it was officially certified on 31 March. There were some tense moments, fears, and minor malfunctions but the aircraft eventually performed as required. The C of A was issued after nine and half-hours of testing over the four days of trials. It was immediately transferred to Channel Air Bridge to begin service. In the tradition started by Freddie Laker it received the name "Chelsea Bridge," which was applied at Southend by Reg Taylor.

Channel Air Bridge

On 02 April 1962, the same day it was delivered, G-ARSD ferried from Stansted to Southend to originate the inaugural service to Geneva. Channel Air Bridge officials were overly optimistic when the inaugural flight was scheduled for 02 April. Their optimism almost spoiled the festivities since it did not have the Certificate of Airworthiness needed for the flight. Fortunately it arrived by motorcycle courier only moments before departure. Captain Don Cartlidge commanded the flight with invited passengers on board. At the same time Captain Bob Langely flew ship one, G-ANYB, with officials and the press. Just four days later G-ARSD and G-ANYB flew the inaugural Basle Service with revenue passengers and cars on board.

The change from Channel Air Bridge to British United did not become official until January 1963 but G-ARSD returned to Southend in October 1962 for maintenance and repainting. Between 24 and 30 October it became the first Carvair to wear the new British United colors. It returned to service with the merger of Silver City Airways into British United and the consolidation of Channel Air Bridge operations at Southend under the Air Holdings group.

The new BUA paint scheme was not that different from the Channel Air Bridge livery. The difference being the red color band was changed to dark blue with a thin yellow border. The Carvair name was added to the fin and "British United" titles substituted for "Channel Air Bridge." It retained the name "Chelsea Bridge" below the cockpit windows.

British United Air Ferries

The new company name and livery was met with great optimism in January 1963. G-ARSD is the only one of the three original Channel Air Bridge ships to be painted in three different liveries. Ship one was scrapped in the original British United livery and ship three crashed at Rotterdam still wearing the Channel Air Bridge colors.

World politics called on British United and Aviation Traders in the March 1963 when the Ministry of Defense jointly contracted by the West German police to transport 29 armored cars from Hannover to Berlin. Aviation Traders engineers built a special platform and ramp for loading. The vehicles were driven onto the platform where the wheels and hubs were removed so the vehicle could be lowered on to a sledge. A jack was placed under the nose of the aircraft to prevent tipping and three winches hauled the armored car into the Carvair. Since only one could be carried at a time, nine were transported on 18 March and the balance from 20–27 June. Access to sanitary facilities were not assured during the operation prompting the installation of a toilet in the nose door in the same fashion as the two Interocean ships.[1]

British United Air Ferries became British United Airways (BUA) late in 1966 adopting a third color scheme for the Carvair fleet. It consisted of a sandstone and blue stripe running the length of the fuselage and up the tail with BUA letters on the forward fuselage and fin. A stylized beige winged bird appeared at the end of the BUA lettering on the forward fuselage. The bird faced forward on both sides of the aircraft. For the second time G-ARSD was the first to receive the new livery on 20 January 1967.

The Long-range service was cut back on 28 February necessitating the grounding of much of the car-ferry fleet. The Bristol 170s were retired to prevent the withdrawing of any additional

Aviation Traders constructed special loading platforms for a British United contract to transport 29 armored cars, one at a time, from Hannover to Berlin in March and June 1963 (courtesy Guy Craven).

G-ARSD in February 1967 two weeks after painting in the "Toppled Mushroom" livery. It is the first of five BUA Carvairs (2, 7, 10, 11 and 13) to be repainted after managing director Derek Platt proposed a uniform color scheme for all British United aircraft (courtesy Ad-Jan Altevogt).

Carvair aircraft. After the summer season, G-ARSD was withdrawn from service and ferried Southend-Lydd to be placed in storage. Two other Carvairs, ship one (G-ANYB) and ship 12 (G-AOFW) were also in storage. It remained at Lydd for a short time, then was returned to service. After only a few weeks of operation it was again withdrawn and stored back at Lydd Kent.

British Air Ferries

Just four days after another corporate restructuring changed from British United Airways (BUA) to British Air Ferries (BAF), G-ARSD made its last revenue flight. On 04 October it was flown to Lydd and on 06 October it was permanently withdrawn from service. It was placed tail to tail with Carvair One and the engines and outer wings were removed. The BUA lettering was blanked out and it was left to deteriorate. The British registration was suspended on 13 April 1968. At one point during the years in storage it was damaged by wind when blown into the hangar. Toward the end it was moved into a grassy field west of runway four and stored along with G-ANYB and several Bristol Freighters.

The second Carvair G-ARSD and first aircraft built at Stansted was withdrawn from service and stored at Lydd Kent in October 1967. The outer wings, engines and main gear doors are removed, awaiting the scrap dealer in May 1969 (courtesy J.W. Bossenbroek).

Broken Up Lydd

The second Carvair built, was broken up at Lydd Kent on 26 August 1970 just eight years after it went into service. The aircraft had been in storage since the last flight on 04 October 1967. It has the distinction of being the last survivor of the original three Channel Air Bridge ships. The first Carvair, G-ANYB, was broken up at Lydd a month earlier. Ship Three, G-ARSF, was destroyed in an incident at Rotterdam in 1963.

9

Carvair Three

c/n 18339-3, C-54B-5-DO, Channel Air Bridge

Carvair three had the shortest operational life of any conversion because of a tragic loss. It was listed on the British Aircraft Registry as number R7367/1 and operated by British United Air Ferries Limited under the Channel Air Bridge colors until lost at Rotterdam. The certificate of airworthiness, A7367, was issued 02 August 1962 and valid until 05 July 1963.

The C-54 aircraft was built at Santa Monica and delivered to the Air Force 10 July 1944 as a C-54B-5-DO c/n 18339 USAAF serial 43-17139. This model is fitted with integral fuel tanks in the outer wing panels. It is also equipped with the double cargo door, cargo hoist, and powered by R-2000-3 engines.

After 15 months of Air Force service the Skymaster was declared redundant. In 1945 it was purchased by Braniff Airways and registered in the U.S. as NC88709. The "C" was dropped from the registration on 31 December 1948 and in October 1950 it was leased to Northwest Airlines. Northwest operated N88709 for nearly four years before sub-leasing it on 10 June 1954 to Transocean Airlines. The same day it was severely damaged by fire on landing at Keflavik Iceland. Transocean was able to salvage and rebuild the aircraft since it maintained a large maintenance facility in Iceland. After extensive maintenance and flight testing it was returned to service.

Transocean, which is reviewed in detail with Carvair 17, owned or operated four Skymasters that were later converted to Carvairs. After only a few months of operation with Transocean it was returned to Braniff Airways and leased to Resort Airlines in October 1954. In less than a year the aircraft was returned to Braniff and placed in storage at Oakland still wearing the Resort livery. It remained stored until purchased by Channel Air Bridge in June 1961.

The DC-4 was on lease the majority of its DC-4 life, yet Braniff remained the sole civilian owner until purchased by Channel Air Bridge for Carvair conversion. This Skymaster had been severely damaged by fire and rebuilt, poorly maintained and allowed to deteriorate further in open storage. Apparently officials at Channel Air Bridge were not aware of the poor condition or were overly anxious to obtain another C-54 at a low cost. This oversight became a major problem resulting in the airframe requiring considerably more repairs than the first two DC-4s selected for conversion.

Carvair Conversion

Production of Carvair three was already underway for Channel Air Bridge when Southend based independent carrier East Anglian Flying Service operating as Channel Airways placed an order for four aircraft. The carrier proposed car-ferry service between Liverpool — Dublin and Bristol — Cork to compete with Aer Lingus. An order was placed on 01 February 1962 for four conversions. The first two were for immediate production. Aviation Traders agreed to a projected delivery date of May 1962 for the first aircraft. The last two were on option for an unspecified future date. The registration of G-ARSF was reserved by East Anglian for the first Channel Airways Carvair. East Anglian purchased DC-4 N33679 from Riddle Airlines on 31 March 1962 with the intent of using it for the first conversion.

Irish authorities assumed Channel Airways would be operating Bristol Freighters when the authority was first requested. When it became apparent that the service would be operated in the Carvair the Irish only granted limited access with authority of only four passengers per flight. Channel Airways immediately placed all plans for car-ferry service on hold. Without route authority the plans for car-ferry service were never completed and the reserved registration of G-ARSF was not taken up. It was eventually re-assigned to British United's Channel Air Bridge for the ship already in production. The DC-4 purchase from Riddle by Channel Airways (East Anglian) for conversion was put on hold and eventually delivered in April and registered as G-ARYY. The DC-4 never went into the Carvair program but was re-configured to high-density seating and put in service in early 1963.

The East Anglian — Channel Airways contract for four Carvairs was never completed. If the Channel Airways Carvair order had been completed, the third ship for Channel Air Bridge would not have been registered G-ARSF. That registration would most likely have been assigned to Carvair four and G-ARYY would have been Carvair five.

Channel Air Bridge DC-4 N88709 was transferred to ATEL in Jun 1961 for conversion prior to the Channel Airways problems. While Channel Airways was sorting out the route authority with Irish officials, Aviation traders took up registration G-ARSF on 17 July. Disassembly of the DC-4 airframe began at Stansted on 22 July and the nose was delivered by road from Southend to Stansted on 31 December. The conversion took 11 months, which is three months longer than the first two ships. The poor condition of the DC-4 airframe requiring retro work accounted in part for the delay.

The Carvair was completely painted at Stansted prior to the first flight and the weight recorded at 40,546 pounds. G-ARSF is the first of only three ships (two were built for Interocean) with a different porthole window configuration in the new nose. There are only two small porthole windows on the right and one at the rear of the hump on the left. There is no explanation for this configuration except this nose may have been originally configured for Interocean.

Carvair three first flew at 17:55 on 28 June 1962 and landed at Southend at 18:17. The

G-ARSF in an early morning haze at Southend on 20 July 1962, one month after the first flight. It was the last Carvair to wear the Channel Air Bridge colors (courtesy Transportraits/Richard Goring).

aircraft required 8:30 hours of testing over the next week to receive the C of A. After successful flight test it was delivered to Channel Air Bridge 07 July 1962, ten days after the first flight. In the tradition established by Freddie Laker it was given an "Air Bridge" name and Christened "PONT DE L' EUROPE." The Certificate of Airworthiness A7367 issued 03 August 1962 arrived at Southend only a few hours before the inaugural flight to Geneva. The third Carvair immediately began scheduled service with Channel Air Bridge between Southend and Basle, Geneva, Rotterdam and Strasbourg.

It was also selected to fly the first schedule service to Calais. The proving flight gave Channel Air Bridge an opportunity to demonstrate the ability of the new Carvair on the short field. Captain Don Cartlidge, assisted by Captain Bob Langley, made the first Calais flight on 07 August 1962 without incident. The crew reported the new aircraft performed well and they were pleased with the performance and its ability to operate on the short field.

G-ARSF is the third and last Carvair to wear the Channel Air Bridge livery. The purchase of Silver City Airways and merger of the two carriers under Air Holdings prompted a new livery. The consolidation was in progress for sometime prior to the 01 January 1963 official date for the new company of British United Air Ferries. It never came under the new company and is the only Carvair to wear a single livery. On 28 December, four days before being transferred to British United, it was lost in a tragic accident at Rotterdam.

Fatal Flight of G-ARSF

G-ARSF departed Southend at 10:02 A.M. on 28 December 1962 with 4 crewmembers and 14 passengers. Captain John Bolton Toothill, affectionately known as "Toots" was in command. The flight was approximately one hour late because of cargo loading and an inspection of the departure runway. The cockpit crew consisted of Captain Tootill, F/O Ronald Joseph Riches, Second Officer Engineer J. Barton. Also on board in the cabin were Flight Hostess K. Wood and the 14 passengers. Commander Tootill assigned co-pilot Riches to operate the radio, flaps, and landing gear. Engineer Barton was assigned the engines, throttles, mixture, props, etc. Actually a flight engineer is not required since the Carvair has a re-designed cockpit for a two-man crew. Channel Air Bridge assigned an engineer for the purpose of reducing the captain's workload since flying the Carvair is a two handed affair. The engineer is positioned in a jump seat between and to the rear of the pilot and co-pilot.

The only irregularity of the flight prior to the landing at Rotterdam occurred shortly after departing Southend. After reaching cruising altitude the cabin heater malfunctioned resulting in it being switched off. This resulted in low temperature on the flight deck and in the passenger cabin. The condition was uncomfortable, however it did not appear to impair the ability of the crew to fly the aircraft.

The flight crew was quite experienced. The trio had considerable Carvair time considering there were only three of this type aircraft in service. Carvairs had only been in commercial operation for 10 months. Commander Tootill was born 09 August 1925 and held license ALTP 51812, valid until 30 May 1963. He was rated in the Bristol 170, DC-4, DC-6, and Carvair. His instrument rating was valid until 04 July 1963. He had a total of 6,534 flying hours of which 450 were in the Carvair. Commander Tootill had considerable experience at Rotterdam airport having logged 176 landings since April 1961. At least 120 of his landings were on runway 24 and 69 of them were in the Carvair. Records show that he had made 10 landings in the last month at Rotterdam. However, he had only made three flights with Riches as co-pilot, two of which were at Rotterdam.[1]

Co-pilot Ronald Joseph Riches born 23 December 1926 was also an experienced airman. He held license CPL 30302 valid until 06 June 1963. He was rated in the C-47/DC-3, Viscount, Bristol 170, DC-4, and Carvair. His instrument rating was valid until 01 July 1963. Airman

Riches had logged 6,014 flying hours, with 201 of those as Carvair co-pilot. He had logged 10 landings at Rotterdam in the last month.[2]

During the days proceeding 28 December weather conditions were poor to marginal. A weak pressure prevailed over Western Europe, which caused a layer of snow of approximately 9 cm. The stratus clouds varied between 750 and 1000 meters and thin broken layers of stratus occurred during snowfall. There was a weak S.S.W. air stream at surface level and a W.S.W. air stream at the upper level. The frontal zone was north of Rotterdam situated above the English Channel across the North Sea from northern Holland to Southeast England. South of this front in the upper air moving N.E. parallel to this air stream were small disturbances. These disturbances caused moderate snowfalls. The 08:45 A.M. weather forecast that morning for 10 A.M. to 7 P.M. weather was as follows: Wind: 200 degrees at 05 knots changing between 1 P.M. and 7 P.M. to 160 degrees at 07 knots. Visibility: 3000 m. reduced to 1000 m. during moderate snowfall.

The Rotterdam airport was covered with considerable snow giving the appearance of a white plain with no contrast with the surrounding countryside. Visibility was restricted by haze and low clouds. G-ARSF was the first arrival to Rotterdam that morning, therefore there were no reports to indicate flight visibility differed from visibility on the ground. Carvair two, G-ARSD, had overnighted and departed Rotterdam at 10:01 A.M. returning to Southend. It did not experience any icing and reported visibility of 600 yards on runway 24. The useable length of runway 24 was 1420 meters or approximately 4650 feet. The runway and threshold had been cleared earlier and was covered with a thin layer of snow. Along the length of the runway the snow had been piled forming a low bank. The runway and threshold lighting was cleared of snow at approximately 10:30 A.M. The red obstruction lights marking a dike, which is about 240 meters in front of the threshold, were turned on. However, the red and white obstruction markers on the dike had not been cleared of snow. The runway and overrun was sanded earlier but still judged to be slippery.[3]

Departing Carvair G-ARSD warned inbound G-ARSF by radio of slippery runway conditions just before it entered Rotterdam air space. At 10:44, forty-two minutes into the flight from Southend, co-pilot Riches contacted Rotterdam tower. At 10:45 he received the following weather transmission. G-ARSF: Wind 200/5 knots, QNH 1015.5 mb, visibility 1500 m in snow, 2/8 at 180 m, 5/8 at 300 m.

The aircraft entered Rotterdam control and was cleared to proceed direct to the R.R. beacon and descend to 1500 feet. As they approached the field. G-ARSF was cleared to descend to 1000 feet. At this altitude they were flying contact with forward visibility of 2 nautical miles.

At 10:55 the aircraft was over the field. The tower advised the following: G-ARSF: Runways and taxiways are covered with compact snow of 2 cm. They are sanded, braking action medium.[4]

As they passed over the field the crew could see the runway. They called the outer marker outbound at 1000 feet and a procedure turn was made at 130 mph, at 2500 rpm. and 30 degrees flaps. Flight Engineer Barton later stated there was no horizon visible, after passing the outer marker they were in "whiteness." The checklist was completed and both altimeters were set and verified at QNH 1015.5 mb.

The manufacturer recommends that threshold speed be computed based on gross weight of the aircraft. The final approach speed should be approximately 15 mph. above the calculated threshold speed. Approach speed should be stabilized while inbound on the localizer and prior to crossing the outer marker. The approach checklist should be completed in the procedure turn or the equivalent. Props will be set at 2250 rpm and manifold pressure set at 25" to 27" hg. At the far end of the downwind leg the flaps are to be lowered to 15 degrees. When the glide slope needle is one dot above the center marker call for "GEAR DOWN." On final, when flaps are lowered to 30 degrees the speed should be stabilized at 115–125 knots (132–142 mph). Flaps are to be lowered to 30 degrees as the glide slope is intercepted; approach speed should be stabilized at 125-mph. and 12" hg. down to touchdown. The nose should be lowered quickly to aid in braking.[5]

Rotterdam tower cleared G-ARSF to land and Commander Tootill began the descent. The ILS glide path was intercepted. After passing the outer marker the runway lights, which were at 100 percent intensity, came in sight to the crew. Between 600 and 700 feet the aircraft was above the ILS as indicated by the needle being deflected to the maximum. The engines were at 2550 rpm. with 27" hg. of manifold pressure. The power was reduced to 2250 rpm. with 18" hg. of manifold pressure. Approximately one half mile from the field Commander Tootill called for 30 degree flaps. There is some confusion but it appears the flaps were at 30 degrees at 10:55 when the aircraft was over the field prior to the procedure turn. Because of insufficient power the aircraft rapidly lost altitude in a horizontal attitude. The aircraft descended below the ILS glide path making the approach and threshold lights poorly discernable in the limited visibility. It was no longer possible for the crew to assess height and distance. The rate of descent was too great in relation to the distance to be covered to reach the approach end of the runway.

At 11:04 A.M. the aircraft main gear struck the dike that formed the main boundary of the airport some 240 meters (787 feet) from the end of the runway. The aircraft bounced and remained airborne for approximately 500 feet further. At that point the right wing dropped and contacted the ground. The forward momentum of the left wing and fuselage caused the aircraft to pivot 180 degrees on the grounded right wing. The right wing broke off at the wing root as the aircraft rolled over while pivoting. The airframe came to a stop upside down on top of the right wing in the reverse direction of approach on the right side of the runway apron. The right wing remained upright even after breaking off. Because of the wet snowy conditions there was no fire.[6]

The forward portion of the cockpit collapsed down to the top of the instrument panel. Commander Tootill was killed on impact and Co-pilot Riches was severely injured. The Engineer and Air Hostess sustained minor injuries. The 14 passengers did not sustain any injuries greater than minor bruises. Notification of the incident and report of Captain Tootill being

The crash of G-ARSF at Rotterdam 28 December 1962, six months after the first flight. Captain John Bolton Toothill was killed when the roof of the cockpit collapsed (courtesy Henk Hartog/BAF).

fatally injured was relayed to ATC controller Ian Callier at Southend for appropriate notification. Rescue crews had to remove the nose door to rescue the injured crew and recover Commander Tootill's body from the cockpit. The vehicles remained secure hanging upside down in the cargo hold leaking oil and fuel into the cargo compartment. It took considerable effort for engineers to upright the fuselage and extract the vehicles.[7]

Records show G-ARSF had 1,081 hours since conversion to Carvair standards. Prior to conversion it had 39,422 total time hours as a DC-4. There had been no serious incidents or recurring defects since conversion. The aircraft landing weight was well below maximum allowable of 63,500 pounds and the CG was within specified limits of 14.2 to 28.4 percent MAC.

The Netherlands Department of Civil Aviation conducted test after the accident in order to evaluate the daytime visibility under such conditions. A DC-3 aircraft was used to evaluate the obstruction marking on the dike and the runway approach lighting when the surrounding terrain is covered with snow. The department's findings were as follows:

1. The dike lies 1.40 m. below the obstruction free zone required by International Civil Aviation Organization (ICAO).
2. The dike did not stand out against the surrounding terrain.
3. The obstruction lighting and the markers on the dike were not effective.
4. The approach lighting between the dike and the runway could not be distinguished clearly during an approach, particularly during a low approach.
5. The threshold lights were not clearly visible and provided the pilot with no point of reference for the runway threshold.
6. The crossbar of the approach lighting was clearly visible the entire approach.
7. It is very improbable that the crossbar can be confused with the threshold lighting.
8. It is possible that the similarity between the upper surface of the runway proper and that of the extension of runway 06 (the underrun of 24), as well as the position of the snow, which had been cleared from the runway, gave the impression that the runway began at a distance of 120 m. (390 ft.) instead of 240 m. (787 ft.) from the dike.

The G-ARSF was uprighted, cars removed, then the airframe disassembled at Rotterdam for transport back to Southend It was eventually melted for scrap (courtesy Richard Vandervord collection).

The Ministry of Aviation accident review board released their findings 30 October 1963. The following is the board's analysis and conclusion.

The aircraft was airworthy, had a valid certificate of airworthiness, and was loaded in the proper manner.

Both pilots were competent to pilot the aircraft. Both had considerable experience with the aircraft type in question and had made a number of landings at Rotterdam airport. They were familiar with the characteristics of the approach area of runway 24 and the conditions as regards the extension of this runway.

The available runway length, although the runway was slippery owing to the snow, was sufficient for a safe landing to be made with ILS approach.

The commander carried out the approach procedure in the proper manner, but during the final stage of approach the aircraft, owing to insufficient power, rapidly lost altitude, thus descending below the ILS glide path.

Once the aircraft was below the glide path the approach lights between the dike and the runway and the threshold lights were not clearly visible to the commander, which made it difficult for him to estimate height and distance. As a result the commander did not realize that the aircraft was losing altitude too rapidly.

The accident was because of the fact that the commander carried out the final stage of approach below the normal glide path with insufficient engine power, as a result of which the speed of descent was too high in relation to the horizontal distance still to be covered to the beginning of the runway. Consequently, the aircraft, at a high vertical speed, hit a dike, after which the right wing broke off and the aircraft came to rest on its back, with its nose facing the direction of approach.

The Board is of the opinion that even if there had been no dike, and it was not an obstruction of any significant height, the aircraft would still have hit the ground a considerable distance short of the runway threshold, although possibly with less fatal consequences than in the present instance. The rate of descent was too great because of full flaps at insufficient power settings.[8]

The British registration was cancelled 05 February 1963. On 23 February the aircraft remains were removed from Rotterdam and taken to Southend for storage and evaluation. The wreckage was moved to Stansted in March and all useable parts were removed. The remaining airframe was scrapped. The aircraft was only in service a few days less than five months. The Carvair fleet was reduced by one before the majority of conversions were built. There were only six Carvairs in existence when G-ARSF was lost. The additional 15 had not been produced or test flown.

Co-pilot Riches returned to flying and was involved in three accidents in his career. In one accident he was a crewmember on Viscount G-AODH. While landing in fog in Germany the aircraft crashed through the ILS Glidepath brick building.

10

Carvair Four

c/n 10338-4, C-54A-10-DC, Intercontinental — Interocean

Carvair four had a relatively short operational life of four and a half years. It has the distinction of being the first of only two Carvairs built as pure freighters and later converted to car-ferry/passenger standards, ship five being the other. Later the three Ansett Carvairs were built as pure freighters and never converted to car-ferry configuration.

The C-54A-10-DC aircraft was built in Chicago and delivered to the Air Force 30 June 1944 as c/n 10338, USAAF 42-72233. This model was built with the corrugated reinforced floor and wing hardpoints for up to 32,500 pounds of internal and external oversized cargo or troop transport. It was fitted with Pratt & Whitney R-2000-7 engines. Only five other Carvairs were converted from older airframes. Ships two and four are the only two conversions from this series C-54.

In November of 1945 after 17 months service with the Air Force the C-54 was declared redundant, converted to DC-4 standards, and assigned civil registration NC65142. It was purchased by Braniff Airways and operated for nine years on U.S. and South American routes. The "C" was dropped from the registration on 31 December 1948 becoming N65142. This DC-4 along with Braniff DC-4 NC59952 (Carvair one) was fitted with Rocket Assisted Take Off (RATO) units for operation out of La Paz Bolivia. Because of the altitude two RATO bottles were fitted to the aircraft for extra thrust at the high altitude field. The bottles have a 12-second burn providing 1,000 pounds of constant thrust. (See Carvair one for more details of RATO.)

The rather weary DC-4 N65142 stored at Oakland in Slick-Logair titles with the orange nose and band on the tail for military contract operations. Its future was uncertain until rescued by Channel Air Bridge for Carvair conversion (photograph by A. Breese/via Jennifer M. Gradidge).

Resort purchased it from Braniff in November of 1954 and operated it until 01 July 1959 when it was leased to Slick Airways. Less than a year passed and it was returned to Resort on 30 June 1960. Channel Air Bridge acquired it on 06 June 1961 and received British registration G-ARSH on 17 June. As with other Resort Airline aircraft it was stored in poor condition at Oakland California. During the lease with Slick Airways it was operated on military Logair freight contracts and abused with the oversize and overweight cargo and minimum maintenance.

Carvair Conversion

The conversion was originally intended for Channel Air Bridge, which had taken delivery of the first three Carvairs. Channel Air Bridge, a division of Air Charter, and Silver City Airways were merged to form British United and it was scheduled to become the first Carvair built for the new company. In reality it never flew with any British carrier.

Aviation Traders received the first order from outside the Air Holdings group in June 1962. In an effort to produce more Carvairs and gain international orders, ATEL convinced Channel Air Bridge to relinquish the delivery position mid-way through production. Although ordered by U.S. based Intercontinental, associate company Interocean was the intended operator to fill an immediate need of a United Nations contract. By adjusting the delivery positions of two ships on the line an acceptable delivery schedule was achieved. Intercontinental most likely would

Carvair four, originally intended for Channel Air Bridge, in stands at Stansted on 15 June 1962. The airframe required considerable retro work before it could be completed for Interocean (courtesy Guy Craven).

not have purchased the Carvair if this arrangement had not been made since the aircraft were required specifically for UN contract work.

G-ARSH was transferred to ATEL in June 1961 and disassembly began on 27 July. The poor condition of the airframe required a considerable amount of restoration work, which contributed in part to the longest conversion time to date of 13 months. It is also the fourth longest conversion of all Carvairs produced. The British registration was cancelled 09 April 1962 during the lengthy conversion. Adding to the construction delay was the request for additional modifications and special equipment and it is the first aircraft that required the cockpit set up for U.S. standards. The cockpit was fitted with double bunks, six seats, one crew jump seat behind the pedestal, abbreviated galley, periscope sextant and chemical toilet.

Intercontinental agreed to the contract that was signed on 27 July 1962 for £127,000 ($355,727). As a condition of the contract Interocean agreed to provide ATL with two other DC-4 airframes less engines to replace those used for Carvair conversion. Both were scheduled to be converted for BUAF at a later date. The two Skymasters scheduled for trade in were LX-BNG and LX-LMK. The registrations are the initials of two of the three owners of Intercontinental/Interocean. The "BNG" was for Bernard N. Goldberg and "LMK" for Lawrence M. Kesselman. LX-BNG arrived at Aviation Traders on 17 August 1962 and eventually became Carvair seven.

The true reason Intercontinental formed Interocean in Luxembourg is not clear but was influenced by several factors, which included international politics, U.S. certification problems and tax purposes. The two Carvairs were never intended for U.S. operation and it was actually purchased by U.S. Transport Corporation of New York for Intercontinental. The two companies were also closely associated. At the time U.S. Transport was also the registered owner of two other Intercontinental Airways aircraft, DC-4 N30045 and DC-6 N12877.

Luxembourg based Interocean was flying the United Nations ONUC contract in Africa with DC-4s and required a front loading aircraft to transport military equipment. The specifications submitted to ATEL for special equipment included a cargo winch located at the rear of the cargo compartment, a toilet mounted in the nose door, double bunks behind the cockpit, plus an enlarged forward bulkhead with self-contained ramp system.

The conversion of ship four was already in progress when the agreement was finalized. In February 1962 Carvair four was months into conversion at Stansted with the nose not yet attached due to engineering changes to the forward bulkhead. The nose arrived by road from Southend on 03 June 1962. Intercontinental specifications required the aircraft be able to transport military trucks. After completion loading test were conducted with Bedford military trucks, which were thought to be similar to those being used in the Congo. The specification required ATEL engineers to build a modified forward cargo door bulkhead. The standard doorframe had been squared and sections removed in order to meet the requirement. The enlarged forward bulkhead was notched at the two and ten o'clock positions and strengthened with stainless steel plates and straps at the top corners. The door lock and alignment pins were moved lower to allow for the notch. The modification was completed before the new nose was grafted to the airframe. No heater ducts were installed from cockpit to the rear of the aircraft.

Carvair four was originally scheduled for BUAF as a car-ferry ship, but it was completed as a pure freighter without the rear passenger compartment. The large square passenger windows and rear cabin emergency exits were not installed.

Only Carvairs three, four, and five have a different configuration of the small porthole windows in the hump behind the cockpit. There are two windows on the right, one behind the cockpit and one at the extreme end of the hump. Only one small window is installed on the left at the extreme end of the hump. All other Carvairs have three windows on the right and two on the left of the hump. To further reduce weight and cost de-icing boots were not installed. It was believed that they would not be required in the heat of the Congo. This assumption was later realized as a mistake when an icing incident on a return flight from the Congo almost ended in disaster.

A self-contained ramp system was developed for the loading of high oversized square vehicles such as the Bedford military truck at remote airfields. Carvair four is the first of only two ships equipped with ramps and notched special enlarged bulkhead. The Ansett freighters 19, 20, 21, were built with a derivative of the bulkhead without the notches. It is also the first of only three ships equipped with crew bunks, cockpit galley, nine cockpit crew seats, periscope sextant, chemical toilet, winch, and ramp system. The number of cockpit crew seats were adjusted related to each mission. On several occasions, as many as six seats were added.

The portable four-ton capacity ramp system folds and stores in the aircraft. When flying empty the ramps were secured in the rear for ballast since the Carvair tends to fly nose heavy. Both ships with ramps have a placard mounted on the lower inside lip of the nose door: "Door Warning To avoid damage to freight door, ramp pickup bars must be positioned inboard before closing." Interocean requested these special modifications solely for the benefit of the ONUC Congo work operation. It appears a bit short-sighted and does not justify the expense to require the special features for a contract that was not long term. Interocean officials did review bidding on contracts to transport car parts between European assembly plants once the Congo work was completed. However, nothing was firm and no contracts were forthcoming. Additionally the transport of auto parts would not require the self-contained ramps.

An embarrassing oversight became evident when the Carvair reached the Congo. Interocean had not checked the dimensions of UN vehicles operating in the Congo. Instead a British Bedford military truck was measured at Southend. In reality this Carvair never transported the military trucks it was modified to accommodate. Once the aircraft got to the Congo it was discovered the UN trucks had the tops welded on and were to high to fit through the forward door.

Repositioning at Southend after first flight from Stansted, Carvair four in Class B registration G-41-2 on 05 September 1962. The aircraft wore this registration for less than a week (courtesy Transportraits/Richard Goring).

Carvair four first flew from Stansted to Southend with Aviation Traders manufactures class B registration G-41-2 at 15:34 on 05 September 1962. The new empty weight was recorded at 40,448 pounds. It was then pushed back into the western bay of the large four bay hangar to complete the installation of the special equipment. It had been painted at Stansted with white upper fuselage and black trim line. The name "Carvair" was applied in small script on the top of the vertical stabilizer.

It was certified after only three test flights. The registration country was in limbo when the aircraft was ready for flight trials. Intercontinental was the new owner, but the aircraft was already scheduled to be transferred to Luxembourg associate company Interocean. On 12 September 1962 the manufactures registration G-41-2 was removed and the U.S. registration N9758F applied. The F suffix indicating a U.S. aircraft being maintained abroad. This rather odd F suffix system of registering United States aircraft was issued in reverse order. The system began in 1950 beginning with N9999F.

The ONUC lettering was applied at Southend prior to leaving for the Congo. On 17 November 1962, the day before it departed it was at Southend with the engine cowlings painted blue in a similar fashion to those of KLM and Eastern. Intercontinental purchased the two Carvairs as airframes only and supplied the engines from an unknown source. The aircraft first flew at Stansted with unpainted cowlings. Records were not located to determine if the engines were changed after it received certification or only the cowlings. The painted cowlings were on the aircraft on 14 February 1963 at Leopoldville (Kinshasa).

Intercontinental/Interocean/ONUC

The Luxembourg registration LX-IOG appeared on the instrument panel data plate but was not immediately assigned to the aircraft. When the aircraft had the Luxembourg registration issued it was LX-IOH. Ship five also had the registrations reversed. The registrations became confused in Luxembourg since Carvair four and five had the U.S. registrations issued in reverse order. Interocean determined it was easier to change the registrations on the actual aircraft rather than try and sort out the paper work. Ship four was N9758F (LX-IOH) and ship five was N9757F (LX-IOG).

Three Interocean pilots arrived on 12 September from Luxembourg in the Interocean Aero Commander. The group included Captains G.R. Morris, A.P. Fairchild and Robert Dedman. Captain Morris, who had been furloughed from TWA, became director of operations for Interocean in the Congo. Robert Dedman left Seven Seas Airline for the better pay at Interocean.[1] Later in the program another ex–Seven Seas pilot, Paul Rakisits, arrived at Southend. After training in England for Interocean, Dedman went to the Congo to qualify other pilots on the Carvair. Fairchild and Rakisits also flew in the Congo and re-appeared in the Carvair story for more than 30 years with at least three other operators.

Most of the aircrews at Interocean were seeking positions with the scheduled airlines. Flying in the Congo was grueling with most pilots logging more than 200 hours a month for months in a row. It was a quick way to build time as Captain in a four engine aircraft with a schedule of two months on and a month off in Europe. Captain Morris encouraged Dedman to build as much Carvair time as possible in order to apply at TWA.[2]

Aviation Traders Captains Cartlidge and Langley began the Interocean training program on 13 September 1962. Carvair four received the Certificate of Airworthiness on 14 September after three certification flights totaling seven hours. The new Carvair remained in white top and black line with bare metal bottom blue trim engine cowlings, and no titles. The last four days of training were in the cockpit. In order to obtain FAA certification N9758F left Southend for Frankfurt Germany on 20 September commanded by Captain G.R. Morris. United States officials from the FAA met the aircraft there for testing and certification.

After U.S. certification Captain Robert Dedman and F/O Ken Windsor picked it up and conducted an acceptance flight with FAA Inspector David Switzer. Dedman and Windsor along with Switzer flew the new Carvair from Frankfurt (EDDF) to home base at Luxembourg. The proving flight continued on to Malta (LMLI) and overnighted. The next day they flew to Kano Nigeria (DNKK) and again overnighted. On the 29th they flew to Douala Cameroon (FKKD) and the final segment to Leopoldville Congo (FCAA) (now Kinshasa). After reviewing the Carvair with company officials, crew rest, and servicing, the Interocean crews began training on 03 October. On 09 October Dedman and John Koontz operated N9758F on the first flight (UN-119) to Albertville (FCRF)(now Kalemie). They continued shuttling UN personnel and equipment throughout the Congo flying 26 legs between 09 and 26 October.[3]

Captain Dedman returned N9758F to Southend for additional maintenance via Leopoldville, Douala, Kano, and Tripoli on 29 October 1962. Dedman was at the controls coming over France empty en route to Southend when giant cumulus clouds were encountered. As an inch of ice built up on the wings, Dedman added more power. As the ice continued to build the Carvair began to sink and even with max power could not hold altitude. French ATC granted a rapid decent and after several tense maneuvers they came out of the clouds at a lower altitude with warmer air melting off some of the ice. They arrived at Southend with ice still falling off. Dedman was almost terminated as he sternly informed Interocean officials that the aircraft could not be flown until de-icing boots were installed. They had been omitted for cost and weight reduction with the belief they were not required in Africa and tropical areas. The assumption was almost a costly mistake that Interocean officials reconsidered. After installation of deicing boots test flights were performed on 17–18 November.[4]

Upon completion of additional modifications, both Interocean ships flew in formation over

Interocean N9758F after unloading 55 gallon drums of fuel at Albertville (Kalemie), Congo, during United Nations peacekeeping operations (photograph by Capt. Paul Rakisits).

Southend on 18 November 1962. Carvair four carrying U.S. registration N9758F commanded by Captain Dedman with F/O Paul Rakisits and Frank Gamble departed that same day to Pisa (LIRP) and Khartoum (HSSS) en route to ONUC operations at Leopoldville (Kinshasa) Congo.

The U.N. operations took the Carvairs from Leopoldville as far south as N'dola Zambia, as far north as Stanleyville (Kisangani) and east to Usumbura (Bujumbura Burnadi). The missions included supplies for the mine at Bakawanga (Mbuji-Mayi); military supplies and troops to Kolwezi, Elizabethville (Lubumbashi), Stanleyville (Kisangani), and Kamina; fuel drums from Albertville (Kalemie) to Kamina; meat and booze to Lulabourg (Kanaga); and produce from Goma.

With the delivery of the Carvairs, Interocean submitted the following crew names to the U.N. for credentials. The list, which included Walter B. McCarthy, General Manager Interocean Congo Division, requested UN identification for the following crew and five un-named mechanics.[5]

W.B. McCarthy	L.M. Flint	M.E. Oberg
M. Ogden	R. Tluszcz	M. Pearce
O.V. Braken	J.M. Harris	M.E. Thompson
R.M. Tracy	D. Fenno	E. Courtemanche
S. Wheatley	G. Pearson	E. Boudreau
J. Meade	J. Johnson	J. Webster
R. King	J. Kent	D. Bolt
A. Cadwell	C. Pollock	C. Wheatley
P. Rakisits	K. Winsor	J. Koontz
R. Dedman	H. Heaney	M. Polski
G.R. Morris	A.P. Fairchild	

Interocean crews became accustomed to primitive loading methods in the Congo war zone. Loading of a boat on LX-IOH at Kamina Base proved to be a challenge for the military winch and local crane (photograph by Capt. Paul Rakisits).

Interocean also submitted a list on 26 November 1962 of seven DC-4s and two Carvairs that would be operating in the Congo. The Carvairs were still recorded with U.S. registrations. This list indicates that both Carvairs operated in the Congo with U.S. registrations for at least a month. Carvair four was registered to Interocean on 10 December 1962 as LX-IOG. The registration was withdrawn and re-assigned to Carvair five on December 17 because of a paperwork problem. The decision was made to repaint the registrations on the aircraft rather than to try and sort out the clerical error.

On 18 December registration LX-IOH was assigned to Carvair four. Still wearing registration N9758F it was officially transferred to Luxembourg based Interocean Airways on 29 December 1962. The purchase transaction from associate company Intercontinental was not signed until January 1963. According to Captain Dedman the Luxembourg registration was applied to the aircraft in January 1963 and it returned to Southend 02 February 1963 as LX-IOH. It maintained dual registration until 27 February 1963 when the U.S. registration was cancelled. Dedman continued flying LX-IOH making his last Carvair flight with D. Fenno on 13–14 June from Leopoldville to Elizabethville and return.

Carvair four was sent to the Congo for military vehicle ferry work under United Nations contract. The aircraft was chartered by the air division of Organisation des Nations Unies au Congo (ONUC) to assume the mission previously operated by Italian Air Force C-119s. Labor and political unrest forced Belgium to grant the Congo independence on 30 June 1960. Eleven days later the Premier of the Katanga Province, Moise Tshombe, seceded from the republic, followed by Kasai province. The Congo was suddenly thrust into civil war. The United Nations sent in a peacekeeping force because of the tension between then President Joseph Kasavubu and rebel leader Patrice Lumumba. Kasavubu in an army coup arrested Lumumba and turned him over to Moise Tschombe of Katanga province who witnessed his execution by a Belgian mercenary.

Tschombe and the Union Miniere Mining Company, based in an area of western Katanga called Bakwanga, had a contract with Seven Seas Airlines to transport supplies to the remote area. The mining company played a strategic role in supplying a major portion of the worlds industrial diamonds.

Seven Seas competed with U.S. based Intercontinental for UN relief contracts while experiencing financial problems. Intercontinental/Interocean reviewed the possibilities of securing additional contracts in the event Seven Seas became insolvent. Intercontinental was experiencing certification problems in the U.S. In order for the company to survive the operation needed to expand, which prompted the forming of Interocean in 1960. Seven Seas paid pilots $4.75 per statue mile for straight line distance flights.[6] This alone created a turn over in pilots and instability in keeping flight crews. Seven Seas ceased operations in 1961 creating a windfall for Interocean. After the collapse of Seven Seas, Interocean purchased four more DC-4s with three of them later becoming Carvair conversions.

United Nations secretary Dag Hammarskjöld was killed in a plane crash while en route to negotiate peace in the Congo. UN secretary U Thant submitted a national reconciliation plan in 1962 and in December Tshombe's troops fired on the peace keeping UN forces. Carvair N9758F arrived in the Congo for the first time on 29 September for trials and training. It arrived the second time about 19 November. It returned to Southend after the Congo trials for maintenance on 29 October. The de-icing boots had not been installed before ferrying to Frankfurt for FAA certification.

The LX-IOH Interocean registration was not official until 10 December 1962, indicating the aircraft was operating in UN military operations by U.S. carrier Intercontinental prior to Interocean ownership or at least Luxembourg registration. Captain Robert Dedman flew DC-4s for Seven Seas in the Congo prior to signing on with Interocean. In 1962 when hostilities in the Congo created additional flying to support the UN mission his Congo experience made him a perfect choice for the Carvair assignment. Although he was the third Interocean pilot to obtain a rating, for a short time he was the only Carvair rated pilot in the Congo.

Carvairs four and five were based at Leopoldville (now Kinshasa) for the Congo ONUC contract work. Details are sketchy, but it is known that the Carvair was used to transport some military equipment such as armored vehicles to Elizabethville (now Lubumbashi). It also transported military equipment for the occupying force in Kolwezi, which was a major stronghold for the rebel forces in Katanga. LX-IOH saw combat when it was reported that over 3000 rebels troops were at Kolwezi. The Carvair was called in to lead a combat assault mission along with four DC-4s and a C-46 loaded with ammunition. After F-86s softened up the area Captain Dedman went in first with LX-IOH carrying two armored cars with crews followed by the troops in the DC-4s.[7]

The Carvair could lift 30 drums (55 gal) of fuel at a time. The fuel drums were transported from Albertville (now Kalemie) on Lake Tanganyika to a base at Kamina Central Congo. The fumes were so strong the aircraft were flown with the emergency exits open to vent the cargo hold. The Carvair was put to the test while in the Congo under wartime conditions of unimproved fields and overloading making the always feared mechanicals common place. While hauling seven tons of supplies to the diamond mining camp at Bakwanga (now Mbuhi-Mayi) LX-IOH damaged a tire landing on the slag runway. Since there were no spares or mechanics Captain Dedman and First Officer John Koontz decided to takeoff with the blown tire. The empty aircraft with 20-degree flaps skipped along the gravel runway for a short takeoff without any other damage to the airframe. Dedman was severely scolded for the unauthorized takeoff and threatened with a reprimand. After review it was obvious he had not other choice under wartime conditions. The two of them gained much experience under adverse conditions in the Congo. Dedman carried his experience to TWA and Koontz later went on to Continental Airlines where he became the pilot for company president Robert F. Six.[8]

Starters were also common failures that had to be dealt with in the adverse conditions. The crews carried a bungee strap with a leather pouch attached. Locals were enlisted to turn the prop through as the crew hit the switch. This was not always successful and most of the time locals were not interested in this dangerous task. The more frequently used technique was the rolling start. If the aircraft was empty the engine was windmilled until it started and the takeoff continued and if loaded it was run to 80 knots. If the engine started the crew taxied back down the runway and made a standard takeoff.

Engine loss under these conditions increased the chances of having to abandon the Carvair. In such an incident LX-IOH sheared a prop shaft at Elizabethville (Lubumbashi). Captain Dedman was forced to do a three-engine takeoff to fly back to Leopoldville (Kinshasa). The prop was secured by tying it off for the 6-hour flight. Dedman was suspended for the incident but after the FAA investigated he was commended for an outstanding job and re-instated. On 26 February 1963, after daily abuse, LX-IOH returned to Southend for much needed maintenance departing back to the Congo on 06 March.[9]

After hostilities stabilized, as part of the reduction in United Nations ONUC strength the two Carvairs took part in transporting an Indian infantry brigade from the Congo to Mombasa Kenya. After the Congo mission Interocean advertising stated that LX-IOH in tandem with the other Carvair uplifted an entire mechanized military division from Leopoldville to Mombasa. This was somewhat public relations exaggeration by Interocean to demonstrate the ability of the Carvair to transport large loads. In fact it was not a division, not mechanized, or heavily motorized. There were some vehicles and considerable infantry support equipment. Mombasa was the loading point for the UN Indian military force ships returning back to India. Almost all the vehicles belonged to the United Nations and were left behind by Indian forces. The Carvairs were utilized to position these vehicles when needed by the occupying forces.

The United Nations forces were in the Congo to keep the peace. Technically they were neutral but in practice they were against Tshombe and his breakaway republic of Katanga. All ONUC charter contracts were terminated in July 1963 for economical reasons. Tshombe who had fought the UN troops surrendered on 14 January 1963. Kasavubu named Tshombe to the position of Premier as part of the agreement to quell the open civil rebellion. Tshombe contracted foreign mercenaries who along with Belgian paratroopers were airlifted by U.S. aircraft. Their objective was to defeat a communist-backed regime in the northeast. They were successful in defeating the communist. The Carvairs had already been withdrawn by this time and were not used in this action.

After the Congo operation ended in July 1963 LX-IOH was operated by Interocean on ad-hoc charter flights until 1965. The ONUC titles were removed and both aircraft were operated without lettering for a short time. Interocean began advertising in a number of trade publications beginning in August. Both aircraft returned to Aviation Traders in March 1964 for maintenance and repainting. Beginning with LX-IOH they were re-configured with the rear passenger cabin for dual-purpose work. The passenger windows were added and were visible in April 1965 promotional photos. The nose door chemical toilet was installed on an as needed basis but the curtain remained.

After return from overhaul in 1963 Interocean began a promotional advertising campaign emphasizing the Carvairs' portable ramp systems. Advertisements stated that Interocean Carvairs had lifted more than 1,012 automobiles, trucks, jeeps, armored cars, trailers, and artillery pieces and moved more than 1,800,000 pounds of cargo, generators, jet engines, and hydraulic lifts.[10]

Limited cargo service was operated under contract to Italian carrier Alisud late in 1963 and early 1964. Alisud ceased operations in February 1964. After the cargo contract with Alisud, Interocean operated limited car-ferry service on the Naples to Palermo routes for Spanish carrier AVIACO.

It is generally believed that the ramp system was removed when the passenger cabin and rear windows were installed. However, the ramps remained with the aircraft and were used as needed until the aircraft was sold to the French carrier Cie Air Transport (CAT). The ramps were removed on or about 28 April 1965 when it was at Stansted for maintenance and repainting for CAT. It is most likely the nose door toilet was permanently removed at this time. The mounting remained in the nose door after the passenger cabin was added in March 1964. The actual toilet was installed for cargo missions.

The ramp system, although very useful, limited the aircraft to very specialized cargo missions. Interocean was not financially flush enough to maintain an aircraft for such limited use. The exception was the UN Congo contract, which required the ramps. The ramps occupied a cargo or car-ferry position reducing the capacity. It was apparent when the Carvair arrived in the Congo that ramps were heavy and required considerable manpower to set up. They were not used at Leopoldville or Albertville because scissor loaders were available. After return from the Congo they were stored for occasional use on specialized cargo work.

LX-IOH operated a charter from Gatwick to Findel Airport Luxembourg on 30 June 1964 and another cargo charter from Gatwick to Athens on 06 August 1964. Interocean continued flying oil contracts, commercial and military cargo charters including one flight for the British military to RAF Changi Singapore. Both Interocean Carvairs flew a number of charters to Helsinki, Rome, and Turin. Other charters flown by LX-IOH included Stansted — Benin City Nigeria on 20 December 1964 and Gatwick — Milan on 06 January 1965.

The certificate for U.S. associate company, Intercontinental, expired in March 1964. The CAB refused to renew it citing questionable practices. Interocean was left to stand alone but began experiencing problems by the spring of 1965. The carrier seized the opportunity to dispose of the Carvairs when French carrier Cie Air Transport tendered an offer. LX-IOH ferried Luxembourg to Stansted on 28 April 1965 for overhaul prior to delivery.

Compagnie [Cie] Air Transport (CAT)

LX-IOH was the first CAT Carvair to receive French registration and the second of four ships it eventually operated. The carrier had been a partner with Silver City Airways since 1961 and closely linked to British United Air Ferries after the Silver City and Channel Air Bridge merger. Three Bristol 32s were originally transferred to CAT for operation in conjunction with BUAF service from Le Touquet and Calais to Lydd and separately operated Nice/Nimes to Corsica. Bournemouth and Cherbourg were added with the arrival of the Carvairs. A separate service was added in the Mediterranean division to Nice and Perpignan.

Compagnie (or, Cie) Air Transport (CAT) purchased Carvair four from Interocean in April and it received a check-4 and repaint at Stansted prior to delivery to Le Touquet on 19 June 1965. Press releases stated it was being converted from cargo to car-ferry but in reality Interocean added the 17-passenger cabin a year earlier. It was being changed from 17-seat 5-car car-ferry to a 55-seat convertible cabin. It was registered in France as F-BMHU and named "Cdt HENRI de MONTAL" after a noted French entrepreneur.

The CAT operation in conjunction with the British car-ferry service for many years with Bristol Freighters made the Carvair an obvious upgrade. When the British wanted to establish air service into Le Touquet for passenger aircraft, transportation to and from the airport was not sufficient. The British requested a two-mile extension of the railroad to the airport. They were met with strong opposition by the French, who were very firm on the negotiations. Since air routes were reciprocal at that time, the French agreed to the rail line and upgrading of the airport only if the British would lease Bristol Freighters giving them car-ferry service. Eventually the British agreed giving CAT authority to operate into Ferryfield (Lydd) and Bournemouth. The French railway system was the majority shareholder in CAT putting it in the airline car-ferry business. The upgrade to Carvairs was necessary to stay competitive and maintain the same level of service with the BUAF interchange agreement. At the time the two Interocean Carvairs became available CAT was operating the older Bristols that were either purchased from Silver City or leased from BUAF. The opportunity to purchase the two Interocean Carvairs was needed to strengthen market share in the Mediterranean.

After French carrier CAT purchased Carvair four (F-BMHU) it was re-named "Cdt Henri de Montal."
Many charters were operated to the Middle East including Sharjah UAE in 1966 (via Jennifer M. Gradidge).

Carvair four is the second of the two Carvairs purchased by CAT in the June 1965 agreement. That same day it ferried from Stansted to Le Touquet with the new French registration F-BMHU. CAT also continued to operate the Bristols to Lydd until November 1969.

The French carrier had a well-planned marketing strategy and focused on interchange agreements for further expansion. It had been the French agent for Silver City Airways prior to being merged into BUAF and had working agreements with Spanish Carvair operator AVIACO. Nice — Corsica service was added in 1964 and routes planned from Nimes to Lydd and Palma. With AVIACO concentrating on the Palma — Barcelona route both carriers were able to expand the coverage of the market. Other CAT routes included Nimes— Corsica, Palma — Ibiza, and Perpignan — Palma.

Cie Air Transport abandoned the option to purchase two more Carvairs in 1965 while waiting government traffic approval on the additional car-ferry routes. The Carvair was utilized for cargo charter work during the winter season to generate additional revenue. The carrier also exploited the long-haul freight routes with ad hoc charters into Africa and the Middle East. F-BMHU was lost on a long haul freight charter after a maintenance stop at Karachi Pakistan.

Crash at Karachi

F-BMHU ferried from Nimes to Le Bourget on the last day of February 1967 to originate a cargo charter. It was loaded with 15,946 pounds (7233 kg.) of cargo at Nord Aviation for transport to Australia and departed Paris on 01 March with a crew of six on board. The Flight was commanded by Captain Dessannaux assisted by a second captain with the co-pilot serving as the navigator. The crew included a flight engineer and two ground engineers (flight mechanics). The aircraft performed well and flights were reported uneventful from Paris-Brindisi-Damascus-Bahrain. After 35 hours en route on 03 March approximately one hour from Karachi Pakistan the Number Two engine developed unspecified mechanical problems. Upon arrival at Karachi it was determine that the engine failure could not be repaired. The flight was delayed at Karachi while waiting for a replacement engine to arrive from Nimes. Captain Don Cartlidge, who had been the senior test pilot for the Carvair program, flew from Stansted to Nimes to pickup a replacement engine, which had 200 hours life remaining. He then commanded the Bristol Freighter from Nimes along the same route to Karachi with the spare. The Number Two engine was replaced on the night of March 7–8. It was run up twice, checked and determined to be performing within limits. Captain Don Cartlidge remained in Karachi and was scheduled to depart in the Bristol after the Carvair.

F-BMHU departed early on the morning of 08 March 1967 with the full freight load at maximum allowable takeoff gross weight of 72,899 pounds (33067 kg) in a high humidity situation, which reduced performance. The Carvair began the roll-out into a light wind with the temperature at 84 degrees Fahrenheit (29 C). The aircraft rotated about one-third down the runway. Witnesses reported hearing backfiring at this point. The aircraft crossed the runway threshold at an altitude of about 150 feet with the Number Two prop feathered. The Number One engine was not giving sufficient power causing the aircraft to lose altitude. The captain banked to the left into the failed Number Two and under performing Number One engine.

The Number One engine is the most critical on the Carvair because of the torque of engine rotation. On all engine out check flights the instructor feathers the Number One engine. This causes the wing to dip requiring considerable skill by the crew to manage the situation. Banking to the left into an under performing or failed engine with a high gross takeoff at low altitude assures disaster. The climb performance data indicates that level flight on two engines cannot be maintained at this weight (72,899 pounds). Additionally one can expect a rate of descent of 170 feet per minute in level flight with maximum continuous power on the two performing engines. The maximum weight at which level flight can be maintained is 58,400 pounds (26,490 kg). With the aircraft altitude of approximately 150 feet impact could be expected within 45 seconds.[11]

The aircraft continued to lose altitude and lined up on the National Highway near the Drigh Road Bridge. The Captain pitched the nose of the aircraft down to increase speed in a possible attempt to bring it down on the highway. It hit a railroad line then an embankment in a nose up attitude as he attempted to flare for contact with the highway. The right engines continued along the embankment while the left engines destroyed several trucks and vehicles on the highway causing seven fatalities and 15 injuries on the ground. F-BHMU burst into flames and airport fire trucks were able to rescue the captain and one flight mechanic who were both seriously injured. The other four-crew members, including second captain, co-pilot, flight engineer, and mechanic, perished. The bodies of the crew were returned to Paris for burial aboard the Bristol that brought the engine in two days earlier.

The post flight inquiry determined that the takeoff was attempted at maximum gross weight in high humidity at a temperature 25 degrees Fahrenheit (14 C) higher than allowable at this weight. The weight and balance manual states that 90 pounds is to be subtracted from the indicated takeoff weight for every degree Fahrenheit above I.S.A. Using this formula with the humidity and temperature the aircraft was approximately 2250 pounds overweight. In these conditions the maximum takeoff weight should have been 70,649 pounds. Furthermore a maximum gross weight takeoff of 72900 pounds at a temperature above 59 degrees Fahrenheit (15 C) is only allowable with an operational fuel dump system. The system was assumed operable, however if inoperable the takeoff gross is reduced to 66,670 pounds. Either way it was not an option because of altitude. The maximum fuel dump rate of 2,280 pounds per minute would require 6.5 minutes to reduce the weight of the aircraft enough to maintain level flight. With only two engines performing at an altitude of 150 feet fuel dumping was not an option. It appears this factor was not considered since the maximum takeoff gross was not reduced.

The Carvair operating manual states: Assuming infinite (sufficient) runway length, the limiting factor which restricts takeoff gross weight will always be the takeoff climb performance of the aircraft. This is the rate of climb in feet per minute with one engine inoperative and its propeller windmilling, takeoff power on the other three engines, landing gear retracted, wing flaps at 15 degrees, and cowl flaps in trail position. This should produce a rate of climb of 250 feet per minute.

The climb performance data shows that with a second engine not delivering takeoff power, level flight cannot be maintained. The Carvair because of a large nose has a tendency to lose altitude more rapidly when banking into a dead engine on the left side. This is covered in Carvair engine out procedures. In the original flight test at Aviation Traders it was noted that the most critical engine loss would be on the left, especially Number One. Test proved that to maintain directional control considerable rudder movement, counteracted by trim is required. A speed of 140 mph. must be maintained to continue climb. The left turn executed by the crew increased the rate of descent.

The survivors confirmed that the backfiring continued after the prop on Number Two engine was feathered. This supports the claim that Number One engine was not performing at takeoff power. The investigation considered the possibility that one of the starboard engines was backfiring. This was discounted considering the course of the aircraft would have been different. In all probability the aircraft weight relative to temperature, and humidity was not taken into consideration.

Under the original Carvair purchase agreement CAT exercised a replacement option and purchased G-ATRV from BUAF on 07 May 1967 to replace the F-BMHU. The French registration of F-BMHU was cancelled 19 December 1968.

11

Carvair Five

c/n 10365-5, C-54A-15-DC, Intercontinental — Interocean

Carvair five is possibly the most well documented and has one of the most colorful histories of all ships produced. It was owned and operated by more carriers and individuals both as a C-54 and Carvair than the majority of ships produced. In addition it is one of only two Carvairs originally equipped with an on-board ramp system with enlarged forward bulkhead for clearance of oversize vehicles. It is the second of five Carvair freighters built without the passenger cabin. Originally scheduled to be built as a car-ferry ship it was changed in production to a cargoliner. After less than a year the passenger cabin was added and it was retrofitted to car-ferry standards. After years of service it was converted back to cargoliner when sold in the United States. Between 1946 and 1960 the original C-54 was owned and operated by nine different airlines and flown in countries from Germany to Australia.

Originally built in Chicago it was delivered to the Air Force 03 August 1944 as a C-54A-15-DC USAAF serial 42-72260. The Air Force immediately transferred it to the Navy as an R5D-1 where it was assigned bureau identification BU50843. On 10 April 1946 it was declared redundant by the Navy and transferred to the Reconstruction Finance Corporation for disposal. The C-54 was assigned civilian registration NC58003 when purchased by postwar start-up carrier Veterans Air Express for $75,000 USD.

Veterans Air Express financed $63,750 of the purchase price and filed documents with the mortgage bank that it would be operated between Newark and Prague. The company was incorporated in New Jersey in July 1945 with Saunie Gravely as President and Harvey G. Stephenson as Vice President. This C-54 (R5D) along with a second Veterans ship were contracted for DC-4 commercial conversion by the Air Transport Division of Matson Navigation Company at a cost of $50,000 each.

Veterans filed for bankruptcy and went into receivership on 19 May 1947. The DC-4 conversion at Matson was partially complete with $31,000 owed. Only about $1800 had been paid on the aircraft mortgage since purchased. The seats and interior equipment were in a warehouse, awaiting a decision when the bankruptcy court ruled that NC58003 would be converted to passenger configuration for disposal. Matson and the Reconstruction Finance Corporation were directed to share equally in the losses for the overhaul. Matson would be allowed to purchase the aircraft with all spares and associated equipment for $61,979.17 USD. After some negotiation and giving consideration for losses the court accepted $90,000 from Matson for both DC-4s NC58003 and NC57777 giving Matson possession on 01 November 1948.[1]

Matson Airlines

Matson Navigation Company, a steam ship company, had financial interest in Pan American Airways until 1941. In 1943 Matson formed an Air Transport Division to acquire Navy maintenance and overhaul contracts. In September 1944 Matson received a maintenance contract for the Navy R5D version of the C-54. When the war ended Matson Air Transport had more than 700 employees at its Oakland base. The company began converting military C-54s to civilian DC-4s for Northwest, California Eastern, Chicago & Southern, Alaska Airlines, Transocean, and many others. The overhaul and engineering service placed Matson in the position to become involved with Veterans Air Express.

The steamship company was planning an airline operation until interrupted by the war in 1941. Matson Airlines was eventually formed in 1946 beginning service in July with two DC-4s on a "For Hire" basis. It had no fixed destinations but operated primarily between the U.S. West Coast and Hawaii. The concept was a luxury air service combined with ship travel. Matson filed an application with the CAB, which was never granted. The carrier operated the Hawaii service for two years until forced to drop it after CAB granted United and Northwest route authority.[2]

The Matson DC-4 N58003 that later became Carvair five operated glamorous charters transporting fashion models to Paris. It also transported not so glamorous immigrants from Puerto Rico to New York. The immigrant flights were described as a miserable business with the Skymasters overcrowded and overloaded beyond limits. The aircraft were grossly abused and eventually sold. In 1956 Matson reviewed the purchase of Transocean, which it had maintained a relationship with for years converting the carriers DC-4s to civilian use.

Matson operated an outstanding maintenance program but one specific incident stands out that has been a point of controversy for many years. After conversion by Matson Air Transport Division back to passenger configuration the Veterans DC-4 registered N57777 was rumored to have been involved in a rather nefarious incident. Because of a clerical error during the Veterans Air Express bankruptcy and subsequent Matson acquisition many records and publications indicated DC-4 c/n 10365 (later Carvair five) was registered as N57777. In fact this aircraft was never registered as N57777. A search of FAA records indicates Matson Navigation Company purchased the aircraft in 1948 and the registration of N58003 was transferred to Matson. The registration of N57777 was the other Veterans DC-4 c/n 3077 also acquired by Matson from the bankruptcy court at the same time.[3]

The airline division of Matson had grand plans for U.S. Mainland to Hawaii service, however a contested CAB ruling, political circumstance, and maintenance problems soon proved otherwise. Public records are not clear and have considerable discrepancies leading to confusion as to the registration and disposition of Skymaster c/n 10365 during ownership with Matson Navigation Company. Many publications and records indicate that Matson DC-4 c/n 10365 was registered N57777 between November 1948 and 03 February 1949. The same sources also show Skymaster c/n 3077 registered as N57777 from 22 May 1946 to 05 January 1949.[4] This indicates that both aircraft were N57777 between November 1948 and February 1949, which in actuality is not true. This single clerical error even fostered the idea that the registration was swapped from one aircraft to the other in a late night slight of hand because of a serious maintenance incident. In an unrelated but very strange coincidence DC-4 c/n 3077 was eventually purchased by Cecil Wroten and was abandoned at Phnom Penh in 1975 along with Carvair nineteen (N33AC) and DC-4 N32AC.

The maintenance incident occurred with one of the three Matson Skymasters. On a night departure out of San Francisco the DC-4 had climbed to 10,000 feet. Legendary aviator and writer Captain Ernest Gann commanded the flight. He related the incident in detail in his book *Fate Is the Hunter*. The crew that night included F/O Drake, Second Officer Hayes as reserve pilot, Engineer L.E. Snow, and a radio operator.

After leveling off the engineer made the power settings for the cruising speed. Within a few minutes the Number One engine backfired. Engineer Snow pushed the control to full rich. The other three engines began to sputter and backfire. The engineer moved all of them to full rich. The fuel flow needles began bouncing off the pegs and the boost pumps had no effect. The crew believed there was air trapped in the fuel lines and after a few minutes the engineer moved the levers back to cruise. The Skymaster smoothed out for several minutes then the Number One engine quit causing the aircraft to dip to the left. At that same moment the other three engines also quit. There was dead silence as the crew began emergency procedures. The engineer managed to get the engines started one by one only at full power settings.[5]

Fog was rapidly moving into San Francisco as the DC-4 turned back for an emergency landing at Oakland with all engines at max power. Captain Gann prepared for a landing at 220 miles per hour without any known procedures. The crew tried throttling back a single engine and each time it would begin backfiring and losing power. This continued down to 4000 feet where each engine began performing normally. The landing at Oakland was performed as if there had been no problem just 54 minutes after departing San Francisco.[6]

A team of mechanics met the flight and after the passengers were moved into the terminal ran up the engines. After the problem could not be duplicated Gann insisted on a test flight. Gann, Drake, Snow and a mechanic took the DC-4 to 10,000 feet and leveled off while Snow set the engines as he had previously done. As he brought the mixture back all the engines backfired and quit. They went through the same experience again. The mechanic, who had been skeptical, was convinced. The next morning Gann and legendary Chief Pilot E.L. Sloniger made another test flight with the same results. The fuel tanks were then drained and checked for contamination. Two days later the DC-4 was taken up and the engines quit again.[7]

Technicians pointed out that this was the only Matson DC-4 with P & W R2000-13 engines. The -13 is more powerful, allowing faster climb and better cruising speed. Unfortunately they tend to starve for fuel when mounted on the early C-54A wing because of the plumbing of the fuel system. After review the engineers pointed out that the -13 engines should be run 30 minutes in full rich mixture after leveling off at cruising altitude and Gann had only waited about 3 minutes. Gann stated that would waste fuel and suggested another test flight. They took the DC-4 with several technicians up to 10,000 feet and waited 30 minutes. The mixture controls were eased back and the aircraft engines ran smoothly. The technicians got rather smug with Captain Gann when suddenly all four engines quit. They were descending at 800 feet per minute and Gann did not get the engines re-started until 3000 feet. Everyone was convinced and the DC-4 was grounded.[8]

The Skymaster originally had the P & W R-2000-7 engines and was recently re-fitted with the -13 engines. After the -7 engines were re-installed the problem did not re-occur. The fuel line plumbing was apparently the culprit. Although the DC-4 in this incident was not converted to Carvair standards, other DC-4s with similar fuel line routings were and presented problems. This is reviewed in depth in the Engineering section of the Carvair.

Engineers at Aviation Traders noted that all DC-4s converted to Carvair standards had the fuel lines in a mixed variety of front and rear spar configurations. The lines, valves, and fittings were not mounted in a uniform system. The fuel line plumbing was changed and standardized on all Carvair conversion and they were all fitted with the R-2000-7M2 engines.

The story of unscrupulous switching of registrations was told for 40 years to justify incorrect accounts from many sources that two Matson DC-4s held the same registration of N57777. A search of FAA records found that only one DC-4 was registered N57777 and the registration was recorded on the wrong form during the Veterans Air Express bankruptcy. The other Matson Skymaster c/n 10365, originally a C-54 converted to a R5D for the U.S. Navy, became Carvair five. In all probability the DC-4 actually involved in the incident was a third ship, N50787 c/n 10278, which was later destroyed by fire when, refueling in Spain in 1961.

Air Ceylon/Australian National Airlines (ANA)

Ceylon, now Sri Lanka, established its first airline and began service on 10 December 1947 shortly after becoming an independent nation. The airline operated DC-3s until 1949 when it entered into an agreement with Australian National Airlines (ANA). Australian National Chairman Ivan Holyman negotiated an agreement to provide pilot training and technical support in exchange for access to the Europe-Australian market. To begin service two DC-4s were purchased from Matson. The first DC-4, c/n 10365 was purchased 20 January 1949 for $260,000. It was delivered on 03 February, registered VP-CBD, and named "Laxapana." It was re-registered CY-ACA on 27 September 1950.

Australian National crews actually flew the DC-4 on the route from Ceylon's (Sri Lanka) capital of Colombo to London with stops at Bombay, Karachi, Tel Aviv, and Rome. The service also extended eastward to Sydney, Singapore, and Jakarta.

Air Ceylon and ANA began to have differences in the early 1950s over how the airline was managed. Air Ceylon accused ANA of skimming the profits since it owned 49 percent of the company. Competing carriers were operating newer equipment and Air Ceylon was concerned about market share. As a result of the dispute the DC-4s were transferred to ANA with CY-ACA being purchased 16 March 1951. It was re-registered in Australia as VH-INY and remained in ANA service for seven years until purchased by Twentieth Century Airlines. KLM purchased ANA's 49 percent interest in Air Ceylon in 1953 ending the Air Ceylon-ANA relationship. As an added twist not only did this DC-4 become a Carvair but Ansett-ANA later purchased three Carvairs and a KLM Skymaster was used for the Carvair mock-up.

Twentieth Century Airlines/Seven Seas Airlines

Twentieth Century Airlines was registered in Nevada with James Fischgrund as president. It operated under the umbrella of North American Airlines until January 1957. North American under the direction of Stanley Weiss combined the certificates of four irregular carriers to operate a low cost charter operation. The regular scheduled airlines opposed the operation through the CAB and accused Weiss of violating flight restrictions placed on irregular carriers. North American was eventually forced to shut down and the following year Twentieth Century operated a DC-4 using a variation of this idea by leasing aircraft to other charter carriers. Four of the C-54/DC-4s owned by Twentieth Century eventually were converted to Carvairs.

Twentieth Century Airlines purchased its second DC-4 c/n 10365 from ANA on 26 March 1958. The purchase price of $198,750 USD included three spare R-2000-D3 engines. The Australian registration VH-INY was cancelled on 29 March and the U.S. registration N5520V was assigned on 30 April 1958. It was immediately leased to Seven Seas Airlines, which was founded by two former Slick Airways employees Erbun L. Drew and Earl J. Drew, for Trans-Atlantic charter work. The carrier began operations in October 1957 with a single DC-4. During the first year of operation a contract was secured to transport immigrants to Europe from Guyana and the West Indies. With the original DC-4 committed in the Pacific, N5520V was leased to fly the Trans-Atlantic charters. The operation only lasted until May 1958 when the DC-4 was returned to Twentieth Century. Even more significant is Seven Seas operation of United Nations contract flights early in the Belgian Congo conflict. Although Seven Seas did not use N5520V on Congo missions, this Skymaster was eventually converted to Carvair five for Interocean Congo work for the UN after the demise of Seven Seas. Adding to the coincidences Seven Seas relocated its European maintenance from Amsterdam to Luxembourg, home of Interocean. It was not unusual for American companies to base operations in Luxembourg as a flag of convenience.

Trans-Avia Flug (TAD)/Lufttransport-Uternehman (LTU)

On 09 May 1958 the aircraft was sold again to Trans-Avia Flug GMBH Dusseldorf and re-registered in Germany as D-ADAL. The U.S. registration was not cancelled until August 13.

A DC-4 at Hannover in 1960 with German registration D-ADAL, only a few months before purchase by British United for Carvair conversion (Roger Caratini via Jennifer M. Gradidge).

Trans-Avia operated it for a year before selling it on 30 September 1959 to another German carrier Lufttransport Union (LTU). The aircraft was named "Dusseldorf" by the new owner.

Lufttransport (LTU) was founded 20 October 1955 as one of the world's first low cost airlines. It operated the first flight 02 March 1956 from Frankfurt to Catania Sicily and later Palma de Mallorca Spain was added. LTU is also the first independent (All Charter) European Airline and was partially owned by two of its Captains, Erns-Jurgen Ahrens and Wolfgang Kraus. The carrier remained a charter airline until 1990 when it joined IATA beginning scheduled passenger flights. The DC-4 remained with the carrier only nine months before being sold again.

British United Airways

The Carvair program was in progress when the decision was made to purchase DC-4 c/n 10365 for conversion. After 16 years of military and commercial service with a number of carriers the DC-4 was quite worn. British United Airways took delivery from LTU on 7 August 1960. In September the British registration G-AREK, that had been reserved for Air Charter, was assigned to the DC-4. Air Charter and ATEL were divisions of the British United Group and former Freddie Laker companies. Ownership transferred to Aviation Traders in September 1961.

Carvair Conversion/Intercontinental

The DC-4 remained assigned to Air Charter until 27 December 1961 when it was delivered to ATEL for conversion. Disassembly began outside by removing the engines and draining the fuel tanks. The front fuselage section was cut away and cross supports added before it was then moved into the hangar. The nose section was delivered by road to Stansted on 22 July 1962.

When construction began it was scheduled to be completed for Channel Air Bridge. Instead Carvair five, c/n 10365, was changed to New York based Intercontinental U.S. Incorporated for delivery to Luxembourg associate company Interocean Airways. The carrier placed an order for two Carvairs to operate on a United Nations contract in Africa after the demise of Seven Seas Airways (see ship four).

British United initially ordered ten Carvairs for the Channel Air Bridge division and had taken delivery of the first three. Because of the reduced traffic during the winter season and no immediate need until spring, the delivery was easily adjusted allowing Intercontinental to assume the line position. It was viewed as an opportunity to gain foreign exposure by selling to a non–British operator.

Intercontinental submitted specifications to ATEL in June 1962 for two special modified cargo version Carvairs. The aircraft were originally designed for BUAF with the rear passenger cabin for car-ferry service. The DC-4 arrived at Aviation Traders in December 1961 but it was apparent that the change in specifications from car-ferry to cargoliner and addition of special equipment would extend the completion date.

Carvair five was originally designed as an ATL-98-A for cross-channel ferry service. The "A" designation is design modification ATL 98/5150 calling for the addition of 250 pounds of lead weights added to each wing tip. This modification increases the maximum zero fuel weight (MZFW) by 1000 pounds. The wing load is more evenly distributed across the wing reducing flex and stress at the wing root effectively increasing cargo payload. It was planned as an "A" model with the weights and reasonable to assume that Intercontinental/Interocean accepted the option. When sold in the U.S. in 1979 it had the weights installed. Adding or removing the wing "Bob Weights" is easily accomplished. It only requires removing the access panel, adding or removing weights, applying the placard, and re-weighing the aircraft.

Time was critical to Intercontinental/Interocean in order to fulfill the United Nations Congo contract. After the delivery positions of two ships under construction was adjusted, Lawrence M. Kesselman, Assistant Secretary of Winston Factors, signed a letter of intent on 27 July 1962 for £127,000 ($355,727). Intercontinental Airlines was actually owned by New York attorneys Kesselman, Goldberg, and Peck. Kesselman and Goldberg signed the purchase contract on 20 August 1962 for two Carvairs less engines. The terms of the contract required £12,700 ($35,572) upon signing, £50,800 ($142,290) upon delivery, and £10,000 ($29,970) per month for six months. The Intercontinental special modifications prompted ATEL to require ten percent of their cost upon signing of the contract and the balance on delivery.[9]

Aviation Traders required two DC-4 airframes as part payment to replace those being converted. Intercontinental traded in two Interocean DC-4s less engines naming LX-LMK as one of the ships. The letters "LMK" were the initials of Lawrence M. Kesselman, one of the owners of Intercontinental. The contract further stated that if -LMK was not available another aircraft of like condition could be substituted. The DC-4 arrived at Aviation Traders on 12 November 1962 and to the surprise of ATEL officials the aircraft was not LX-LMK c/n 7469 but LX-BBP c/n 10382.

The substituted Interocean Skymaster LX-BBP had been damaged in an incident on 27 October 1962. The carrier decided to keep the better aircraft and take advantage of the option to substitute another DC-4. No details of the incident are known. It appears that it was hastily repaired with the reasoning that it was to be disassembled anyway so the damage was not an issue. It was later converted to Carvair ten and purchased by BUAF.

There are 10 older and 10 newer airframes used in the Carvair program. The Skymaster had a total of 34,931.5 airframe hours on 02 September 1962. The conversion, which began 27 December 1961, took ten months to complete with the additional special modifications. The new empty weight was recorded as 40,750 pounds without the passenger cabin. Because of the extra options the weight is within 200 pounds of the first three aircraft built with passenger cabins. Unlike Carvair four for Intercontinental, the enlarged, notched, and reinforced forward bulkhead was installed in the nose assembly during construction at Southend prior to being transported to Stansted. Like ship four the special equipment included a self contained ramp system, cockpit crew bunks, four crew seats, cockpit galley, periscope sextant, chemical toilet mounted in the nose door, and cargo winch system. De-icing boots were installed on the wings, dorsal fin, and horizontal stabilizer after a serious incident involving Carvair four.

The British registration of G-AREK reserved for BAF was cancelled 04 September 1962. The U.S. registration was requested on 28 August and assigned to Intercontinental on 12 September. Testing and first engine runs were performed 25 October 1962. The upper fuselage was painted white with bare metal bottom. It flew for the first time on 02 November 1962 carrying the U.S. registration N9757F. The "F" indicating U.S. registered aircraft operating in foreign countries. The first flight departed Stansted and flew for several hours over the North Sea and Southend returning to Stansted. The C of A was issued at Southend on 13 November 1962 after 11 hours 20 minutes of flight-testing. Interocean crew training was conducted at Manston, Southend, and Stansted.

The British C of A for export was issued on 13 November. The total hours recorded at that time were 34,943.40. The ONUC lettering was applied at Southend prior to hand over. On 20 November honored RAF pilot Captain Patrick Barron delivered the new ship to Intercontinental at Luxembourg where a street is named after him. Two days earlier on 18 November the two Intercontinental/Interocean Carvairs N9757F and N9758F flew in formation over Southend.

Although Carvair five did not receive a Luxembourg registration immediately, LX-IOH was applied to the instrument panel when completed. When the aircraft had the Luxembourg registration issued it was LX-IOG. Ship four also had the registrations reversed. The registrations became confused in Luxembourg since Carvair four and five had the U.S. registrations issued in reverse order. Ship four was N9758F and ship five was N9757F.

Interocean

Intercontinental transferred it to Interocean Airways of Luxembourg on 17 December 1962. The companies shared the same management making it a paper purchase. Sometime in December the registration was sorted out and LX-IOG was assigned to the aircraft. Carvair five carried dual registrations for a short time either for the purpose of operating in politically sensitive situations or to mask the country of origin and registration. A letter from Bernard Goldberg to the FAA dated 25 February 1963 can verify this. The exact date the registration was changed on the aircraft is not known but strong evidence suggest it was in December 1962. An Interocean memo to the UN dated 26 November 1962 shows the registration of N9757F with only 65 hours on the aircraft since conversion and 65 hours on engines one, two, and four and 64 hours on Number Three. The U.S. registration of N9757F was cancelled on 23 February 1963 and the Luxembourg registration LX-IOG had already been applied to the fuselage.

Labor and political unrest forced Belgium to grant the Congo independence on 30 June 1960. The Italian Air Force supported the United Nations operation early on with C-119s. When the Italian C-119s were recalled in the summer of 1962 for financial reasons, the air division of Organisation des Nations Unies au Congo (ONUC) was put in desperate need of aircraft capable of transporting vehicles. The Carvair was able to fill the vehicle transport requirement.

Like ship four N9757F was built as a pure freighter without the passenger compartment making it quite different from previous British Carvairs. It featured a strengthened floor to complement the self-contained ramp system and enlarged forward bulkhead. The ramp system had a four-ton capacity to accommodate heavy military vehicles. The ramps on the two Inte-

Carvair five wearing U.S. registration N9757F at Kamina Base, Congo, in 1962 with the nose door open. The deicing boots are installed, indicating it is November or December. The cargoliner had no large windows in the rear (courtesy Carl Gustaf Wesslen/Leif Hellström collection).

rocean aircraft were not the same. Ship four ramps had guides on the inside edges but the unit built for ship five did not. The legs on one set of ramps had cable supports and on the other rods with reinforcement plates. The ramps were switched between aircraft from time to time. When needed they would be loaded from the storage area without regard as to which set. The ramps were not used when the aircraft was staged at Leopoldville (now Kinshasa) because cargo loaders were available.

Interocean press releases in March 1963 cited the Congo work and the ruggedness of the Carvair. It did perform quite well in Africa but the advertising was a bit exaggerated in an effort to demonstrate the ability to transport large loads. It did transport considerable military supplies, fuel drums, vehicles, artillery pieces and Argentine armored cars. Before leaving the Congo Indian troops were flown to Mombasa to return home by ship. It has not been determined why Aviation Traders Captain Don Cartlidge along with Ken Smith were dispatched to ferry LX-IOG from Leopoldville via Kano and Tripoli to Luxembourg. It only served with Interocean for slightly more than three years, operating in the Congo until July 1963 where it logged more than 2500 flying hours.

After returning from Africa it was repainted in the Interocean livery and operated for a short time without the passenger cabin. In the fall of 1963 Captain Jessie Meade and F/O Paul Rakisits flew a charter from Luxembourg to Bari Italy where two 70 foot antennas were loaded destined for Comodoro Rivadavia Argentina. The flight took them to the Azores, Dakar, Natal with a stopover at Recife and São Paulo. This was the last time they flew for Interocean. Rakisits later flew with two other Carvair operators during his 53-year career and Meade was tragically killed in a crash at Annabell airstrip Biafra.[10]

At the end of August 1963 LX-IOG was returned to Aviation Traders to be retrofitted to car-ferry standards. The passenger cabin, galley, and cabin lavatory were installed at Southend and it was ready for service on 28 September. The emergency exit on starboard side of the rear passenger cabin was installed with a round window making it the only ship with this feature. It was most likely fabricated or engineers used what was available. All other Carvairs fitted with passenger cabin emergency exits have a square window.

When first delivered to Intercontinental, ATEL engineers included along with the manuals the passenger cabin specifications of the first Carvair G-ANYB. Even though a lavatory was added with the addition of the passenger cabin the chemical toilet was not immediately removed from the nose door.

The Interocean passenger cabin contained only 17 seats while the British Carvair offered 22 seats. The portable ramp system was not removed with the installation of the passenger cabin although it reduced the car transport capacity to four. Interocean advertising photos from 1963 show the aircraft with passenger cabin and ramp system. The carrier launched an advertising campaign emphasizing the merits of the ramp system. The ramps although very useful, limited the bulk capacity and added weight to the aircraft. Interocean was not financially flush enough to maintain two Carvairs for limited use. The decision was made to transport the ramps on an as needed basis. If facilities were available at the destination the ramps were left behind at Findel Airport Luxembourg.

During the time that Intercontinental/Interocean owned the aircraft it was painted and lettered in three different variations. It was delivered to Intercontinental with white upper half, thin black cheat line below the windows and bare metal bottom with Carvair title on the fin. During UN operations ONUC lettering was added in large letters on the lower forward fuselage. It carried the ONUC titles both registrations N9757F and LX-IOG. After the Congo

After returning from the Congo, Interocean embarked on a major advertising campaign demonstrating the portable ramp system. The notched and squared bulkhead for loading military trucks is quite visible. The legs of the ramps have steel rods rather than the cable cross supports of the first set produced (courtesy Interocean/Leif Hellström collection).

mission, Interocean Airways letters with the blue and white striped arrowhead logo with a red lion was added to the forward fuselage. A variation of the arrowhead in a circle appeared in print ads but was never applied to the aircraft.

During 1964 and early 1965 LX-IOG operated a number of charters transporting general cargo, oil drilling equipment and British military contracts. On 06 August 1964 LX-IOG in tandem with LX-IOH operated a charter from Gatwick to Athens. On 16 August it operated Frankfurt to Prestwick. There were other erratic bookings to Helsinki, Malmo, Rome, Turin and the Azores. On 21 December 1964 it operated Stansted — Benin and on 06 January 1965 Southend — Athens. One of the last Interocean Carvair charters was 14 April 1965 from Gatwick to Naples. In March 1964 the U.S. operating certificate for Interocean parent company Intercontinental was suspended because of alleged improprieties. Interocean was already seeking a buyer for the Carvairs and seized the opportunity when French carrier Compagnie (or Cie) Air Transport (CAT) made an offer to purchase in May 1965. Interocean had operated some service for Spanish carrier AVIACO that maintained a marketing agreement with French carrier CAT in the Mediterranean.

Compagnie [Cie] Air Transport (CAT)

LX-IOG ferried from Luxembourg to Stansted 28 April 1965 for a check-4 overhaul and repainting. ATEL performed a zero-fuel modification on the aircraft. Like ship four the passenger cabin was increased from 17-seats to 55-seats. The nose-door toilet was also removed. It was test flown on 30 May pending sale to Compagnie [Cie] Air Transport of France (CAT). After completion it was the first of four Carvairs eventually operated by the carrier. It ferried from Stansted to Le Touquet on 31 May 1965 as F-BMHV officially recording French registration (F-BMHV) on 24 June.

The carrier formally named it "Cdt Max Geudt" in honor of World War II French Ace Max Geudt. He was the commander of a Banff Strike Wing Squadron of the Royal Air Force and was killed in Norway in 1945 while leading an attack on a German convoy. The two Carvairs were purchased by CAT to replace the Bristol MK.32 Super-freighters that were obtained from BUAF. The interchange agreement originally began with Silver City Airways. The CAT livery was only a minor alteration of the existing color scheme by adding a wide dark blue band along the windows and Cie Air Transport titles in white over the line on the forward fuselage. A square logo with two aircraft profiles and the name "Cie Air Transport" was applied on each side of the forward fuselage. This is a variation of the "Silver City" logo that appeared on Bristols, validating old market agreements between the two carriers.

CAT operated Bristol car-ferry service between Le Touquet/Calais and Lydd in a marketing agreement with Silver City before the merger with BUAF. The Mediterranean division was also operating from Nimes to Corsica and Palma. The decision was made to increase the network by purchasing the Carvairs and adding Bournemouth and Cherbourg to the cross channel service. F-BMHV remained with CAT for five years, which is longer than any of the carriers other Carvairs.

Transports Aeriens Reunis (TAR)/British Air Ferries

Transports Aeriens Reunis (TAR), a French charter airline, purchased F-BMHV in 1970 along the two other CAT Carvairs. One of the other two aircraft, F-BOSU, was ferried to Nimes France and used for spares. The other was moved to Nimes to be repainted with TAR lettering. Carvair five retained French registration F-BMHV although in a unique arrangement it was leased to British Air Ferries. The aircraft ferried from Le Touquet to Southend 29 April 1972

F-BMHV appeared rather plain in 1972 when owned by French carrier TAR and leased to British Air Ferries. It is the only Carvair to receive this bland color scheme (courtesy VIP/Nicky Scherrer).

F-BMHV as "Porky Pete" in a one-of-a-kind BAF livery late in 1972 while still owned by French carrier TAR. One month later it was purchased by BAF and re-registered G-AREK (courtesy Zoggavia/Paul Zogg collection).

to begin service and made the first revenue flight on 03 May from Southend to Le Touquet-Ostend.

The CAT logo and letters was replaced with BAF painted on each side of the forward fuse-lage above the stripe and "ON CHARTER TO BAF" added above the passenger door. Sometime after August 1972 F-BMHV was overhauled and repainted in a one-of-a-kind variation of the two-tone blue stripe BAF livery. The aircraft "Bridge" names were replaced with character iden-tities. The French registered British Carvair was renamed "Porky Pete." It was applied on the new livery of white upper half and two-tone stripe similar to other BAF Carvairs. The stripe ended at the rear passenger windows and did not continue up the tail. The letters "ON CHAR-TER TO" was moved from above the passenger door to above the British Air Ferries titles on the fuselage and TAR was added above the passenger entry door.

On 18 February 1973 British Air Ferries purchased Carvair five from TAR prompting a liv-ery change. It was temporarily registered G-41-1-73 for two days for transport then re-regis-tered G-AREK in Britain on 23 February. This is the same British registration the airframe carried in 1960 as a DC-4 prior to conversion and for the first flight as a Carvair. Shortly after repainting T.D. "Mike" Keegan proposed a new BAF color scheme of white top with blue stripe and dark blue bottom. For the second time in a year it was scheduled for repainting but retained the character name "Porky Pete."

The aircraft continually experienced mechanical problems and was considered unreliable. The situation became serious enough that BAF operations avoided scheduling it for charters. When the Paulings lease was negotiated in 1974 Keegan was very pleased and took it as an opportunity to get G-AREK out of his scheduled operation. It would be out of the BAF fleet and operating far from England.

Pauling Middle East

G-AREK remained in the last BAF color scheme with British Air Ferries only 16 months. Pauling Middle East Limited wet leased the aircraft on 20 June 1974 for support on construc-tion projects in remote areas of the Gulf region. The Paulings lease required a partial repaint in a new livery. The large blue stripe was changed to yellow with the Pauling logo on the tail and behind the cockpit window. Officials at BAF believed that the Carvair was no longer prof-itable and often stated they were only marginally profitable at their best. In addition G-AREK had been considered unreliable and the worst aircraft in the BAF fleet. The British car-ferry era had come to an end but the Carvair was still very useful for oversize cargo missions. Possibly this aircraft would have been more profitable if the expense of repainting it so many times were taken into account.

The Pauling lease called for four engineers, two based at Abu Dhabi and two at Thamarit (Midway). The field at Thamarit was a rolled sand strip, which wasn't good for the aircraft. The crews would fly every other day 3.5 hours Thamarit—Abu Dhabi and overnight then 1.2 hours Abu Dhabi—Muscat (Seeb Airport) and 3 more hours to Thamarit. Pauling operations were confined to Bahrain, Oman, and United Arab Emirates. The pilots and engineers took it in turns, alternate crews for alternate days. Once G-AREK settled into the Pauling, cycle it was reported to be more reliable. Almost immediately the crews considered this Carvair to be the least likely to bite you with an engine failure. This is contrary to the opinion of BAF crews in the U.K. and Keegan who had been very anxious to get rid of it. It would appear that by this time most of the maintenance bugs had been worked out on the longer segments.[11]

When G-AREK first arrived in the Gulf it was flying the Pauling Gulf circuit from Abu Dhabi—Muscat (Seeb)—Thamarit (Midway)—Salalah without HF radios. ATC at Seeb airport became irritated with Pauling because communications were limited until the aircraft was within range. BAF dispatched engineers Brian Mees and Sam Barrington to Abu Dhabi with a spare

Keegan saw a way to get G-AREK out of the BAF fleet in 1974 by leasing it to Pauling for work in Oman and Abu Dhabi. Once it started flying long segments it became reliable (courtesy Richard Vandervord).

Collins 618S-4 double sideband radio and antenna. After this first trip Engineer Mees was assigned to the aircraft and returned to the Gulf in March 1975 to keep G-AREK flying. The Collins was a crystal type radio and proved not to be reliable since the locals often changed frequencies. BAF made up another system with a Sunair digital type with the TR and ATU on the same board, which was installed late in 1974.[12]

According to Engineer Mees going out of Seeb on a hot day (ISA +30) was an adventure. Taking off over the ocean wasn't too bad but over the mountains (Jabal Akhdar) was critical. Although very forgiving, the aircraft was always overloaded and would not climb. The Seeb airport presented a number of problems. Because of the high mountains the Carvair always requested to takeoff toward the sea sometimes circling for 45 minutes to get enough altitude to clear the "Jebels" (mountains). It was standard procedure for the engineer to sit with his hands on the fuel dump levers until the aircraft cleared the mountains. The theory was that if an engine failed he could immediately dump enough fuel to buy enough time to recover and return to the field.

The contract-loading agents always fudged the weights and Paulings claimed everything was "priority" and had to move. It was extremely hard to make schedule since the cargo handlers were less than dependable and took breaks without notice in the middle of loading. Pauling officials had no concept of weight restrictions or lift in the desert heat and did not care as long as the cargo arrived at its destination.[13]

On one occasion G-AREK departed Abu Dhabi commanded by senior BAF Captain Vic Surrage with a cargo of two large generators. The aircraft used the entire 10,000-foot runway and mushed into the air. Engineer Mees stated, "We struggled and I mean struggled to climb"! Upon arrival at Seeb Captain Surrage ordered the generators off and the cargo load weighed. The Carvair had taken-off two-tons (4000 lbs.) overweight. Captain Surrage ordered one of the generators left behind and the other re-loaded for the second leg to Thamarit Oman.[14]

The Pauling contract called for transporting almost anything from machinery, generators, tools, spare parts, frozen food, liquor, beer, and large amounts of cash. Passengers were also transported in the 22-seat cabin which had not been removed from the aircraft. A Pauling employee was carried as a flight attendant-purser. One of these flight attendants almost got the crews thrown into prison, the aircraft confiscated, and Pauling kicked out of the Middle East.

The aircraft was required to clear customs each time it transited Muscat (Seeb) Oman inbound from Abu Dhabi. To continue the flight a non-objectionable customs certificate was required. The flight attendant always volunteered to take care of the customs papers. The crew often joked and wondered why he carried a large briefcase when he went inside. They assumed he was a little eccentric and carried all his forms in the case. However, since he took care of the papers they would meet him in the cafeteria and buy his coffee. This went on for some time until suddenly the authorities took him away and confronted the BAF crew. It was discovered that he was leaving the customs office and buying liquor at the duty free shop before re-joining the crew. He transported the liquor in the large brief case and sold it at exorbitant prices to locals in the desert at Thamarit.[15]

The Seeb airport was new and catered to shinny new jets. The authorities were not accustom too or liked the old leaky Carvair. Immediately upon landing the engineer was required to put drip pans under the engines to catch the dripping oil or be assessed an exorbitant fine. Because of the many restrictions imposed by local authorities it was impossible to keep on schedule.

Eventually a new captain was rotated in who was determined to show the company that he could stay on schedule in spite of the extra procedures and slow loading. After departing Seeb and climbing to a little over 2000 feet the captain turned east. The Carvair struggled to gain altitude as it approached the mountains. Not being accustomed to the on going problems of overloading and falsified weights he didn't have enough altitude. The Carvair was so heavy and low that he ended up flying through passes and valleys trying to maintain altitude. The engineer who had his hands on the dump valve and the co-pilot were scared out of their wits and protested to the company. The Captain was soon rotated back out and never returned.[16]

On 07 May 1975 G-AREK made a special car-ferry flight to Bahrain. The Pauling International local manager was Mr. Nevelli who drove a silver Rolls-Royce. The nearest dealer where it could be serviced was in Bahrain. The loading at Muscat was not a problem since Paulings had a pallet loader on station for the Carvair. However, it was not taken into account that there was not a car loader at Bahrain. The local cargo handlers devised a loader by taking a huge steel plate and positioning it on opposing forklifts. The Rolls-Royce was unloaded safely although it was a rather tricky maneuver.[17]

This same day Norwegian Carvair 14, LN-NAA arrived at Bahrain with Number Three engine shut down. It was en route to Southeast Asia for Red Cross humanitarian work in Cambodia. Brian Mees stayed behind to assist the other BAF engineer change the engine on LN-NAA and G-AREK returned to Abu Dhabi. Mees returned to Abu Dhabi on 13 May only to learn that his aircraft, G-AREK, was stranded at Thumrait with engine problems. The Norwegian Carvair was now called on to come to the rescue of G-AREK.[18]

BAF crews were also called on to fly the Carvair under combat conditions. Each month G-AREK was dispatched to extend the stage length from Thamarit to RAF Salalah to pick up wages for Paulings employees and ex-patriot mercenary types plus four to five tons of beer for a bar known as Charlie's. The bar was the local meeting place for Pauling employees and the British and Commonwealth mercenaries employed by the Sultan of Oman. The money was shipped in an extraordinarily heavy container and always the last item loaded on the aircraft by forklift. Anytime the Carvair was dispatched to Salalah the crew was forced to wait around for hours and generally wrote off the entire day. The Sultans military would come up with a Land Rover and take the crew to the beach. Ahead of the crew they used a Bedford military truck with a large roller on the front to clear the mines from the beach road.

The flight to Salalah was always an extension flight from Thamarit. The distance by road was only 50 kilometers, however it was mined. The aircraft was required to climb for 30 to 45 minutes to quite an altitude to avoid the guns of a rebel group known as "Adoo." This group was very active and could be heard at night in small arms skirmishes. The rebel group had Chinese weapons and was reported to have crude Chinese guided missile, which prompted the crews to be very cautious. The rebels continued their long lasting conflict for years. The approach to Salalah was always made from the sea to avoid any ground fire. The crews proved their resolve and the durability of the Carvair making many trips without sustaining damage.[19]

After 19 months in the Gulf the mission changed and Paulings had little need for the Carvair and scheduled it to return to BAF. Periodic maintenance and inspections were overdue when it left the Middle East. It was quite weary and needed major work because of the abusive mission transporting construction equipment and almost anything else needed in the remote desert outpost. After departing Saudi Arabia for the ferry back home to Southend, it suffered an in-flight engine failure and diverted to Athens Greece. It was flown by a British crew but still under lease to Paulings. The crew deadheaded home and Paulings abandoned it at Athens and refused to pay for servicing. BAF maintenance engineers were dispatched to Athens on 11 January to access the condition of the aircraft. They had problems with the local authorities at Athens because of Paulings unpaid invoices and returned to England. A second recovery crew was dispatched consisting of Captain Vic Surrage, F/O Tom Burt, and engineers Brian Mees, Doug Wall, and Frankie Briggs. After two weeks of considerable work, the engines were run up and the team got enough systems working and the aircraft airworthy enough for the ferry flight. The de-icing boots were toast and the crew was concerned about the weather. It had been damp and cold at Athens and the working conditions were so bad the crew would attempt anything to get home.[20]

Once airborne, G-AREK climbed out from Athens and over the Strait of Otranto. Brindisi control instructed the Captain to climb to 14,500 feet. It is not a good idea for a Carvair to attempt this altitude even when empty and it was heavy with seven hours of fuel onboard, two spare engines, miscellaneous parts, tools, and a crew of five including the two engineers. Captain Surrage advised Brindisi he would attempt to comply. Ice began to build up on the wings at 9,000 feet and the de-icing boots were intermittent. At 11,000 feet the captain advised they could not climb any higher and were picking up heavy ice. He requested a lower altitude as the Carvair began losing altitude even at full throttle. The situation became critical, as Surrage was unable to control the aircraft and continued losing altitude. First Officer Burt was busy with ATC while Captain Surrage had his hands full. After falling about 4,000 feet, they got G-AREK under control enough to maintain altitude. After just 2.5 hours of flight, five hours of fuel had been consumed, prompting Captain Surrage to declare a fuel emergency and divert to Nice. Once on the ground the first priority was to find the nearest pub. The crew agreed that if not for the outstanding flying ability of Captain Surrage they would not have survived.

Captain Surrage was known to be friends with BAF Chairman "Mike" Keegan. During the three days of repair he drove down to see Keegan who was on holiday near Nice. He told Keegan that they were there ferrying an aircraft back to Southend and offered a ride back on the Carvair with his car. Keegan was not aware it was G-AREK, which in his opinion, was the worst aircraft in the fleet that he had happily passed off to Paulings. The plan was to depart the fourth day at 12 noon after Keegan came up with his Jensen Interceptor convertible car. Captain Surrage did not realize that the two spare R-2000 engines in non-mobile stands had been loaded on the aircraft. They were in a position blocking the loading of Keegan's car and were not on rollers. The engineers put the car on the "Hylo" loader and used the bumper to push the engines back in the cargo bay so the car would clear. Keegan was not amused.[21]

The next day they went out and re-adjusted and fine-tuned the engines. Conditions in Nice were far better than Athens and the local FBO provided stands and assistance. It took three days to get the Carvair in a condition they felt safe to fly. The U.K. weather was considered excellent on 07 February and the decision was made to try for it. The next morning they departed Nice for England arriving at Southend without incident on 08 February at 16:30. This was the last time a BAF crew flew G-AREK other than training and test flights.[22]

Upon arrival at Southend it was parked next to the cannibalized Carvair eight, CF-EPV. Within a few days after returning to BAF, G-AREK was moved to a parking area on the north side of the field. After a review of the condition and repairs required it was removed from service and stored on an apron roadway at Southend, Essex. After 14 years of hard service throughout Europe, Africa, and the Middle East it was back at Southend. It appeared that this would be the end of G-AREK as it sat waiting for authorization and payment from Pauling for contracted maintenance agreed upon in the wet-lease contract.

As time passed waiting for the settlement BAF began removing parts for use on other Carvairs. Three propellers were removed almost immediately followed by the good engines that were installed at Athens to get it home. Eventually only one engine remained. It sat derelict during the first half of 1976 with a single engine in position four and later in position three and remained in this condition until purchased by Uni Air of France.

UNI Air

After the five months of storage at Southend G-AREK was brought back to flying status for Toulouse based UNI Air, the third French carrier to own it. The British C of A for export was issued on 22 Jun 1977 for purchase by Uni Air on 24 June. The aircraft was flown to Toulouse receiving French registration F-BYCL on 06 July 1977 and canceling the British registration G-AREK the same day. The Pauling logo was removed but the aircraft still carried the bright yellow stripe, white upper fuselage and navy blue wings and lower fuselage.

Uni Air purchased F-BYCL for proposed cargo charters from Europe to Africa and the Middle East. The contracts did not materialized and the aircraft remained stored at Toulouse. It is not known with absolute certainty, but there is no record of Uni Air ever operating the aircraft since public transportation authority was never requested. It remained in storage at Toulouse for two years until purchased by Ruth May of Dallas Texas for lease to Falcon Airways. A study of records and photographs indicate that very little was done to the aircraft during the two years in storage. The call sign G-AREK was still on the panel when purchased by Ms. May.

Ruth May/Falcon Airways

British Air Ferries advertised three Carvairs in January 1979. Only two were airworthy since G-AOFW had been stored without engines since early 1977. Falcon Airways purchased BAF Carvairs, G-ASDC and G-ASHZ, in April 1979 and agreed in principal to purchase a third. It was obvious that G-AOFW was not an option, since it would require considerable expense to make airworthy. The only other available Carvair was F-BYCL that belonged to Uni Air in France. It was complete and although stored required less maintenance to become airworthy. Lewis May, President and majority owner of Falcon Airways along with Vice President Jim Griswold, were reviewing aircraft for a very ambitious expansion. In July 1979 Griswold was interviewed by *Aviation Week* and presented his very aggressive ideas and overly optimistic view of the market potential of the aging Carvair.

Carvair five was not purchased by Falcon as the other two ships. The aircraft was purchased by Ruth May, wife of Falcon president Lewis May, and leased to Falcon. Ms. May owned other aircraft in the past and had considerable experience in aviation prior to her marriage to Lewis May. Her career in aviation began during World War II when she convinced the U.S. Weather Bureau to allow her to operate a weather station. Until this time few women had been allowed to work in aviation but with most men in the military she got the job becoming the first woman in the state of Texas to work for the U.S. Weather Bureau. A graduate of Scripps College for women in Claremont California, Ms. May was almost immediately taken by the thrill of flying and the young "Fly Boys" as she called them. She is truly one of the most gracious, fascinating and colorful individuals in the Carvair story and can weave a story of her aviation experience in such a way that you are poised on the edge of your seat.[23]

Ms. May participated in the negotiations and purchase of the first two Falcon Carvairs and flew to London on behalf of her husband to inspect them at Southend with Jim Griswold. There had been some friction between Ms. May and Griswold since she was a primary financial backer and he preferred her not in the operation. Ms. May had placed a substantial amount of personal assets as collateral for the Falcon Carvairs. Griswold arrived in England first and convinced the BAF officials that as Vice President of Falcon he would be acting on behalf of Lewis May. However, Lewis May was relying on Ms. May to participate in the negotiations on his behalf and she had her own interest to protect. This confused the staff at BAF who were hesitant to negotiate the aircraft sale with a woman that was not an officer of Falcon Airways.[24]

Ms. May recalled seeing Carvair 12, "Big John," which was stored at Southend. It had been removed from service in 1977 and used as a static display. Ms. May and Griswold were taken out to look at it as part of the sales pitch and to consider the possibility of it being put back in service as a third Falcon ship. At this point officials at BAF were still not clear as to who would be making the decision to purchase Carvairs seven and nine. They were not aware that Lewis May, who was back in Dallas Texas, would make his decision based on the impression and recommendation of Ruth May.

Ms. May stated that as they were walking out to the aircraft she overheard one of the BAF officials say to Jim Griswold. "Would you be so kind as to assure me that the lady is not going to climb a ladder into the aircraft." Ms. May stated that although she was wearing a dress and expensive shoes she was determined to scale the ladder to board the aircraft. This was her first encounter with the Carvair. It was not established at this time that she would purchase F-BYCL and it would become part of the Falcon fleet secured by her personal loan. Uni Air dispatched a representative to Southend to offer Carvair five, which was in storage at Toulouse France. During the negotiations for the first two ships for Falcon, a tentative sale was negotiated with Jean-Pierre Sauval of Uni Air for F-BYCL.[25]

Sauval agreed to coordinate the maintenance and painting to be performed at Toulouse. His letter to Ms. May of 07 July stated he had given the spares list to V.J. La France, who was an official at Falcon Airways. Sauval stated that La France advised him that he would be at Toulouse in August to take delivery of the aircraft. This did not sit well with Ms. May who was actually purchasing the aircraft.[26] Ship five remained registered in France while the sale was pending. Application for U.S. registration was filed with the FAA on 15 July 1979. The registration of N83FA was granted on 04 August for the transfer to Texas. The French registration F-BYCL was cancelled on 24 August 1979.

While processing the registration it was discovered that the FAA did not have a release on file from Aviation Traders giving clear title of the aircraft. The previous U.S. registration N9757F held in 1962 by Intercontinental was never cleared. Apparently during the Intercontinental/Interocean shuffling with the double U.S. and Luxembourg registrations for Congo work and the request for British registration requested prior to conversion, the proper release was never filed.[27] This oversight put the aircraft registration in question. Ms. May was livid that her bank did not catch the discrepancy and would put up the money and allow her to purchase an aircraft without clear title. The bank had either not made a title search or they had not informed her of any title problem. Eventually the title was cleared and arrangements were made for Ms. May to take possession.

The Falcon Carvair livery is a variation of the last BAF cargo scheme using the paint left from the last car-ferry colors. Only Carvairs seven and nine were ever painted in the BAF Cargo yellow and blue "sash" colors and both became Falcon Carvairs. Carvair five was originally intended for BUAF before Intercontinental took the delivery position in 1962. The Carvair had worn the Interocean livery; three French liveries and the colors of a number of other carriers, plus several one-of-a-kind BAF paint schemes. Once again it was returning to a variation of BAF colors. A portion of the cargo livery was still visible on the vertical stabilizer in 1997 when the aircraft was lost in a tragic incident.

Ruth May purchased the N83FA for $200,000 USD in August 1979. The purchase included two spare engines, miscellaneous other spares, and one cargo loader all valued at approximately $120,000.[28] A security agreement, FAA document A99619, was executed 30 August 1979, filed 12 December, and recorded 15 January 1980. The discrepancy of title was finally cleared and the U.S registration was granted 15 October 1979.

Estimates were as high as an additional $40,000 USD above the purchase price to make the Carvair airworthy and fly it back to Dallas Texas. This included seven days of expenses, air fares to France, 7200 gallons of fuel at a cost between $10,800 and $14,400 USD, painting, maintenance, and miscellaneous expenses. The Carvair actually returned to the U.S. as N83FA with the title problems, which had not yet been cleared. The purchase appeared to be in the favor of Ms. May. The aircraft had Cardex; a considerable list of spares valued at $120,000 USD, including two engines. Mr. Sauval noted that the spare engines were not QEC and would require about $2500 USD each to fit the kits. Also included in the deal was an ATEL "Hylo" scissors loader. The first sign of problems came in a letter from J. P. Sauval of Uni Air to Ruth May dated 10 July 1979. Mr. Sauval stated that he did not think the aircraft would be ready before 20 August. He also stated that La France had inquired four days earlier about a delivery date in order to plan his arrival at Toulouse.

The Carvair had to be repainted to match the other two Falcon Carvairs. The repainting ultimately cost $5,000 USD. Accounting records show the original cost estimate was $1,500 USD since Falcon had adopted a variation of the BAF cargo color scheme. The BAF livery consisted of a yellow and navy sash around the forward fuselage and a block and stripe pattern on the tail with the letters "CARGO." The only alteration necessary on those ships was to change the BAF yellow stripe to light blue. Unlike the other two Falcon Carvairs it was no easy task to repaint Carvair five. The entire aircraft was still painted in the old Pauling colors. Someone failed to take into account that Carvair five was not already in a "BAF Cargo" color scheme and would have to be stripped to bare metal on both fuselage and wings and repainted, resulting in a $3,500 cost over run.

Ruth and Lewis May arrived at Toulouse with a group of 19 people including V.J. La France, and Falcon business associates. Jim Griswold had remained in Dallas to oversee Falcon operations. Griswold made it obvious to Ms. May that he did not want anything to do with the third Falcon Carvair because it belonged to her. Most of the group that traveled to France were not essential to the ferrying of the aircraft but tagged along through the generosity of Lewis May. Ms. May took delivery on 18 August 1979 and affectionately re-named her Carvair "Creme Puff." The spares were loaded along with many cases of wine that were a gift from the French. Ms. May warned the group that they would not be able to get the wine back into Texas without problems with customs. She tried in vain to get them to leave it behind. The Carvair departed France on 20 August for England with Ruth and Lewis May, three-cockpit crew, mechanic, loadmaster, 13 merry maker passengers and all spares on board.[29]

After leveling off at cruising altitude the crew experienced erratic oil pressure on Number Two engine. Just before crossing the English Channel they were forced to shut it down. The decision was made to continue on to Southend where maintenance was available. Upon landing an inspection found a blown cylinder on Number Two, s/n 100576. It appears that Mr. Sauval was right when he wrote Ms. May and said it would not be ready by 20 August.

While the Number Two engine was being repaired the band of merry makers went to a hotel. Lewis May felt some responsibility and generously opted to pay for the hotel. Ruth May was not as generous when she discovered some of the group making frivolous charges to their rooms. While waiting for repairs one of the passengers purchased an old touring car and had it loaded on board without consulting with her.[30] V J La France had previously purchased a Triumph Vitesse from a Southend Customs Agent and transported it back to Texas on the delivery of N80FA. There had been some friction about that incident since the car was transported at Falcon Airways expense.

The crew expressed concern about Number Three engine, which had also been performing erratically on the flight from Toulouse and was leaking oil. After some adjustments and run up test it appeared to be performing within limits. The group remained at Southend until 23 August completing maintenance and waiting for the paperwork to be processed to export the aircraft.

N83FA departed England on 23 August 1979 with the band of merry makers, wine, car, and spares on-board. Only a short time into the flight the Number Three engine, s/n 27094 with 471 hours, failed over Lockerbie Scotland. The decision was made to return to Southend where heavy maintenance was available rather than divert to a field in Scotland. Many years later Ms. May stated that this was only the beginning of a disastrous financial time for her. She was already experiencing a cost over run from the original estimate. Servicing and painting were over budget. The band of merry makers had charged excessive amounts to their hotel rooms in France and England and in addition the scissors loader had been left behind. Because of size it could not be loaded and there was no one available to disassemble it for transport back to Dallas. The cylinder change on Number Two engine and now an engine failure on the first day of actual ferry operation caused her concern.[31]

The Number Three engine had oil pressure fluctuations and lacked power from Toulouse to Southend. The problems had been addressed at Southend and the engine run-ups and adjustments were deemed adequate for the flight back to the United States. After arriving back at Southend the second time the inspection of Number Three revealed a master rod failure. The decision was made to have BAF Engineering replace it with s/n 107555, which was one of the spares on-board with only 32.1 hours on it. Ms. May took this opportunity to unload the unauthorized automobile and non-essential riders who had been frightened by this second engine failure. She instructed BAF to unload the car with the spare engine.[32] It should be noted that if the aircraft had been converted to all cargo as the other two Falcon Carvairs there would not have been a passenger cabin and the 19 passengers to deal with. Lewis and Ruth May along with the crew decided that rather than spend an unknown number of days at a hotel they would return to the U.S. while BAF changed the engine and completed additional maintenance. The engine was changed on 31 August and run up with no problems.

Ruth May along with the crew returned to Southend on 13 September to prepare for a second attempt. The ferry flight was planned again and the export paperwork completed. The removal of the touring car and the passengers that returned to the U.S. on commercial flights reduced the takeoff gross and fuel consumption. Ms. May was concerned with ferry cost after receiving the second invoice from BAF Engineering totaling of $5,365 USD for the engine change, fuel, oil, and maintenance.[33]

Ms. May wrote on the back of the day five page of the ferry log. "Power project at Churchill Falls, Carvairs carried passengers and equip. One may be in the bag at Twin Falls airport gravel strip, Robert Scott/Eastern Prov. Bert Lambert, Churchill Falls." This entry is in reference to Carvair six, which had been destroyed at Twin Falls Labrador. The plan had been to possibly acquire the spares since the ferry flight to the U.S. would pass close to the area. It appears that

Ms. May was given incorrect information. The Carvair lost at Churchill Falls had long since been salvaged and the remains plowed under. A photo from Ruth May's files shows a rudder with the EPA lettering being loaded. It is the rudder from CF-EPX, which was included with the spares acquired and was actually on board.

On their second departure from Southend on 17 September the crew decided to stop at the Outer Hebrides Islands off the coast of Scotland to top off the fuel tanks before the first leg over the North Atlantic. The FBO there was not interested in taking U.S. currency. After some negotiating Ms. May was allowed to purchase fuel on credit and send the bill to Falcon Airways in Addison Texas. Once again problems arose with the replacement Number Three engine, s/n 10755. The crew noted erratic oil pressure before landing at the Outer Hebrides but after an inspection and run-up with no major problems, the decision was made to continue. At this point more than $40,000 USD had already been expended, which was the total estimated expenses, and they were still in Scotland.[34]

After departing the Hebrides for Reykjavik Iceland the crew began to experience fluctuating oil pressure on Number Three engine again. After several hours it began leaking oil, which became a growing concern since there was no central oil system to replenish the supply. The DC-4 central oil tank had been deleted when the aircraft was converted to Carvair standards. The co-pilot who had little or no sleep in several days became exhausted while constantly monitoring the pressure and quantity gages. Ms. May volunteered to take a turn flying the right seat while he lay down in the rear passenger cabin, which offered some degree of comfort over the flight deck. After arriving at Reykjavik Iceland the crew became concerned about the weather. Ms. May had considerable experience from operating a weather station in Texas. She reviewed the forecast and decided it would not be a problem. The crew was convinced and decided that maybe they should continue on to Greenland after attempting to fix the oil leak and refueling at a cost of $1065.00. This would allow them to get ahead of the front if her assessment of the weather was incorrect.[35]

They arrived in Greenland exhausted and the decision was made to overnight. Ms. May stated she had never stayed in a hotel room that had one bare light bulb hanging down in the middle of the room. She thought they might have trouble when the innkeeper told her to stay in her room and block the door. In the interim the crew was having something to eat in the bar/restaurant and made the mistake of speaking to one of the local girls. Shortly afterward a large local fellow they described as "Nanook of the North" decided to take on the entire flight crew. Once again the wisdom of the innkeeper prevailed when he shut off all the lights allowing the crew to escape.[36]

The next morning the crew decided it would be best to leave as soon as possible. After refueling and oil service at a cost of $2,873 USD N83FA departed en route to Goose Bay Labrador and again experienced problems with Number Three engine. They were forced to shut it down and continue on rather than return to Greenland where the crew had escaped physical harm. Ms. May was to sit in the rear and be prepared to kick out the life raft if they had to ditch. She recalled that the ocean looked very rough with deep troughs. She was very concerned with the situation and became sick from watching the water. She did not realize how rough the seas were until they reached Goose Bay. Upon arrival the crew was told that the swells were 30 feet. Ms. May kissed the ground and threw up.[37]

The weather was so bad they had to rent a hangar in order to make repairs to Number Three engine. Ms. May spent the time sitting in the back of her aircraft knitting while the engine was repaired and checked out. After another day of waiting and repairs, gas and oil costing $2,607 USD plus hangar rental N83FA was again airborne en route to Ottawa. At Ottawa the aircraft was refueled at a cost of $1850.46. The balance of the trip was less eventful with one more stop at St. Louis before arriving home at Addison Texas on 21 September. After arriving in Texas records dated 29 September indicate the total time on the aircraft as 47,399 hours.

The crew poses with N83FA after arriving Dallas from France without the Merry Makers. From left, Mike Robinson, Tom Leatherton, Ruth May (owner) and Stuart Dawson (courtesy Ruth A. May).

The planned seven day trip of 32 hours of flying that began on 21 August had taken a month and nearly twice the flying hours. Records of Ruth May's personal journal shows a cost over run of $11,575 USD, considerably above the liberal estimate of $40,000 to ready the aircraft and ferry it from Toulouse to Addison. The amount does not count the use of one of the spare engines and the value of the scissors loader left behind. The estimate to return from England was based on the cost to ferry back each of the other Falcon Carvairs. Both of those aircraft were in better condition in England and N83FA was stored for several years and was in France.

Carvair five finally received clear title to N83FA on 15 October 1979 allowing it to be leased to Falcon Airways. Once in Dallas the wing "Bob Weights" were re-installed to allow for an additional 1000 pounds of cargo. The nose door was upgraded with the di-electric patch for radar and gill-liners were installed in the cargo bay as required for U.S. operation.

Because of the cost overruns the loan on the aircraft was renewed at $265,000 on 10 December 1979.[38] Mechanical problems continued for some time after joining the Falcon fleet. It would appear that the reputation of being the worst aircraft in the BAF fleet had come back to haunt it. The Number Two engine, s/n 100576, failed with only 459 hours during a test flight in March 1980. It was replaced with the other spare engine s/n 100400 brought from France, which was considered good and only had 13.8 hours on it.

Falcon used N83FA to uplift oil-drilling equipment to North and South America. Items such as pipe and directional drilling equipment, explosives, helicopter blades, aircraft engines, and ship props were also transported. The cargo carrier had an established record for transporting explosives and hazardous materials. Ms. May had a habit of riding along on some of these exotic and dangerous charters. Once she showed her quick dry sense of humor while accompanying a dynamite charter to Aruba. She rode in the cargo compartment playing Pinochle on the boxes of dynamite. One of the flight crew came down from the flight deck and told her that it would probably be much safer if she rode in the cockpit. She casually commented. "The way I see it is that it doesn't really make any difference. The front half of this airplane is not going to get there without the back half."[39]

N83FA did not remain with Falcon Airways very long. By March 1980 Falcon was in financial trouble and bankrupt. Ruth May stated that she did not think that her aircraft ever turned a profit at Falcon. Since the aircraft belonged to her, after the demise of Falcon, it was moved to Fort Worth and she drove from Dallas weekly with a mechanic to run it up and keep it airworthy. Letters, notes, and telephone records from her files indicate an attempt to operate ad hoc charters in order to generate cash flow.

Ms. May's hand written notes indicate an inquiry on 18 July regarding the transport horses. The charter did not materialize and the Carvair remained idle. By October 1980 it had only operated a few charter flights. All of the efforts by Ms. May did not pay off and late in October 1980 it was removed from service and stored at Ft. Worth.

Gifford Aviation/Kodiak Western Alaska Airlines (Air Fortynine)

Gifford Aviation Inc., formerly Air Fortynine, a third level cargo operator based at Merrill Field in Anchorage had grand plans of expansion. Gifford Aviation agreed to lease the Carvair N83FA from Ruth May on 17 February 1981 for a period of six years ending in 1987. John Gifford and his wife Darlene were the majority stockholder, with 59 percent of outstanding shares. The other officers and shareholders were Howard G. Fowler 10 percent, David M. and Debra Pickworth 21 percent, and Ralph Stemp 10 percent. Gifford Aviation was the parent company of Kodiak Western Alaska Airlines and Starbird Incorporated.[40] Gifford was leasing N83FA to operate under Kodiak Western transporting general cargo and fish.

Ms. May negotiated the contract with Gifford from her home in Dallas. She had not visited Fort Worth for several months relying on a local mechanic to run up the engines periodically. After signing the contract with Gifford Ms. May learned N83FA had been damaged on 15 January while parked at Meacham Field. The damage to the nose gear and doors was serious enough to make the aircraft not airworthy. Gifford was notified of the delay in delivery, which was not well accepted. Ms. May had to sustain the cost of repair until the responsible party could be contacted for settlement. Gifford turned to Aero Union and leased Carvair 17 until N83FA could be repaired.

The details of the incident were never clear. A Beech 18 registered to Edward B. Harp of Arlington Texas ran under the nose of the Carvair impacting the nose gear, doors, and lower fuselage. The FBO did not advise Ms. May and authorities were not notified. Unofficially it was alleged that the Beech was under power at the time and the pilot lost control. After some pressure the owner of the Beech 18 accepted responsibility for damages. The Statement in Proof of Loss dated 24 February 1983 stated "A Beech 18 aircraft was blown into insured aircraft causing damage." Although N83FA was repaired immediately for delivery to Gifford the settlement was a full year after the damaged had occurred.

The financial status of Gifford Aviation was never very strong. However, on 22 February 1982 a year after leasing N83FA, Gifford opted to purchase it for $254,095.82 with the lease payments credited against the purchase. The monthly payments were recorded as $5,912 per month with John Gifford personally guaranteeing the loan. Numerous letters of insurance cancellations and late notices plus warning letters from attorneys for lack of insurance coverage indicated financial problems for some time and pointed to Gifford's demise. During 1982 Gifford only made five payments on the aircraft which were frequently delinquent. Mrs. May's attorney, Frederick Zimiring, stated in a letter of 18 May 1983 that he had spoken to Gifford's attorney about the problem. He was told that Gifford Aviation was being sold and the new management would settle the loan or return the aircraft.

Gifford signed an agreement on 16 May 1983 with James B. Burr, president of Rocky Mountain Helicopters, a Utah corporation, to purchase the company for $1.5 million. The deal with Rocky Mountain Helicopters was already too late. In a letter to Ruth May dated 19 May 1983 Gifford wrote. "I have sold almost everything I own, including my house to keep Gifford Aviation afloat. The court may have told you that we had $290.000 in receivables; however, January, February and March are the worst months of the year and it requires more than this to cover expenses. The cash required to stay current with everything is $330,000 per month — so there is not enough cash to go around." ... "I am sure you realize that 83FA has not had a revenue flight since October of 1981. I realize that I personally guaranteed this note, but as I mentioned above, I have no assets to contribute. Therefore, I believe our only alternative is to abandon the 121 program that we have tried so hard to hold together. We have placed the DC-4 on the market and expect to sell the aircraft in order to inject the cash required to correct the debt to you."[41]

Prior to receiving the letter Ms. May had been contacted by a friend and advised that N83FA was for sale in TRADE-A-PLANE. Gifford had not advised her of their financial problems and lack of business or that he was trying to sell N83FA for which she had not been paid. Ms. May's outrage prompted John Gifford to write his letter. The end to Gifford Aviation was at hand. Gifford filed for chapter 7 bankruptcy 23 May 1985. N83FA was impounded in Alaska with 47,984 hours on the airframe and it had not been flown since October 1981. Ms. May filed suit in 1984 against John Gifford, case 3AN-83-5736 in Alaska Third Judicial District for nonpayment. The suit was in the amount of $285,220,56 for all monies owed her through 03 January 1984. From that time on she was incurring aircraft parking expenses at $339.45 per month plus insurance, attorneys fees and the interest on the loan was compounding. The unpaid balance now due with penalties and associated cost was more than $340,000, which was twice the value of the aircraft.

Ruth May was agreeable to sell N83FA for $221,779 and give full credit against the principle for payments that Gifford had made less the insurance premium. Gifford had purchased N83FA on 22 February 1982 but because of Gifford's erratic payments Ruth May was shopping for a new owner by July 1983.

Ms. May was not aware that Carvair 17 was leased in April 1981 to Kodiak Western Alaska Airlines, a subsidiary of Gifford Aviation. This was during the same period Gifford was negotiating with her for Carvair five. Carvair 17 belonged to Aero Union at that time and was involved in a dispute over lease payments. It is not clear what role Charles Willis Jr. played in that dispute but he was somehow involved in a lease with Aero Union. At the same time Willis was also making inquires to purchase Carvair five from Ms. May.[42]

Willis, who later appeared in Honolulu as an officer in Hawaiian based Pacific Aerolift indicated to Ms. May that he had an immediate need for a Carvair for work in Panama. He told her that he had a contract in the amount of $350,000 USD to operate a charter or series of charters to Howard AFB in Panama. Willis proposed a lease purchase-agreement with payments of $15,000 per month for a dry lease with an option to purchase and credit for payments made. Ms. May offered to sell N83FA to Willis for $250,000 on 22 July 1983 but the negotiations did not go well. A hand written entry in her files stated. "Stay away from Pacific Aerolift, bad news." In addition she noted in her journal, "Dick Foy," "Charles Willis is bad news he had Dick's plane on lease, Just got it back. Dick used his own crew." This was in reference to Aero Union Carvair 17 when there was some dispute about the terms and payments resulting in Aero Union demanding the aircraft returned.[43]

Ms. May flew to Anchorage on 21 July 1983 to inspect N83FA and determine what was needed to get it airworthy. On 22 July she sent a letter to Lewis Khem of Pacific Air Express offering it for $225,000. Pacific Air Express was shopping specifically for Carvairs for its Hawaii operation and had purchased another Falcon ship, N80FA, in January 1983. A month after these negotiations Khem purchased Carvair 17 from Dick Foy at Aero Union.

There were also inquires from Warren Basler, Air Resorts Airline, Atorie Air, Aviation Enterprises, and Cary Cargo based in Barbados. Cary operated cargo service to Trinidad, Tobago and Barbados and expressed a strong interest even inquiring about crew training and modifying its certificate. There was also discussion with Dick Foy of Aero Union. This is rather unique since Aero Union owned Carvair 17, which had been leased to Kodiak Western Alaska Airlines, a subsidiary of Gifford Aviation. Ms. May was attempting to sell Carvair five to Aero Union after repossessing it from Gifford, who owned Kodiak Western.[44]

Ms. May even offered N83FA to the Confederate Air Force on 02 September 1983. She stated in her letter that N83FA flew the same as a standard DC-4 and the Cardex and flight logs were up to date with 4451 hours left on the airframe before the next scheduled overhaul. She mentioned the historical significance of the aircraft and the original DC-4 airframe of the World War II era. It appears officials of the Confederate Air Force did not realize how rare the aircraft was, being only one of 21 Carvairs built. Robert Griffin's response of 20 October 1983 stated.

"After very careful consideration of your kind offer, the acquisitions committee has recommended we decline the offer. The reasoning of the committee was primarily one of cost of restoration of the aircraft to full C-54 configuration in museum display condition. It was felt this restoration process would be beyond the financial capability of the museum."[45]

It was a missed opportunity to recognize the rarity of the Carvair and a turning point in history of the type. N83FA operated until lost in 1997 leaving only four in existence at that time. It could have been an opportunity to save one example for future generations but the CAF committee was thinking of re-converting it back to a DC-4. While waiting for a decision from the Confederate Air Force Ms. May also contacted BAF on 01 October 1983. They could not offer any assistance since the Carvair was long past its time and there was no demand for the aircraft in the jet age.

Another interested party of note is Jim Blumenthal who had purchased the other two Falcon Carvairs (N80FA) (N89FA) and was operating a standard DC-4. He considered N83FA for speculative resale and spares to support N89FA and his DC-4. He contacted Ms. May on 28 March 1984 and urged her to return the aircraft to airworthy status for a test flight. He advised her that it would be easier to sell and she could add the cost to the selling price. He also informed her that he bought spares from the Gifford bankruptcy sale and many of the items that were originally with N83FA were missing and offered suggestions as to where they might be found.[46]

Ms. May aggressively sought cargo contracts to put N83FA back to work. She contacted a group in Jacksonville seeking cargo support for a $25 million contract to supply ARAMCO. There is also notation of a contract of 40,000 pounds a week between Guyana and Supervalu of Minnesota to transport produce. While the aircraft was parked in Alaska deteriorating, other offers came in but none were close to the actual value. The net payoff of Ms. May's loan to the Mercantile Bank of Dallas in 1985 was $76,712.44 as noted in her files. She signed an agreement of consignment with aircraft broker John Woods Inc. of Dallas on 19 February 1985. The agreement contained a condition that if Stephen Wong of Hong Kong, Carlos Bustamente of Puerto Rico or Lewis Khem of Honolulu purchased the aircraft no commission would be paid. It is of note that Lewis Khem was still a potential buyer after several years of negotiations.

The aircraft remained parked in Alaska unsold. The Anchorage Aircraft Exchange wrote Ms. May on 24 May stating that they would contact her immediately when a sale was put together. As of April 1985 Ms. May's attorneys were still continuing their litigation against Gifford and the bankruptcy trustee regarding the disposition of N83FA. She assumed this indicated a sound offer was forthcoming, however months passed. Lambeth Aerospace Group of Mesa Arizona offered $65,000 on 11 July, which was declined. Ms. May had taken such a loss by this time she would entertain any offer that would pay off the $76,000 balance and allow her to walk away. A hand written note in Ms. May's files stated, "Show Bob McSwiggan airplane 07 August 1985." The Mercantile National Bank of Dallas filled for repossession of N83FA on 08 August 1985 with $239,007.30 unpaid balance by Gifford Aviation.

Custom Air Service

Bob McSwiggan, Director of Operations of Custom Air service, purchased the aircraft for $79,000 on 17 October 1985. He had been a major operator of DC-3's for schedule and ad hoc charters and was looking to increase lift. The Carvair would hold an entire forty-foot trailer of freight and the straight in loading made it the perfect upgrade to his fleet.

Custom Air Service a FAR Part 125 cargo only carrier had operated for freight forwarders such as Emery and on call to automobile manufactures. Custom also flew occasional emergency AOG charters for the major airlines. McSwiggan was pleased that large commercial jet engines would fit in the Carvair with only three-quarters of inch clearance. The auto parts were the bread and butter flights and the aircraft AOG spares charters were the bonus flights.

The Carvair was not operated by "Common Carrier" Academy Airlines as often reported. Instead it was operated by Custom Air Service in compliance with FAR Part 125, which is a Private Carrier that does not hold itself out for transport. Custom Air Service did not advertise, post any signs, or employ any agents or salesmen. It relied strictly on ad hoc charters primarily from repeat customers Chrysler, Cummins Diesel, Ford, General Motors, and Textron with no long term contracts, only verbal agreements. Some of these companies had been long time

Bob McSwiggan saw the cargo potential of N83FA. The 80-foot cargo hold could transport an entire 40-foot trailer of auto parts. The low ceiling of the early model Carvairs is quite noticeable (author's photograph).

customers of the Academy DC-3s. The larger Carvair could transport more volume such as a 40-foot trailer of auto parts. Custom Air Service declined offers from DHL to carry its overflow since that would constitute common carrier status.[47]

Custom Air Service also provided support for other carriers with AOG situations such as a flap for a Kitty Hawk 747. On one particular flight of 02 December 1996 Air Aruba paid $37,000 for a 3,400-mile trip to pickup a spare jet engine at Oklahoma City and deliver it to Aruba then return the bad engine. Bob McSwiggan was not aware until May of 1996 that he had purchased the only Carvair still in existence that had an enlarged forward cargo entry that could handle a large aircraft engine. Carvair five was one of only two built with the enlarged forward bulkhead. The other was Carvair four, which was tragically lost in 1967. Mark McSwiggan stated that when the nose door was damaged and required changing the spares on hand would not fit. They were not aware that the nose door on ship five was not interchangeable with the 19 other Carvairs. He was also not aware the aircraft was once fitted with the ramp system.[48]

N83FA was still painted in the Falcon "sash" livery of two-tone blue, which was a color modification of the old BAF cargo livery. Soon after purchasing the aircraft McSwiggan had the two-tone blue sash around the forward fuselage removed. He never removed the word "CARGO" and stripes from the tail.

A major cigarette manufacture approached McSwiggan with a proposal to transport cigarettes. They wanted to repaint the aircraft like a large flying cigarette billboard. The forward fuselage would be the lighted end with smoke swirling up the sides of the large bulbous nose. McSwiggan declined because of a long-term contract would constitute "Common Carrier" status and he was not interested in promoting cigarettes. The transporting of auto parts was quite lucrative and the aircraft was still available for the very profitable oversize and AOG charters.[49]

Some of the charters operated were missions of considerable length and in many cases

The Falcon Airways two-tone blue sash stripes were removed from N83FA as soon as Custom Air Service of Griffin, Georgia, acquired it (photograph courtesy Mark McSwiggan).

operating an empty ferry to pickup a critical shipment. The Carvair has a considerable range empty that is only restricted by oil consumption. Each engine is equipped with a 22-gallon tank with 1.8-gallon held in reserve for propeller feathering. The R-2000 engines are considered efficient if they consume less than three gal/hr. It is fairly obvious that if an engine has high oil consumption it is necessary to make unscheduled landings for service. McSwiggan conceived a clever solution to this potential problem. A converted 40-gallon prop alcohol tank was mounted in the rear belly compartment and converted to a fuselage oil tank. Lines were run to the DC-4 engine oil tank manifold, which was capped off when converted to Carvair, reactivating and providing in-flight oil service.[50]

The fuselage tank is serviced through an access cap mounted flush in the cargo compartment floor. In addition, four five-gallon "Jeep Cans" of oil were secured at the rear cargo compartment bulkhead for refill in flight. The system eliminated individual tank service and range was no longer restricted by oil consumption. The payload weight was reduced by this modification. On the positive side the tank and extra oil reduced ballast problems with the nose heavy Carvair.

The aircraft received a tri-annual weight check required under Title 14 CFR Part 125.91(b) on 28 January 1997. The records show the aircraft dry weight of 42,376 pounds at that time. The CG was also calculated at that time to be 366.67 inches aft of datum.

Fatal Flight of N83FA

Custom Air Service based N83FA along with Carvair N89FA at Griffin-Spading County Airport. The airport is an uncontrolled public facility located one mile south of Griffin, Georgia. Airport elevation is 958 feet with a 3700-foot runway designated 14/32. The runway is 75 feet wide and has 200-foot threshold on each end. The southeast end threshold drops off an embankment with a public road, traffic light, and intersection below and to the south of it. There is low intensity preset runway lighting from dusk to dawn.

On the night of 03 April 1997 night visual meteorological conditions existed.[51] The (METAR) report at 23:56 was as follows: visibility 10 statute miles; clear sky; temp/due point 56/37 F; altimeter setting 30.16 "Hg. The conditions were ideal for night flying.

Shortly before midnight of 03 April 1997 N83FA was being preflight inspected for a ferry flight operated by Custom Air Service. It was being positioned to Americus Georgia to pick up a cargo of auto parts destined for Rockford Illinois. This was a standard charter that the aircraft had flown many times before.

The aircraft had been serviced for the flight and was to be dispatched with 2000 gallons (12000 lbs.) of fuel. This Carvair was equipped with four main and two auxiliary fuel tanks. The number one and four main outboard tank capacity is 495 gallons each and inboard main tank capacity is 508 gallons each. The auxiliary tanks were not serviced and contained the minimum of 15 gallons each. The four main tanks had been fueled to capacity. Added to the minimum in the aux-tanks, there was a maximum of 2036 gallons on board. The Carvair fuel burn is 250 gallons the first hour and 220 each additional hour at 65,000 pounds gross weight. The first hour is increased by 25 gallons at max gross of 73,800 and the first two hours are increased by 30 gallons each.[52] Depending on the weight of the cargo, the aircraft was fueled for seven to eight hours of flight.

A note of interest on the Carvair is that even though all the aircraft appear the same no two of them are alike. The fuel selector valves on this ship are located on the forward face of the control pedestal. The positions are, OFF, MAIN TANK ON, and AUX. TANK ON. There are three positions on each lever except for Number Two and Three because these inboard tanks were not installed. The flight manual states: "Pilots are warned of the need to correctly locate the fuel cock control levers by the feel of the detent in the intended position. Failure to do so even by only a small amount will allow all ports of the cock to be open to each other which may lead to feeding engine(s) from unintended tank(s) or unwanted tank to tank fuel transfer. If not noticed and corrected, this can lead to power loss from unexpected fuel exhaustion in the tank feeding the engine(s)." Although the fuel tank selectors are not considered a factor in the loss of N83FA it was noted in the NTSB report.

The flight crew consisted of Captain Larry Whittington and co-pilot Ralph Josey. Captain Whittington had more than 12,000 total flight hours and more than 1,000 in the ATL-98 Carvair. He had flown 45.4 hours in the last 90 days with 27.7 in the Carvair. He also had in excess of 32 flight hours at night and more than 10 hours of actual instrument flight hours, within the last 90 days.

Captain Whittington was issued an airline transport certificate on 25 May 1995. He was multi-engine rated and type rated in the DC-3 and DC-4. Also he held a commercial rating with single engine land rating. In addition he also held a mechanic's certificate with airframe and power plant ratings. He was issued a first class medical certificate on 25 November 1996, which specified the holder shall wear corrective lenses for near vision. Whittington received a DC-4 rating on 11 March 1995 and had successfully completed a proficiency check in the DC-4 on 29 October 1996. The record validates that Captain Whittington was a well-qualified airman.

The co-pilot, Ralph Josey, had more than 5,000 total flight hours with 1,500 hours in the ATL-98 Carvair. It is noted that he had no pilot in command experience in the DC-4/ATL-98 aircraft. Records indicate he had logged 78.2 hours as co-pilot in the last 90 days all in the ATL-98 Carvair. Airman Josey held a commercial pilot certificate with airplane single and multi-engine land, and aircraft instrument ratings. He was a retired Delta Air Lines mechanic with a mechanic's certificate with airframe and power plant ratings. He also held a flight engineer certificate for reciprocating engine aircraft. His second class medical certificate dated 23 July 1996, indicated the holder shall wear glasses, which correct for near and distant vision. A waiver was issued to him on 16 July 1993 for defective distant vision 20/200 corrected to 20/20 bilaterally.

Captain Whittington and airman Josey boarded the aircraft sometime after 23:00 and are believed to have completed their pre-flight checklist. Although there was no one to observe the pre-flight routine, Whittington was known to be methodical in following procedures and checklist.

The Carvair has a cable operated gust lock control located in the cockpit just behind the left seat and toward center of the aircraft. This allows the flight controls to be locked manually. The control handle is made of magnesium and is locked with a stainless steel pin. The pin is attached to a wide red lanyard that is stowed on a spring-loaded reel in the cockpit ceiling. When engaged the pin is in the control handle and the lanyard extends down from the ceiling and across the captain's seat preventing access. It was a common practice for crewmembers to tuck the lanyard behind the captain's seat to allow easy access without removing the pin.[53] The gust lock control is on the pilot checklist along with fuel tank selectors, flaps, and trim tab positions.

Just before midnight 03 April 1997 Mark McSwiggan observed engine start, and dispatched N83FA. The run-up was completed on all engines with no problems noted. The aircraft taxied into position at the far end of the field for departure on runway 14. Mark McSwiggan remained on the southeast end of the field, at the end of runway, as was his habit on all departures he dispatched. He stated later, to the investigators, that he routinely looked at the tail and elevator when he dispatched an aircraft as it turned away from him. He stated that the elevator was in the down position. Tonight for some reason Mark did not stand at his usual spot adjacent to the apron, instead he stood further back on the parking ramp near the corner of the hangar.[54]

The flight manual states that elevator trim should be correctly set prior to takeoff. The chart in the manual indicates at 14 percent MAC, flaps 15 degrees; the trim position should be 1.5

Tab Wheel Divisions Nose Up at 12 percent MAC 2 Tab Wheel Divisions Nose Up. This flight had a takeoff weight of 56,345 pounds with a 14 percent MAC as stated by Custom Air Service. The calculated zero fuel weight was 43,225 pounds.

At approximately 00:10 N83FA was in position at the end of the 3,700-foot runway 14. Mark McSwiggan commented that he heard the engines power up and the aircraft started roll-out. The lighting at Griffin-Spalding as stated is barely adequate. The aircraft landing lights were visible, however silhouette was not clear until it was a considerable distance down the runway. At approximately 1500 feet into the takeoff the main gear partially extend as the aircraft became light. Approximately 2000 feet into the roll, McSwiggan observed the color of Number One exhaust turn from blue to yellow. There was an audible change in power level. At 2700 feet the nose of the aircraft yawed left and the left wing dipped. Skid marks began at 2900 feet indicating the crew locked the brakes, however power was still on engines two, three, and four. The tires blew as the aircraft passed McSwiggan's position and he noted the tail was high. McSwiggan realized that if he had been standing in his normal observation position the left wing of the aircraft would have struck him as it veered off the left side of the threshold.[55]

The skid was very heavy at the runway end and continued about 15 feet into the grass over run. The terrain drops abruptly and the aircraft was airborne and trying to fly for about 75 feet. The left gear impacted the ground again as the aircraft reached the bottom of the embankment and went through the perimeter fence and a wooden sign mounted on 4 × 4 post.

The aircraft was now on a heading of 95 degrees carrying debris and skidding across a municipal road. The left wing severed a utility pole sparking a fire. The fire was spread along as the wing hit a wood privacy fence of an apartment complex. The gear continued to skid across a parking lot as it clipped a fire hydrant and a metal light pole. The aircraft was now 1300 feet from the end of the runway as it crashed through the front wall of an abandoned super market. The building's steel girders severed the top of the cockpit at the windshield level as the building roof collapsed on top of the aircraft and exploded.

The aircraft came to rest with about thirty feet of the fuselage and the empennage remaining outside the building. The intense fire of the 2000 gallons of av-gas extensively burned the aircraft and building. The crew did not survive.

Post Crash Investigation (NTSB ID ATL97FA057)

The cockpit instruments recovered were badly burned and generally not readable. The left fuel-actuating fixture, which corresponds to the Number One engine fuel selector, was not aligned with the three other fuel selector fixtures. This would verify McSwiggan's account of power being pulled back on Number One engine.

The Number One engine s/n BP700946 was overhauled by Genair Corporation on 26 April 1982 and was repaired by them at 11 hours on 17 November 1987. It was originally installed on 09 October 1994, removed several times for repairs and last installed on 16 January 1996. The engine was extensively damaged by fire in the crash. The rear of the engine was burned with the supercharger section burned away exposing the crankshaft. Rotation of the engine was difficult until most of the cylinders were removed. There was evidence of heat distress to the connecting rods and the master rod. No mechanical failure was found during disassembly. The Carburetor was too burned to be tested, both magnetos were removed and tested. They both produced a spark for each cylinder. The cowling and the propeller were found outside the northeast wall of the building. One blade of the propeller was bent forward and the other two blades were bent aft. The propeller pitch control mechanism was on low pitch, high rpm. mechanical stop.

The Number Two engine s/n BP702224 was overhauled by Precision Airmotive Corp. 13 January 1995 and installed on the aircraft at that time with zero hours. The forward section, nose case and engine cowling were burned away in the crash. The rear of the case was burned away exposing the accessory drive shaft. The shaft was found beneath the engine. All of the cylinder heads except those on top were burned away exposing the pistons. Both magnetos were too badly burned to be tested. The propeller was separated from the engine.

The Number Three-engine s/n P105712 was installed on 16 January 1996. It was overhauled by Garside Aircraft & Engine Service and reinstalled on 16 March 1996 with zero time hours. This engine was burned similar to Number Two with cowling and nose case burned away. Most of the cylinder heads were burned away as well as the rear case. The carburetor and magnetos were too badly burned to be tested. The propeller remained attached to the engine with gearing exposed. The blades were partially melted.

Firemen fight the intense fire of 2000 gallons of AV gas at Griffin, Georgia, just past midnight on 04 April 1997. The author and Ruth May were meeting in Honolulu at the same time to discuss the aircraft (courtesy R.D. McSwiggan).

The Number Four-engine s/n 702373 was overhauled by Precision Airmotive Corp. on 09 September 1994 and installed on the aircraft on 21 October 1994 with zero time hours. In the crash the forward section was burned away. Most cylinders had melted cooling fins. The carburetor was burned and both magnetos were missing. The propeller was separated from the engine; one blade was bent forward and twisted into the hub. The other two blades were burned away.

The most alarming find in the wreckage was the steel flight control lock pin. It was found in the wreckage inboard and aft of the pilot's seat in the melted control lock assembly. The pin should have been with the reel in the cockpit roof debris that was severed by the building girders and found some distance away. This find supports the theory that the co-pilot was attempting to remove the pin by leaning over behind the center pedestal. The captain was braking and pulled back power on Number One engine. It is suspected that he could not pull back two, three, and four because the co-pilot's right arm, shoulder and torso were blocking movement of the throttles. The captain being decapitated further supports this and the co-pilot was not because he was lying across the pedestal attempting to remove the pin, which held the elevator in neutral position.[56]

The Georgia Bureau of Investigation Forensic Divisions and toxicological examinations conducted post-mortem examinations of the pilots by the FAA Toxicology and Accident Research Laboratory. The examination of Capt. Whittington was negative for alcohol and drugs. The examination of Co-pilot Ralph Josey found no alcohol, however 0.148 (ug/ml, ug/g) of Paroxetine was detected in the bloodstream. Paroxetine is a prescription antidepressant that has been shown to have little effect on performance. This level is consistent with therapeutic dosage. Also detected in the blood of the co-pilot was 1.110 (ug/ml, ug/g) of Diphenhydramine. This level is approximated at ten times the level that would be present following a dosage at twice the recommended strength. Diphenhydramine is a sedating antihistamine often found in over the counter allergy medications. Test also found Pseudoephedrine in the blood and Ephedrine, Pseudoephedrine, and Phenylpropanolamine in the urine. All of these are commonly found in over the counter decongestants.

Post crash review of log forms used to record hours and in-flight discrepancies were generally in order. The three page preprinted sequential forms from 05 February 1996 to 02 April 1997 did have three pages missing accounting for 17 hours. The last log page 83055 was for the flight prior to the incident flight and only had the portion completed with flight hours and crew names.

The co-pilot of the fatal flight, Ralph Josey, was also the co-pilot on the 02 April flight. That flight was a ferry from Griffin to Americus Georgia and a cargo flight to St. Louis. The captain of the previous flight stated that he and Josey had commented on how well the aircraft had performed and particularly the engines. When asked he also stated in his interview that the control-lock pin was normally removed in sequenced checklist. He related a past incident when he had pulled on the runway and discovered the control lock pin had not been removed. He stated that the checklist was reviewed again. He also stated that he had discussed with Josey that in the event of a malfunction at Griffin he would continue takeoff. Most takeoffs at Griffin were at 59,000 pounds, which can be done, on three engines since the field is too short for an abort after 50 knots.

N83FA List of Statistics

Custom Air Service 1996 year end totals for N83FA show that it was averaging 4.6 revenue flights per month. The following listing indicates that although it was a 36-year-old Carvair converted from a 53-year-old airframe, it was still producing a profit.[57]

Total Flights	56
Total Miles	111,090
Total Gross	694,312
Average Gross/Flight	12,398
Average Trip Miles	1,983
Total Hrs. Operated	555
Average Trip Hrs.	9.91

N83FA, Carvair five c/n 10365 had 50,558.31 total flight hours. It was powered by four Pratt & Whitney R2000-7M2 radial engines. The total engine hours were as follows: #1 engine 688.1 hrs. #2 engine 936.3 hrs. #3 engine 399.1 hrs. #4 engine 1,448.6 hrs.

In addition to the tragic loss of life a part of aviation history ended with N83FA. This aircraft flew millions of miles covering the globe for 53 years. It survived wars, weather, rebels, overloading, engine failures, and financial hard times only to be lost by an overlooked lock-pin.

12

Carvair Six

c/n 7480-6, C-54-A-5-DO, Aer Lingus

Carvair six has the distinction of being the only Carvair converted from a C-54A-5-DO. Only two older airframes were used for conversion. It was built at Santa Monica with all the features of the C-54A and fitted with R-2000-7 engines, twin boom cargo hoist, and fuselage fuel tanks. The aircraft was delivered to the USAAF on 11 April 1944 with military serial 42-107461. It is unique to have a 1942 military registration yet delivered in 1944 because the serial reflects the fiscal year ordered not delivered. After exactly two years with the Air Force it was declared redundant and transferred to Reconstruction Finance Corporation for disposal. On 10 April 1946 it was sold to American Airlines for $90,000 USD and named "Flagship Philadelphia."

American received registration NC90431 and operated it for more than eight years before it was declared surplus and sold in 1954 to the aircraft broker Airplane Enterprises. The "C" was dropped from the registration 01 January 1949 becoming N90431. Within a month it was purchased by Pan Am affiliate AVENSA. It was registered in Argentina as YV-C-AVH and operated for four months before being sold to Resort Airlines and returning to the U.S. as N75298. The DC-4 operator flew it on contract work for five years before leasing it to World Airways for one year. Shortly after returning to Resort on 27 June 1961 it was parked with other DC-4s at Oakland. It remained idle until purchased by Channel Air Bridge in May 1962 for the express purpose of Carvair conversion. The British registration G-ARZV was assigned on 08 June 1962.

Channel Air Bridge had two Carvairs in service and a third within days of the first flight. Conversions four and five were diverted to Interocean. Aviation Traders was quite pleased to receive an order from a foreign flag carrier. Not only was it a morale boost but in the best interest of sales and quite easy for ATEL change the line position from Channel Air Bridge to Aer Lingus. The change in delivery prevented Channel Air Bridge from ever having more than three Carvairs. The company became British United Air Ferries under the Air Holdings consolidation before the next Carvair was completed. The DC-4 is the fourth purchased for conversion from Resort just prior to the company going into receivership. Like the others, it had been abused and stored in the open at Oakland. It was not disclosed when purchased but later discovered during tear down that it had suffered an unrecorded belly-landing requiring additional work.

Carvair Conversion

Under pressure from the Irish government Aer Lingus announced in January 1963 it planned Carvair service with a start-up date of 20 June 1963. This was partially prompted by the attempt of Channel Airways (East Anglian Flying Services) to fly the Carvair on Irish routes a year earlier. Aer Lingus officials estimated the Carvair would only be marginally profitable yet had

N75298 parked at Oakland on 07 November 1962. The cabin windows are blocked out, indicating it had been operated in cargo configuration before storage (courtesy AAHS/Robert Hufford).

little choice since not purchasing it could cost market share. Carvair six was scheduled to be a British Carvair, but in order to make the start up deadline and secure a new contract ATEL made G-ARZV available to Aer Lingus.

The purchased agreement was signed for £175,000 ($490,175) including spares and support equipment. The total purchase price for the first two Aer Lingus Carvairs was £350,000 ($980,350). The contract was viewed with considerable optimism by ATEL since this was the first order from a national flag carrier. Up until this time there had been great interest in the Carvair but no orders outside the Air Holdings group except charter carrier Intercontinental — Interocean based in Luxembourg.

The Channel Air Bridge DC-4 was transferred to ATEL and conversion began on 18 June 1962. The nose arrived by road from Southend on 09 September and the fitting to the airframe was completed on 28 October 1962. The full conversion took approximately six months to complete becoming the shortest to date. It first flew as a Carvair from Stansted to Southend unpainted on 21 December 1962. The British registration was cancelled on 21 January 1963 after the Irish registration was assigned. The Certificate of Airworthiness number 317 was issued 05 February 1963 after 15 hours and 15 minutes of testing.

The new Carvair ferried to Southend for painting since Aviation Traders pushed it out ahead of schedule to make the promised delivery date. It has been reported that it was painted in BUAF colors then repainted for Aer Lingus. This is totally false since the conversion was assigned to Aer Lingus before conversion and months before it was painted. It appeared in full Aer Lingus livery on 23 February. The extra emergency exit on the starboard side was installed at Southend in January prior to the painting.

After conversion it retained the short ram-air carburetor intake scoops that were fitted to this early model DC-4. Most C-54/DC-4 aircraft and majority of Carvairs have the later non-ram long fairing extending along the nacelle. However, this DC-4 arrived with the short scoops and possibly set the standard for the other Aer Lingus aircraft. It most likely arrived for conversion equipped with the R-2000-D3 engines, which commonly had the short carburetor intake scoops.

The Aer Lingus specifications were quite different from previous Carvairs. The carrier ordered a different galley plus a 34-seat four-car twin cabin configuration. A separate passenger cabin with 12 seats was installed forward of the rear cabin occupying the fifth car position. An additional emergency exit was installed at the second to last round window on the starboard side on the Irish Carvairs. The 34-seat cabin was ordered specifically for Dublin-Cherbourg service.

As the only C-54A-5-DO used for Carvair conversion it required additional modifications to comply with Aer Lingus standards of uniformity. The Irish specifications required all Aer Lingus Carvairs to have the same fuel tank configuration. This C-54A aircraft was originally built with four wing tanks and optional fuselage tanks, but no outer wing tanks. The outer wing tanks and fittings were added making the six tank fuel system similar to the C-54B. The other Irish aircraft were converted from different series C-54s validating the need for fuel system uniformity. The Irish also specified uniform avionics on its fleet. The Carvairs were delivered with the Collins FD 101 flight director, which was standard on the Aer Lingus F-27, and Viscount 808.

The Aer Lingus livery of white upper fuselage, bare metal bottom and green trim was applied at Southend between 11–23 February 1963. The blue and white passenger cabin was identical to BUAF with only a slight difference in trim. After an additional two months of finish work and testing the Carvair made a second first flight on 26 February 1963 wearing the full Aer Lingus livery. The new empty weight recorded as 41,127 pounds and it was officially transferred to Aer Lingus on 28 February.

The aircraft was named "St Albert," which appeared on the port side of the cockpit and the Gaelic name "Ailbhe" appeared on the starboard side. Crew training began immediately and continued until delivered to Dublin. It was flown to Boscombe Down for compass checks on 01 March and received the Irish C of A on 08 March 1963. Aviation Traders test pilot Captain Don Cartlidge ferried the aircraft to Dublin on 14 March 1963. In a delivery ceremony Aer Lingus Captains McCarty and Tooke and Air Hostess Downes accepted the first Irish Carvair.

Aer Lingus

Training resumed at Dublin on 15 March 1963 and continued into April when limited cargo service began. On March 21st the Minister of Transport and Power, Mr. Erskine Childers, inspected the Carvair and witnessed a loading demonstration using the "Hylo" loader. The first flight delivered two and a half tons of mushrooms from Dublin to Liverpool and eight pianos to Manchester. It did not officially begin cross channel car-ferry service until May. The aircraft was chartered for the first Aer Lingus car-ferry revenue flight from Cork to Bristol 05 May 1963 to transport four vintage automobiles for the annual London to Brighton race. The first scheduled revenue cross channel flight operated from Dublin to Liverpool via Holyhead on 08 May 1963 and returned via Isle of Man. It was commanded by Captain J.J. Sullivan and assisted by F/O John Howe with flight attendant Ms Brown. The Carvair cruised at 200 mph at 6000 feet. Although it was a cloudy day the flight was uneventful.

Aer Lingus put the Carvair in service with some skepticism but after it came on line and produced positive results a decision was made to order a third aircraft. Reliability was not as good as had been expected because of engine failures. Initially the engine problems were placed on the crews not being familiar with the characteristics of the R-2000 engines. It was not until 1966 just before the Carvairs were phased out that the engine overhaul contractor was changed

Irish Minister of Transport and Power Erskine Childers inspected the Carvair and witnessed a loading demonstration using the Hylo loader on 21 March 1963. The Rolamat floor system is conspicuously absent, confirming that it was installed on EI-AMP and EI-AMR along with the raised ceiling in the winter of 1963–64 (courtesy Aviation Traders Ltd.).

from Scottish Aviation to Air France. The engine failures were drastically reduced with the change vindicating the flight crews.[1]

Another option considered by Aer Lingus was a 55-seat three-car configuration, which would allow the operator to quickly convert from the 34-seat four-car configuration depending on bookings. An additional 21 seats could be added ahead of the 12 seat forward cabin. Although considered, use of the 55-seat configuration was not utilized. All the Aer Lingus Captains interviewed stated they never commanded a Carvair with 55-seats.

Car-ferry services were not operated during the 1963–64 winter season because of lack of traffic even though the summer market was moderately successful. Aer Lingus took advantage of the winter down time to return EI-AMP to Southend on 08 January for service and additional modifications. After the first season Carvair six became the earliest conversion to be fitted with the "Rolamat" floor system, which gave the carrier the option of car-ferry or palletized cargo service. It was believed the additional option would increase cargo bookings by reducing loading and turnaround time.

BKS Air Transport operated car-ferry service during the summer of 1960 and 1961 between Liverpool and Dublin in the Bristol 170 freighter. Aer Lingus studied the BKS operation and noted that the carrier was the primary operator in the horse ferry (bloodstock) charter market. BKS discontinued car-ferry service in 1961 because of financial problems. In an effort to increase profit on the marginal Carvair it was reasoned that with no car-ferry competition it may be possible to develop the horse ferry market bringing additional pressure on BKS with the larger aircraft and increase revenue.

Aer Lingus officials were concerned that the Carvair would never be profitable unless the aircraft was totally utilized. While reviewing the transport of horses, it was pointed out that the dead space in the hump behind the cockpit could be utilized. The Carvairs were scheduled to return to ATEL for installation of the "Rolamat" floor and to raise the ceiling in the cargo bay behind the cockpit to accommodate horse bookings. The modification begins under the hump area behind the cockpit and gives a cathedral like effect in the forward section of the cargo compartment. EI-AMP is the first Carvair returned to the factory to receive this modification to increase the volume of the cargo hold. As a result seven other aircraft had the raised ceiling.

Aviation Traders was successful in selling the raised ceiling and "Rolamat" floor system to other carriers such as Ansett. Years later in 1997 when the two ex–Ansett aircraft 20 and 21 were ferried to Griffin Georgia the raised ceiling option was reviewed. Bob McSwiggan of Custom Air Service saw the option and was so impressed he reviewed the possibilities of modifying Carvairs five and nine, which he owned. After considerable study it was concluded raising the ceiling would not be possible without major structural changes that would be cost prohibitive.[2]

Upon arrival at Stansted on 08 January 1964 for the ceiling modification the nose wheel of EI-AMP partially collapsed on landing. Although there was no major structural damage to the airframe there was considerable damage to the nose wheel well and mounting. External damage was confined to the nose doors and surrounding skin.

The investigation of the incident determined that the fault was human error and not the aircraft. The aircraft was towed at Dublin prior to departure. The prescribed procedure before towing is to remove the toggle pin from the torque links to prevent damage to the hydraulic steering. The lower scissors of the torque link usually drops down freeing the lower end of the strut to turn freely. After towing the pin is to be replaced back in the torque link. The pin was removed at Dublin prior to departure. Either the torque link scissors remained together because of the weight of the aircraft on the strut or the ramp serviceman did not push the lower half down. Possibly after towing the torque link was engaged and the pin was not re-installed. Either way the aircraft had steering at Dublin for taxi and takeoff.

When EI-AMP departed Dublin the oleo strut extended when airborne allowing the scissors to separate, disconnecting the steering. Upon landing the nose gear shimmied violently spinning around and forcing the upper mounting through the cargo floor. The captain radioed on arrival at Stansted that he had lost steering and requested a tow. The ground crew inspected the nose gear and found that it was damaged but still supporting the aircraft. The crew was able to attach a tow bar and cautiously tow the aircraft off the active runway to the hangar.

This was an unfortunate event for EI-AMP but because it occurred at Stansted while in for maintenance recovery time was reduced. It was repaired and the ceiling modified returning to service 15 February ready to begin the 1964 summer season.

The Aer Lingus Carvair operation peaked in the summer of 1964. The car-ferry flights increased 12 percent over the previous year and the Carvair took over the cargo and mail service from the Aer Lingus DC-3s. The success was short lived because of the reliability issue of chronic premature engine failures. EI-AMP was used solely on the all cargo routes to Birmingham, Glasgow, Liverpool, London, and Manchester during the winter of 1965–66. Captain Dible and F/O O'Callaghan flew EI-AMP to Bristol on 24 April 1966. As they prepared for the return a starter burned and failed. Fortunately the Irish Carvairs carried an extensive spares kit on board. The starter was changed for the return flight causing a six-hour delay.[3] Just seven days later again with Captain Joe Dible in command on an overnight freight circuit the Number Four engine went in to over-speed on takeoff. The prop was feathered and the engine seized. The aircraft landed safely on three engines. The following morning Captain Dible was authorized by Telex to carry out a three-engine ferry back to Dublin.[4] This type of problem so plagued the Aer Lingus Carvair operation that it was a factor in phasing them out. Michael O'Callaghan made the last scheduled cargo flight with EI-AMP from Dublin-Liverpool-Manchester-Dublin on 30 October 1966. One week later EI-AMP was removed from service and stored at Dublin.

Eastern Provincial

The Irish Carvairs remained stored until purchased by Gander based Eastern Provincial Airways (EPA). EI-AMP was purchased 10 January 1968 with spares and support equipment for approximately C$167,000.00. Eastern Provincial officials realized a need for the Carvair with the rapidly expanding market because of major projects in eastern Canada. The primary reason EPA purchased the Carvairs was to transport heavy bulky cargo into the airfields at Churchill Falls, Twin Falls, and Wabush. The field at Churchill Falls (CZUM) was 5,500 × 150 feet with an asphalt surface. The Twin Falls field was graded with hard packed earth surface. The mammoth Churchill Falls hydro-electric project in eastern Canada, the smaller Twin Falls project and the iron ore mining complex at Wabush required regular air service. EPA needed a large volume aircraft to transport crews and equipment to remote sites from Newfoundland and the Maritime Provinces.

Eastern Provincial advertised the Carvair as being capable of transporting 46 passengers or 15,000 pounds of freight at a rate of $2.00 per mile. Newspaper advertisements of the time also displayed photos of the Carvair flying over remote Canadian forest and waterfalls with the following copy.

> As in other progressive undertakings in Newfoundland and Labrador, Eastern Provincial Airways is lending vital aerial support services of many kinds to the rapid realization of the Churchill Falls dream. Bushline and mainline planes and aircrews of Eastern Provincial Airways fly the wilderness skyways daily — around the calendar — in transporting men, machines and materials in ever-increasing quantities from Newfoundland and the Maritimes to the Atlantic Region's economic "Sparkplug" at Churchill Falls. This is just one more manifestation of EPA's determination to fulfill its frequently-expressed ambition: "To contribute everything possible to the growth and prosperity of the Atlantic Region."[5]

Carvair six was to receive the first Canadian registration CF-EPU for the EPA trio. All three Aer Lingus Carvairs were pulled from storage and work began to bring them up to flying status. They had been stored in open damp weather, which proved to be a problem. Some deterioration was blamed on the moisture as well as some of the chronic mechanical problems throughout the time with Eastern Provincial.

EI-AMR was selected as the first to ferry to Southend in February 1968 for overhaul, repaint, installation of a nose entry hatch, twin high capacity heaters, and nose door insulation for the cold weather operations in northern Canada. On 01 May 1968 AMP was scheduled for a performance flight before hand over to Eastern Provincial. Aer Lingus training Captain J.J. Sullivan was in command of this test flight. The test involved a three-engine climb with the critical Number One engine feathered. The scheduled procedure was conducted at Dublin. An engineer observer was seated in the jump seat noting the engine readings and performance. Captain Sullivan performed the engine out procedure and shut down Number One.

During the three-engine performance climb over County Meath with the Number One engine feathered EI-AMP suffered an in-flight emergency. Upon passing through 4000 feet the Number Three engine fire warning came on. There was no drop in oil pressure (normal 80 psi, warning light below 48 psi) or cylinder head temperature light prompting a visual inspection. Another maintenance engineer was in the passenger area and observed the fire. He rushed into the cockpit to tell the crew at the same moment the fire warning sounded. The fire drill was successful and Captain Sullivan prepared to restart Number One engine when the feather light for Number Three came on again. The feathering button was pulled and Number Three propeller went into over-speed. The noise got higher and higher until it was at a whistle pitch and exceeded 3000 RPM. Captain Sullivan made more attempts to feather with no results. Realizing he would have to shed the propeller he gently lowered the nose and pulled up abruptly. The propeller separated marked with silence.[6]

Sullivan stated he was enjoying the relief when he discovered that the severed propeller had struck Number Four engine damaging that propeller. I was now down to one operating engine, Number Two. He said, I was out over County Meath. I called Dublin Airport and declared an emergency. I started a slow descent and began to un-feather and restart Number One. This was successful, however, after the long shutdown in the climb the engine was very cold. I had to delay opening up this engine awaiting cylinder head and oil temps. I had enough height in hand to continue the descent on Number Two engine and avoid damage to Number One. Captain Sullivan landed EI-AMP on runway 17 at 17:10 with full fire escort. The aircraft safely coasted to a stop when the Number Three engine, which was extinguished by all indications, flamed up. This time the airport fire department had to extinguish the blaze.[7]

After arrival at the service hangar Captain Sullivan was viewing the damage when he was informed that a farmer in County Meath had called saying he saw a propeller fall from the sky onto his farm and was inquiring what he should do with it. The subsequent investigation showed that the flexible hose line from the propeller feathering pump had come adrift where it passed through the bulkhead between the power and accessory sections. The gradual separation sprayed oil on the hot engine causing the fire. Consequently the oil supply to the prop-governor was interrupted causing the runaway propeller and over-speed separation. A review of other Carvair over-speed and engine failures to date revealed that three of the four incidents were on Number Three engine.[8]

After repairs Captain Sullivan test flew EI-AMP on the 16th, 17th and 19th of May. On 20 May it became the last of the three Aer Lingus Carvairs to depart Dublin to Southend for the Eastern Provincial modifications. This is also the last flight of a Carvair in Aer Lingus livery. After the engine fire incident the fate of the aircraft had been placed in limbo. While being evaluated and awaiting repairs EPA officials expressed concern that the damage was severe enough that it may not be worth completing the transaction. As a result the planned Canadian registration of CF-EPU was not used. When it was determined that the Carvair was fully repaired and airworthy it was ferried to Southend. After overhaul began a new Canadian registration of CF-EPX was assigned and the Irish call sign EI-AMP cancelled on 29 May 1968.

When EI-AMP arrived at Southend Carvair eight, EI-AMR, had been completed for Eastern Provincial and was registered as CF-EPV. Carvair 14, EI-ANJ, was in the final stages of overhaul and modifications as CF-EPW. Like the other two EPA aircraft, the new high capacity heaters were installed. New dual intake scoops were added on both sides of the upper fuselage behind the cockpit. A second exhaust was added on the port side. A smoke stain track was soon visible on each side of the nose hump. The nose door fresh air intake was sealed, ducting removed and it was insulated to reduce heat loss in the severe cold weather operations. A hatch was added in the bottom of the fuselage into the nose wheel bay. The hatch allowed mail to be dropped to villages without airfields. Flight and ground crews soon began using it for easy access. The three EPA aircraft are the only Carvairs to receive this modification.

Carvair six made the first flight after overhaul in EPA livery on 01 July 1968. The color scheme of CF-EPX was the same as Carvair 14, CF-EPW. It was painted white with a gray bottom separated by a red band. Eastern Provincial titles were applied above the red stripe and the EPA logo on the rudder. The black "EPA" letters were not applied in the red band on each side of the nose as on CF-EPV, Carvair eight.

It departed Southend on the delivery flight 05 July 1968 stopping at Dublin to pick up spares with intermediate stops at Prestwick and Keflavik (Iceland) en route to Gander. Aer

CF-EPX receiving final work at Southend in July 1968 after overhaul and fitting of cold weather modifications for Eastern Provincial (courtesy Transportraits/Richard Goring).

Lingus Captain J.J. Sullivan was present when CF-EPX stopped at Dublin. As he inspected the Carvair he stated. "This is the end of an enjoyable five years." It arrived Gander with 47,799:46 TT hours.

Crash at Twin Falls

After four months in Eastern Provincial Service CF-EPX operating as Flight 81 departed Goose Bay at approximately 12:15 P.M. on Saturday 28 September. Captain Ivan Delong was in command of the one hour ten minute flight assisted by F/O Dave McNeil. There were two flight attendants on-board, Carol Emslie and Pat Laviolette. The fifth member of the crew was not identified but believed to be Loadmaster Byrne. The aircraft originated at St. Johns that morning with interim stops. The flight with 33 passengers and a crew of five was uneventful until a landing was attempted at Twin Falls.

As the flight approached the gravel field Captain Delong attempted an early touchdown on the short runway, which was bordered on each end by water hazards. Because of his limited experience on the Carvair and a high rate of decent the main gear struck and embankment approximately eight feet from the threshold. The left main gear was severely damaged by the impact, debris punctured the Number Two main fuel tank and the Number Two engine lost power. The controls to Number One were damaged preventing power adjustment. The aircraft bounced and overshot the runway. As Captain Delong applied power to go around, Number Two engine did not respond. The prop was feathered and a three-engine climb procedure was initiated with Number One under performing.[9]

After gaining altitude the crew assessed the damage to the fuel tanks and hydraulic system. They then reviewed emergency procedures while circling for about 45 minutes. Fuel was burned off before attempting a second approach. The right main and nose gear indicated down and locked but the crew could not determine the extent of damage to the left main gear. Captain Delong made the decision to attempt a second landing on the right main and nose. At approximately 3 P.M. the aircraft made a long slow approach touching down on the nose and right gear. The captain was able to carry the aircraft for some distance before it settled on the left wing. The left wing tip was ripped off as the wing struck the dirt and gravel runway. The damaged left main gear separated from the aircraft. The nose and right main gear collapsed after the aircraft slid off to the left hitting large rocks in the soft ground. It came to rest perpendicular to the runway digging in the gravel and dirt. The left wing remained in tact with the right outer wing buckling outboard of Number Three engine. The partially collapsed right main gear caused the outer wing and Number Four engine to pivot down burying the prop-hub in the soft earth.[10]

There was no fire and all passengers were evacuated without injury. The aircraft was declared a total loss. Because of the remote area of the airfield salvage was not considered and the Carvair was cannibalized for parts. The airframe was burned and plowed under. The salvaged parts were eventually sold to British Air Ferries for spares. The Canadian registration of CF-EPX was cancelled four months later. The field at Twin Falls was eventually abandoned in favor of the more improved Churchill Falls, which still exist today.

Conclusion

The pilot's inexperience with the Carvair and the high position of the cockpit combined with his distraction with the landing hazard and short gravel runway led him to attempt a touch

Less than three months after delivery to Eastern Provincial, CF-EPX crashed at Twin Fall, Labrador. It was stripped of parts before the airframe was burned and plowed under. Many of the spares were used on other Carvairs (courtesy Bert Lambert/Rick Gaudet collection).

down to close to the runway threshold. The pilot was assigned to perform a landing on a marginal landing area in an aircraft that he had limited time commanding. The Captain's proficiency in emergency procedures and engine out climb procedures allowed him to recover and land the crippled aircraft on an already marginal field without loss of life. Examination of the maintenance records revealed no discrepancies that would indicate the aircraft not airworthy.[11]

The Canadian registration was cancelled on 31 January 1969. The salvaged parts were purchased by BAF and transported to Southend. The remains of the airframe were plowed under at Twin Falls.

The rudder eventually became part of the spares that were purchased by Ruth May of Falcon Airways when Carvair five passed through Southend in 1979. It eventually came to Griffin Georgia after Bob McSwiggan purchased two of the ex–Falcon Carvairs.

13

Carvair Seven

c/n 10273-7, C-54A-1-DC, British United Air Ferries

The C-54 converted to Carvair seven was delivered to the USAAF on 06 November 1943 as a C-54A-1-DC and assigned serial 47-72168. The Skymaster was built at Chicago in military configuration for 50 troops or 32,500 pounds of cargo and fitted with Pratt & Whitney R-2000-7M2 engines. This is the same series of engine specified after Carvair conversion. It was equipped with cargo hoist and reinforced floor. Aviation Traders determined that the corrugated floors on most aircraft were not suitable for automobile and cargo handling. They were damaged on many of the DC-4s prompting the decision to replace them on all Carvairs. Ship seven is converted from the second oldest C-54 airframe used in the program with the Carvair 13 being the only older airframe.

California Eastern Airways

The C-54 served with the U.S. Air Force until 03 May 1946 when it was transferred to the War Assets Administration for disposal. California Eastern Airways, which was formed by Jorge Carnicero in 1946, applied for civilian registration on 15 May 1946 purchasing it for $75,000 USD. California Eastern president J.J. O'Brien and vice president W.J. Hoelle signed the purchase documents. It received civilian registration NC54373. The "C" was dropped due to a change in U.S. registration policy effective 31 December 1948. Four of the airframes converted to Carvair standards were owned by California Eastern or one of its associated companies.

The Air Transport Division of Matson Navigation Company removed the carburetor intake filter system on 18 July 1946. This non-ram carburetor intake installed for the military had hot, cold, and filtered air intakes positions. Blocking plates were installed for the filters and the actuator controls were removed reducing aircraft dry weight by 123 pounds.

California Eastern operated the C-54 from company inception in 1946 until bankruptcy reorganization in mid 1948 and did not resume service until December 1950. All the carriers aircraft were leased out during this period. Some records indicate N54373 was registered to Arrow Airlines in July 1949. A search of FAA records does not indicate any application to transfer the registration, but there is a distinct possibility Arrow did operate the aircraft. FAA records do indicate N54373 was leased to Twentieth Century Airlines on 18 May 1951. Jorge Carnicero founder of California Eastern along with Stanley Weiss of North American Airlines formed associate company Aero Leases. Weiss acquired Twentieth Century Airlines along with four other carriers in order to circumvent the limitations of his operating certificate. The bankruptcy of California Eastern was a windfall for Weiss since more DC-4s were need to expand. The Skymaster was leased out most of the nearly seven years it was owned by California Eastern. This is the longest ownership of the carriers four DC-4s that were eventually converted to Carvair standards. (See Carvair two for additional on California Eastern.)

During the years it was leased out and operated on various contract missions, a number of modifications and upgrades were performed. De-icing boots were added in 1948. On 18 April 1949 the seating configuration was changed from 59 to 66. The cabin bulkhead was removed at station 250 on 20 October 1949 and five more seats were added. Two fuselage fuel tanks were installed on 17 September 1950 to increase the range and once again the seating capacity was changed. The tanks were the same equipment configuration as the military C-54 fuel system.[1]

Seaboard & Western

Seaboard & Western purchased N54373 along with another DC-4 and seven spare engines for $610,000 on 15 February 1953. The registration was transferred on 17 February 1953 and it was named "Wake Island Airtrader." Seaboard & Western was established as an international air freight company on 12 September 1946 in New York with offices at 150 East 35th Street. The company was the brainchild of two brothers Arthur V. and Raymond A. Norden. Both were former Air Transport Command pilots. Their staff consisted of former military staff officers and pilots, which included Wallace P. Neth and Captains Carl D. Bell and Warren H. Renninger.

In May 1947 the government cracked down on small charter companies for a host of infractions. Established companies such as Seaboard & Western were allowed to continue to operate since they were already established in 1946. After the government regulations were imposed the company was re-registered in Delaware in May 1947. At that time the company had only $20,000 in operating capital and contracted all maintenance to Lockheed Air Services at MacArthur Airfield. After flying the first commercial pure freight charter across the Atlantic, Seaboard & Western filed with the CAB for scheduled cargo authority on the North Atlantic route as an Irregular Common Carrier.

Seaboard & Western owned two DC-4s that later became Carvairs. N54373 was added to the fleet after two other Skymasters were involved in separate incidents. One crashed and burned at Pisa Italy and the other was severely damaged in a hard landing at Haneda Airport, Tokyo. As Seaboard added more Constellations to the fleet the Skymasters were surplused resulting in N54373 being leased to Transocean in 1954. It remained with Transocean nearly two years until 04 November 1955 when Air World Leases purchased it. Andre de St Phalle who headed up Air World immediately leased N54373 back to Seaboard for $15,000 per month with an option to purchase between 01 August 1956 and 31 January 1959. The aircraft registration was transferred to Air World on 02 December 1955.[2]

Maintenance records indicate that the overseas equipment, which included the fuselage fuel tanks, were removed on 13 June 1957 and 80 passenger seats were installed along with a galley at the rear entry door. After almost three years of Air World ownership, it was sold back to Seaboard & Western on 06 October 1958 and the registration transferred on 12 November.[3] The Skymaster had transported record amounts of cargo. After considerable maintenance work the DC-4 began a second tour with Seaboard & Western by being positioned in Europe. It took over the routes that were serviced by a C-46 Commando at Shannon Ireland. The connecting cargo from Seaboard Trans-Atlantic Connies was broken down for distribution to Prestwick, Amsterdam, Hamburg, Paris, Geneva, Basle, and Zurich. By 1959 only three of the 13 DC-4s operated by Seaboard were still in service. Two of those remaining three, N54373 and N1221V, became Carvairs seven and ten respectively.

Captain Arvid Fairchild flew N54373 while with Seaboard & Western from 1955–1961. He became Flight Director for Interocean from 1961–1963 when the DC-4 was with the carrier as LX-BNG. In 1983 he became Vice President of Pacific Air Express and flew the same airframe as a Carvair. Over a 30 year span he flew Carvairs with three carriers (Interocean, PAE, HPA) and was involved with a fourth carrier (PAL) that attempted to operate them.

Seaboard & Western (name change to Seaboard World in 1961) is not only tied to the Carvair and BAF by two of its DC-4s but also through the purchase of five CL-44s swing-tails beginning in 1961. Many of the same individuals involved with these CL-44s were a major part of the independent British air and car-ferry operators. The CL-44s were flown by Seaboard until sold to Transglobe in 1968. Several weeks later Transglobe ceased operations and the CL-44s reverted back to Seaboard & Western. An arrangement was made for Tradewinds to take over the CL-44s under Bobwood Limited, which became Tradewinds Airways. Wealthy London Insurance broker Charles Hughesdon financed the new company in order to continue the operations of Transglobe. There were three minor shareholders in the venture. George Townsend and Ted Parker were former directors of Transglobe and Cyril Stevens was the former Chairman and Managing Director of BKS Air Transport. Mike Keegan who purchased BAF and the Carvairs in 1970 co-founded BKS with Stevens and Barnby. BKS operated car-ferry service with Bristols and also competed with Aer Lingus Carvairs on Bloodstock charters.

Trans Mediterranean Airways (TMA) of Beirut leased two of the CL-44s from Hughesdon. TMA had seriously considered Carvairs in 1962 and continued to review them with ATEL through 1966. The carrier wanted DC-6 Carvairs and believed they were forthcoming. Aviation Traders purchased a TMA DC-4 in 1963 that became Carvair 15. Realizing that the DC-6 Carvair probably would not be built, Trans Mediterranean leased CL-44s in 1967. In 1969 British United Airways parent of British United Air Ferries strongly protested the Tradewinds license application with the Air Transport License Board (ATLB). By 1970 Tradewinds was successful in acquiring proper license.

During the same period in 1970 when Trans Meridian Airlines (TMAC) was operating CL-44s, Tradewinds flew two aircraft in the Ugandan airlift to Entebbe in concert with Mike Keegan's TMAC operation. In 1971 Keegan purchased Carvair operator British Air Ferries (BAF) and operated it under Trans Meridian. This DC-4 was already converted to Carvair standards and in the BAF fleet at that time. Carvair 10, which was converted from the other Seaboard & Western Skymaster N1221V and was leased by BAF in 1972.

Air-World Leases sold N54373 to Marshall Landy, a Miami based aircraft leasing and purchasing firm on 20 May 1959. The registration was transferred to Landy on 29 June and it remained on lease to Seaboard & Western until 01 November 1960. It was idle for a short time before Interocean, one of the original Carvair operators, leased it with an option to purchase on 09 March 1961. Within days Interocean secured financing and exercised the purchase option. Aero Factors on behalf of Interocean requested U.S. de-registration of N54373 for export to Luxembourg. Interocean applied for Luxembourg registration on 07 March 1961 requesting the assignment of LX-BNG. The "BNG" is the initials of Bernard Nathaniel Goldberg, one of the officers and financial backers of Intercontinental and Interocean.

Luxembourg based Interocean successfully bid on United Nations relief missions contracts in Africa. Because of an oversight in the records of ownership regarding a lien on the aircraft the Luxembourg registration was not forthcoming. The U.S. registration was cancelled on 03 April 1961. Landy purchased N54373 from Aero Leasing but the FAA was not notified of the Seaboard & Western lease agreement being transferred. The FAA showed Landy as the owner requiring Seaboard & Western to submit proof there were no liens on the aircraft. The financial release was not cleared until 31 July 1961.[4]

Once the records were corrected Interocean put the newly acquired aircraft into service. It was obtained specifically for the Interocean UN Congo contract while the carrier was waiting for the badly needed Carvairs. Aviation Traders required two DC-4 airframes as part of the contract for Carvair four and five. Interocean opted to trade in Skymaster LX-BNG delivering it to ATEL in August 1962 as part of the purchase agreement. It was placed in the Carvair program as conversion seven. Interocean could not wait for -BNG to be converted and agreed to purchase ships four and five less engines on 02 August 1962.

Carvair Conversion

LX-BNG departed the Congo to Luxembourg on 17 August 1962 and flew for the last time as a DC-4 on 22 August to Stansted. It is the only C-54A-1-DC converted to Carvair standards. The disassembly began after arriving Stansted and the R-2000-7 engines were removed and returned to Interocean. It was purchased by ATEL in early August but it retained the Luxembourg registration during conversion.

The fuel tanks were drained and the engines, outer wings, and empennage were removed outside as the airframe was stripped. The DC-4 nose was cut away and the fuselage placed in the jigs by October 1962. The Carvair nose was delivered to Stansted on 11 November and fitted on 21 December. The conversion was relatively straightforward and completed in seven months becoming the second fastest conversion to date.

G-ASDC was built as an ATL-98-A for cross-channel service. The "A" designation is to identify the design modification 98/5150 of the 250-pound wing bob-weights. The adding of the lead weights in each wing tip increases the maximum zero fuel weight (MZFW) by 1000 pounds. The more even distribution of weight across the wing reducing flexing and stress on the wing roots thus allowing additional cargo payload.

Conversion seven received British registration G-ASDC on 11 January 1963 and was painted in the new BUAF colors in March. It is the first new conversion delivered in the BUAF dark stripe livery. The first three Carvairs were delivered in Channel Air Bridge colors. Two of them were repainted after the merger with Silver City Airways. Because aircraft four, five and six were delivered to foreign customers, G-ASDC was the first aircraft to emerge from conversion in the new British United livery. The new empty weight was recorded as 40,787 pounds.

The DC-4 was in continuous service for 19 years, flying mostly in the Pacific, before going to Europe and Interocean as LX-BNG. It arrived on 22 August 1962 from ONUC work in the Congo as a trade-in for Carvair four, which is being completed behind it. A Carvair nose cradle sits near the nose as work begins at Stansted (courtesy Terry Leighton/Aviation Traders Ltd.).

Captain Don Cartlidge commanded the first flight from Stansted to Southend on 19 March 1963. The seven hours and 35 minutes of test flights were completed in seven days with the C of A issued on 26 March 1963. It was handed off to BUAF the same day and christened "Pont du Rhin" before making a scheduled flight to Rotterdam. G-ASDC holds the distinction of the only Carvair receiving three different names while in British service.

British United Air Ferries/Pont du Rhin

During the period that G-ASDC was still in car-ferry configuration prior to being converted to a pure freighter it achieved celebrity status. It was on a short-term loan to United Artists in 1963 for the filming of the James Bond thriller *Goldfinger*. It appeared in the movie at Southend being loaded with the gold Rolls-Royce of Auric Goldfinger. The aircraft was also used for several promotional demonstration flights. On 13 December 1963 it became the first Carvair to visit Lydd Kent and was eventually transferred to Ferryfield (Lydd) on 21 March 1964 to inaugurate deep penetration service. The Bristols continued to operate out of Lydd-Ferryfield until 1965. The last Carvair car-ferry flight operated from Lydd on 03 October 1970.

On 05 December 1964 G-ASDC was damaged in a bizarre incident at Rotterdam when the nose wheel collapsed sideways while taxiing. The aircraft was not traveling at high speed or being handled in a manner exerting excessive sideways weight and momentum on the nose gear. Inspection and investigation could not determined if any previous incident had stressed or cracked the mounting. After repairs it was returned to service.

G-ASDC was leased to United Artist in 1963 and appeared in the James Bond movie *Goldfinger*. The Rolls-Royce of Auric Goldfinger was loaded at Southend in practice sessions before filming the movie scene (courtesy Aviation Traders Ltd.).

British Air Ferries/Big Louie/Plain Jane

G-ASDC was repainted in the BUA blue and sandstone stripe with flying bird logo in the fall of 1966. A year later on 01 October 1967 British United Air Ferries became British Air Ferries (BAF). The new two-tone blue stripe livery was not immediately adopted and an interim BAF livery was created by extending the blue and sandstone stripe over the "toppled mushroom" and adding the BAF titles. After the BAF change to the new two-tone blue stripe livery it maintained the name "Pont du Rhin" into the 1970s.

It was severely damaged while landing at Rotterdam on 17 February 1970. The incident was similar to the Channel Air Bridge crash of G-ARSF on 28 December 1962 that destroyed the aircraft. Carvair G-ASDC departed Southend with a cockpit crew of three and only three passengers on board. The aircraft is designed for a two-man crew, however BAF continued with the practice initiated by Channel Air Bridge of having an engineer on board. The engineer was usually assigned the engines, throttles, mixture, props, etc. to reduce the workload on the captain.

Weather conditions were marginal at Rotterdam and the airport was covered with a light layer of snow. Light snow was falling and low clouds limited visibility contributing to hazardous landing conditions. The runway threshold was cleared of snow and the approach lights were at high intensity. The approach checklist was completed and the props were set at 2550 rpm. with the manifold pressure at 27" hg. as the glide slope was intercepted.

After passing over the outer marker the approach lights, which were at 100 percent intensity, reflected off the falling snow and low ceiling. The glare temporarily blinded the captain who called for 30-degree flaps and the power was reduced to 2250 rpm and 22" hg. Because of lack of visibility the procedure was implemented too early in the approach sequence. With the landing gear down, full flaps, and the power reduced, the rate of decent was too great in relation to the horizontal distance to be covered placing the aircraft short of the threshold. The nose gear struck the cross bar on one of the approach lights separating the nose wheel from the strut. The captain pulled up the nose of the aircraft as the main gear contacted the runway. The aircraft carried on the main gear a short distance before the oleo leg of the damaged nose strut contacted the runway. It formed a skid until the aircraft settled, then began plowing up the runway. At the same time the props of engines two and three struck the ground. The Carvair remained straight until it came to a rather abrupt stop.[5]

None of the passengers or crew was injured. G-ASDC sustained considerable lower forward fuselage skin damage from the impact with the lights and flying debris. The nose gear doors, strut and upper mountings were severely damaged as well as the inboard props and engines. G-ASDC remained at Rotterdam for weeks while repairs were carried out. Once it was determined to be structurally sound and repaired it returned to regular service.

After "Mike" Keegan acquired controlling interest of BAF in 1971, a new marketing campaign was implemented and the Carvairs were given character names. On 04 October 1972 G-ASDC was re-named "Big Louie." It continued in car-ferry service for two more years before returning to Stansted in January 1974 for maintenance and re-painting in the last BAF livery. It remained out of service until 08 April while the interior was upgraded with piped in music and the seating configuration changed to 55 passengers. After completion the empty weight was recorded at 41,307 pounds. Most BAF Carvairs had already received the new livery. The repainting of G-ASDC was rather short sited since it was withdrawn in the fall and stripped of all paint for conversion to cargo only.

Cargo Conversion/Plain Jane

By the end of 1974 it was obvious the Carvair era was long passed. The roll-on roll-off sea ferry business was rapidly bringing an end to its usefulness. The idea of getting the last bit of service out of the aging aircraft as freighters was enhanced by a London dock workers strike. Officials at BAF under Mike Keegan believed that a stripped down version would allow an increase payload and the additional cargo weight would increase the yield. It is known in theory that any aircraft collects weight over years of service from repairs and repainting. As the empty weight increases upon each periodic weight check the cargo payload is reduced. Engineers at BAF decided to experiment with G-ASDC in a weight reduction project.

The project began on 29 January 1975 when it was returned to Southend for conversion to all cargo. After detailed inspection all the paint, which was quite thick, was removed reducing surface drag and weight. To further reduce weight the passenger cabin and all fittings were removed. All the windows on the starboard side including the large rear passenger windows were removed and skinned. All the windows on the port side were removed and skinned over except a single rear passenger window. The only windows left were those on the emergency exits. The rear cabin and lavatory were removed, leaving only five seats against the rear cargo compartment bulkhead. A chemical toilet similar to those on Interocean Carvairs was mounted in the nose door. The exercise was a success removing 2000 pounds from the interior and 1000 pounds from the exterior increasing the payload to an 8.5 tons.

The overhauled Carvair made its debut in cargo configuration on 11 February 1975 in polished bare metal skin. It was re-named "Plain Jane" becoming the only BAF ship to have three names. The titles "BAF Cargo" was applied on the nose and tail complete with telephone numbers. BAF handed out advertising flyers to potential customers. The ad read in large letters above

Stripped of all paint in 1975 to reduce weight and converted to all cargo, G-ASDC lands at Southend. It was the only British Carvair to wear these titles, making it easy to identify. More than once it was reported for spraying oil on the fresh laundry drying in the countryside (courtesy Transportraits/ Richard Goring).

a night photo of the stripped Carvair "FLY BAF Plain Jane." The copy gave capacity details and stated that five seats were left. These were for grooms on bloodstock charters, guards on treasury shipments or any cargo that required an escort.

During the spring and summer of 1975 freight demand was extremely high and G-ASDC was in almost constant use. In full cargo configuration it could transport up to 13 racehorses even though it did not have the raised ceiling. It also transported factory goods, cigarettes, machinery and even the instruments of the Vienna Philharmonic Orchestra.

Invicta/Return to BAF

Invicta wet leased G-ASDC for one month beginning 28 October 1975 for cargo charters between Manston and Rotterdam. The contract flights were booked prior to the planned shut down by the carrier. Management tried to resurrect Invicta but the fate of the airline was already decided. After less than two weeks the Carvair was returned to British Air Ferries and officially back on line on 09 November 1975.

British Wing Commander Hugh Kennard and his wife Audrey founded Invicta. He became managing director and obtained financial backing from Commodore Taffy Powell and one of his wartime colleagues W.C. (Bill) Franklin. Kennard was previously with the Britavia group that owned Silver City Airways and founded Air Kruise Limited in 1946. Audrey Kennard was co-editor of *Ferry News*, which was a publication that promoted travel and Ferryfield (Lydd). Through his contacts Kennard was able to secure Ministry of Defense charters to Europe and North Africa with the bulk of them to Germany. Invicta International was a family operated Manston based British Air Charter operator for many years. After an unsuccessful attempt to merge with British Midland in 1969 it was re-established as a freight charter company with three DC-4 aircraft. Because of the success of the cargo charter business Kennard decided to re-introduce the passenger charters in 1971 with two Vanguards.

The carrier operated holiday tour charters to the Mediterranean, Ministry of Defense trooping flights to Germany, pilgrim flights to Tarbes, and ad-hoc charters from Luton and Stansted. During this time Invicta added Vickers Vanguard aircraft to their fleet. This was all part of a lease arrangement with Air Holdings, the parent of Aviation Traders. Air Holdings had a surplus of Vanguard aircraft, which were acquired in a not well thought out deal with Air Canada. The volume of traffic encouraged Invicta during the summer of 1971. In 1972 after BAF was acquired by Mike Keegan it began operating charters on behalf of British Air Ferries from Stansted and Southend. By the fall of 1972 Invicta was operating five Vanguards acquired from Air Holdings with ATEL performing maintenance at Southend.

Invicta was again in financial trouble in November 1972. Failing to make lease payments, Air Holdings repossessed the Vanguards. This time European Ferries who purchased 76 percent of the company saved Invicta. The revival was short lived and the fate of Invicta was sealed on 10 April 1973 when one of the Vanguard aircraft crashed on approach to Basle Switzerland. Of the 145 passengers and crew on board 108 perished. When the investigation was complete it was generally accepted that the captains license and ability was questionable and he had possibly falsified his qualifications. European Ferries owned control of Invicta and announced that unless a buyer could be found it would cease operations by the end of 1975.

Invicta continued to operate in a normal manner during the 1974–75 season as if there were no problems. Operations ceased in October and BAF Carvair G-ASDC was leased for the cargo work. It was returned to British Air Ferries 08 November 1975. After the return to BAF it remained in the bare metal until the spring of 1976 when BAF adopted a new cargo color scheme of a yellow and black. It became the first of only two Carvairs repainted with a yellow and black sash around the forward fuselage. The vertical stabilizer was painted in a stripe and bar pattern with "Cargo" titles. In June 1976 only six months later G-ASDC was offered for sale.

While waiting for a buyer it continued in an all cargo role until December 1978 when it was withdrawn from service and stored at Southend.

In 1979 BAF placed the last three Carvairs, G-ASDC, -ASHZ, and -AOFW up for sale or lease. G-ASDC continued in limited cargo service picking up overflow and ad hoc charters. Captain Eric Sears flew the last BAF Carvair revenue flight on 04 April 1979. The overnight flight from Brussels arrived Southend at 03:55 with a cargo of auto parts. The first Carvair, G-ANYB, had made the first revenue flight on 17 Feb 1962. In the early hours of dawn, after 17 years and 46 days the Carvair era was over with no fanfare or acknowledgement.

Falcon Airways

Falcon Airways of Addison Texas submitted a letter of intent to purchase three Carvairs on 11 April 1979. G-ASDC was one of the last two airworthy BAF Carvairs and the first of two BAF ships converted to all cargo configurations. The repainting of the aircraft for Falcon was completed on 22 April at Southend. It was an easy livery change from BAF to Falcon by substituting blue for the yellow bands of the BAF Cargo livery and adding the Falcon name to the side of the fuselage.

Texas based Falcon Airways purchased the Carvair on 01 May 1979 for $166,667. The aircraft was turned over to Falcon Vice President Jim Griswold the week before for training and crew preparation for the ferry flight to Addison Texas. The U.S. registration N80FA was requested on 11 April and the British registration G-ASDC was cancelled on 30 April. Also arriving early with Captain Griswold was Falcon Airways Captain Stewart Dawson, F/O Gary W. Poplin, and Engineer/Flight Mechanic Tom Littleton. BAF fleet Captain Don Willis made a number of check flights with the group to prepare for the 33 hour four day ferry back to Texas.[6] The British certificate to export was issued on 02 May and the FAA issued registration N80FA to Falcon Airways the same day. While the crew was preparing for the ferry flight Falcon official V.J. La France purchased a Triumph Vitesse automobile from Southend Customs agent Richard Vandervord and arranged to have it loaded on board for the flight to Texas.[7]

Carvair seven, with new U.S. registration N80FA departed Southend at 13:15 on 03 May with Jim Griswold in command. The aircraft performed well without incident at 8,000 feet on the 5-hour flight to Lisbon Portugal. After a one-day layover the Carvair departed Lisbon on 05 May for the second five-hour leg to Santa Maria Azores with Stewart Dawson in the left seat and Griswold in the right seat. Because of changing weather conditions the decision had been made not to take the northern route via Iceland and Greenland. After stopping at the Azores the decision was made to continue on across the Atlantic. N80FA departed Santa Maria at 17:30 for Canada. After a grueling 11 hour flight alternating between eight and 10 thousand feet to get above the weather the Carvair landed in low ceiling and heavy rain at Stephenville Newfoundland. After bucking headwinds and dodging weather the crew was exhausted and almost didn't make it. There was less than 200 gallons of fuel on board when it arrived. Griswold was able to purchase only 980 gallons of fuel at Stephenville and decided to fly to Moncton to get additional fuel. After resting the day at Moncton and taking on more fuel the Carvair was back in the air at 23:35 for the final leg to Dallas. The flight to Dallas is approximately 1700 miles and was estimated to take about 11.5 hours.[8]

N80FA with 71,000 hours on the airframe and 21,500 cycles on the gear arrived at Love Field in Dallas just after 10:00 A.M. on 07 May 1979. Lewis and Ruth May were there to welcome home the crew and inspect the first Carvair addition to the Falcon fleet. Ruth May had participated in the inspection and purchase of the first two Falcon Carvairs in April and personally purchased Carvair five in August for lease to Falcon Airways. She verified a story that the Falcon crew wanted to change the name of N80FA from "Plain Jane" to "Large Lewis" after Lewis May. It had been previously named "Big Louie" while serving with BAF.[9]

N80FA during a test flight at Southend in April 1979 after repainting in Falcon livery. Traces of the BAF letters are clearly visible on the nose (courtesy Richard Vandervord).

After arrival in Texas Falcon added Bendix RDR-150 radar and replaced the old Rockwell Collins avionics with King Gold Crown navigation—communications equipment. It was not equipped with the di-electric patch nose door and was still fitted with the original unit with only two lights in the lower section. Falcon acquired a number of spares with the purchase of the two Carvairs that included several nose doors with the radar nose patch. They were transported back to the U.S. aboard the second Falcon Carvair N89FA. The radar nose door was installed on N80FA on 17 July 1979. The wing "Bob weights" that had been removed at Southend prior to the purchase by Falcon were re-installed at the same time.

Falcon Airways officials were very optimistic the Carvair would generate new business. Unfortunately the expected contracts did not materialize for the Carvairs and Falcon suffered financial trouble by early 1980. Mercantile Texas Corporation repossessed N80FA on 21 March 1980. Ownership was transferred to the Mercantile National Bank of Dallas on 07 April.

Nasco Leasing

Carvair N80FA remained stored until 10 October 1980 when it was purchased by aircraft dealer Nasco Leasing of Incline Village Nevada. FAA records indicate the registration was transferred to Nasco on 22 January 1981. However, Nasco defaulted and the Mercantile National Bank was already in the process of repossessing from Nasco when it was transferred. N80FA remained idle and the registration was transferred back to Mercantile National Bank on 19 February for the second time in a year. It was put up for sale and conflicting records indicate it was at Dallas in June 1981 when it was actually at Long Beach.

James Blumenthal

Established DC-4 operator James Blumenthal purchased the two ex–Falcon Carvairs N80FA and N89FA on 01 December 1982. N80FA was stored at Long Beach after being re-possessed from Nasco. Blumethal had been instrumental in Falcon Airways originally becoming interested and eventually acquiring the Carvair in 1979. He was flying support on the cleanup of the Amoco Cadiz disaster and inquired about the two BAF Carvairs that were idle at Southend and available. After the collapse of Falcon in 1980 several attempts by poorly financed individuals were made to purchase them. Blumenthal eventually purchased the pair opting to re-sell N80FA and retain N89FA to operate in tandem with his low time DC-4.[10]

Calm Air

Calm Air a regional carrier established in 1962 in Manitoba is owned by the Morberg family. During the 70s the carrier operated many types of aircraft including two DC-4s. During a period of expansion Calm Air reviewed the purchase of inexpensive aircraft to capture more of the cargo market. When Carvair seven, N80FA, became available it was considered and even rumored to have been purchased by Calm Air. The purchase was considered late in 1981 but did not materialize instead the carrier opted for two Air Express Australia DC-4s in November 1981. That carrier also attempted to purchase the two Ansett Carvairs eventually acquiring two surplus QANTAS DC-4s after the demise of Air Express Australia.

Pacific Air Express/Kemavia Inc

Pacific Air Express (PAE) was founded in 01 June 1982 and incorporated in the state of Nevada after DHL and Air Distributors ceased operations in Hawaii. PAE operated DC-4s, and Beech 18s within Hawaii in an attempt to gain a portion of the $30 million a month passenger / cargo market of the Honolulu-Maui corridor. Pacific Air Express began as a financially marginal carrier with high expectations in the very competitive Hawaii cargo market. Pacific Air Express operated under the umbrella of Kemavia Corporation, which is an acronym for Khem Aviation (Kemavia). Louis Khem was president of both Kemavia and Pacific Air Express. The carrier added the Carvair to its DC-4 inter-island operation because of its high bulk capacity. Khem purchased N80FA at the urging of A.P. Fairchild, who was Vice President of PAE. He pointed out to Khem that during early Carvair development in 1962 the State of Hawaii approved and recommended the use of the Carvair for inter-island service. Fairchild had flown this same DC-4 airframe when he was a Captain with Seaboard & Western in the 1950s. He also flew DC-4s and Carvairs with Interocean Airways in the 1960s. He had most recently been involved in an attempt by Pacific Aerolift to operate Carvairs.

In 1982 just seven days prior to PAE purchasing Carvair N80FA, James Turner of Turner Aviation purchased the two ex–Ansett Carvairs in New Zealand. The registration of N406JT and N407JT was requested and Pacific Aerolift (PAL) was incorporated to operate the Turner Carvairs with Charles Willis and A.P. Fairchild as corporate officers. In March 1983 two of the PAE DC-4s were re-registered as N300JT and N301JT indicating Turner was providing aircraft for both companies. (See Carvair 17, 20, and 21 for more details on PAE, PAL, and HPA.)

Fairchild claimed to have taken part in the original Carvair program in England and held the first type rating.[11] In reality he trained in England on the Carvair while with Interocean. He held the second FAA type rating and was Interocean manager of flight operations in the Congo for a short time in 1962. He eventually became Director of Operations for Hawaii Pacific Air, the third Carvair operator in Hawaii. In a strange twist the two Carvairs slated for the failed

Pacific Aerolift (the first attempted Hawaii Carvair operator) later became the Hawaii Pacific Air fleet under Fairchild in 1990.

Louis Khem purchased the first Pacific Air Express Carvair N80FA on 29 December 1982 from James Blumenthal. Khem leased it to PAE in January and the registration was transferred to Kemavia on 08 February. The aircraft had been moved to Tucson for maintenance and painting, which was completed on 05 March 1983. Maintenance records indicate that three 556-gallon KC-97 fuel tanks were installed in the cargo compartment for the ferry flight to Hawaii.

Khem requested the FAA reserve registration N103 for the "C-54A" Carvair on 23 March 1983 to replace registration N80FA. The registration of N80FA was then transferred from Kemavia to Pacific Air Express on 10 April. He also requested that reserved registration N302JT be assigned to a PAE Beechcraft.[12] The registration of N103 was not applied to Carvair seven until 12 July 1983.

The ownership of aircraft and registrations are somewhat cloudy during this period. James Turner of Honolulu based Turner Aviation was the financial backer of the failed upstart Pacific Aerolift with two Carvairs registered N406JT and N407JT. Turner was also president of Love's Bakery in Honolulu. Several aircraft at PAE were registered N300JT, N301JT, and N302JT indicating the involvement of James Turner. Fairchild, who was Vice President of PAE and PAL, would become Flight Director of Hawaii Pacific Air (HPA) seven years later in 1990. That carrier would purchase the financial interest of the two Carvairs N406JT and N407JT from James Turner.

By 1985 PAE was Hawaii's largest air cargo operator. The crews of Captains Larry Nelson,

After the demise of Falcon Airways, Carvair seven eventually went on to Pacific Air Express as N103. After an early morning bread delivery it departs Maui for the return to Honolulu (courtesy Bob Murdock).

Randy Tucker and F/O Kirk McBride and Denny Turner were rated in the DC-4, Carvair and Beechcraft. Later Paul Rakisits who flew the Carvair for Interocean and SEEAT joined PAE at the request of Fairchild. PAE transported flowers, pineapples, and papayas from the outer islands to Honolulu for both consumption and re-shipping to the U.S. mainland. Dry goods, bread, and automobiles were transported outbound from Honolulu to the outer islands. There are no major bakeries on the outer islands and all bread is baked in Honolulu and flown daily to the other islands. There is fierce competition for this lucrative contract among the cargo carriers in Hawaii. In addition to the Carvair and DC-4s PAE was also operating three DC-6s in inter-island cargo service. One of the DC-4s, N300JT crashed at Kahului Maui on 29 June 1983.

Pacific Air Express was already beginning to feel the pressure of competition when Mid Pacific Air introduced YS-11 cargoliners. Aloha and Hawaiian were also moving into the cargo business and seeking the bread contract. By early 1986 Pacific Air Express grounded the three leased DC-6 cargoliners and freight traffic was not sufficient to pay operational cost of the Carvairs.

Hondu Carib/Great Southern Airways

With the demise of Pacific Air Express in 1986 Carvair seven was grounded at Honolulu. Frank Moss and Lee Mason entered into a joint agreement and purchased N103 from Louis Khem. During the negotiations Khem also offered the second PAE Carvair N55243 to Moss (see Carvair seventeen). In July N103 was transferred to the Hondu Carib operating certificate. After 1985 many of the remaining Carvairs operated under Hondu Carib Cargo certificate, which is a short name for Honduras Caribbean founded by Frank Moss of Port Charlotte Florida. Moss paid his half of the purchase in cash and Fred Lee Mason elected to make payments. Mason felt it was in his financial best interest to have the aircraft registered in his name even though he was only half owner. The Carvair operated under the corporate name of Great Southern Airways based in Orlando. The arrangement ultimately resulted in a dispute that grounded the aircraft.[13]

Carvair N103 ferried to Tucson in October of 1986 for temporary storage until the new registered owner Lee Mason could arrange financing and complete the deal. Frank Moss had already paid his half of $35,000. Over time it was registered or owned by several other companies until lost in Alaska but Moss was the final owner. It remained at Tucson until ferried to Georgia on 27 April 1987. During the ferry flight it was forced to divert to Memphis because of a major oil leak.

Frank Moss has built a reputation flying old prop workhorse aircraft. He began his career flying as a corporate pilot on light twins and graduated to DC-3s when he flew with Air New England. After leaving Air New England he established Hondu Carib registering it in the Cayman Islands. The company was actually based at Tela Honduras, which is a better location for cargo operations between Central America, Miami and Caribbean destinations. Moss operated a pair of DC-3s and DC-4s for six years before moving to La Grange Georgia in 1983.

Great Southern Airways owned by Fredrick Lee Mason operated ad-hoc cargo charters under Hondu Carib and depended on regular calls from automobile factories in Georgia to transport parts. Lee Mason of Williamson Georgia and Frank Moss purchased N103 from Kemavia on 08 September 1987. Mason applied for transfer of the registration on 18 November and secured a loan against the aircraft for $25,000 on 06 January 1988. It was ferried from Tucson to La Grange Georgia on 27 April 1988. Frank Moss and Lee Mason had an inter-working business relationship. Mason was having problems making his payments on both the Carvairs seven and seventeen. As a result tensions developed between the pair. Great Southern Airways was grounded and ultimately out of business. The Carvair became a casuality of the conflict and was parked at La Grange because it was registered to Mason but Moss had put up all monies paid to date. During this time, another ex–Falcon Carvair, N83FA, owned by Bob McSwiggan was flown by Custom Air Service under the Hondu Carib certificate.

With Great Southern grounded N103 became the center of the dispute while parked in Georgia. Frank Moss became a contract pilot back in the Caribbean flying DC-4s, -6s, and -7s for Haiti Air Freight, Jamaica Air Freighters and many others. The FAA restricted many of these operations in 1993 because of numerous accidents and Moss opted to move to Alaska to fly for Brooks Fuel. All the while Carvair N103 was still parked at La Grange Georgia deteriorating.

The ownership dispute over N103 was finally resolved in 1994 in favor of Frank Moss. Yesterdays Wings was formed as a Delaware corporation with Moss as president. Fred Mason settled by paying all monies owed on the Carvair and full ownership went to Frank Moss when the registration was transferred on 31 August. Moss only had $9600 but decided to go back to Georgia to rebuild it. The aircraft had been parked for years and was in need of major work.[14]

Soon after arriving in Georgia Moss realized the aircraft was so deteriorated that he did not have the funds to make it airworthy. In an incredible gesture of friendship the people of LaGrange Georgia pitched in to help. Not only did they come out to turn wrenches and make repairs but the women brought covered dishes and made ice cream. Moss was overwhelmed with their generosity when he would go to town to buy wiring or hardware. The vendors would tell him "Your money is not good here" and donate the parts. After 10 months of living in the aircraft, while working on it, Moss was able to make it airworthy.[15]

Just as the project was nearing completion disaster struck again. This time it was a Tornado, which slammed the nose door against the side of the aircraft causing structural damage. Stringers 17, 18, and 19 had to be replaced. In order to gain access to them three skin panels were removed between stations XN57 and X14. Moss was financially exhausted by this time. The sheet metal dealer donated materials for repair and the group of local supporters and friends from Brooks Fuel helped repair the damage.[16]

N103 was registered to Yesterdays Wings of Wilmington Delaware on 20 December 1995. It departed La Grange for Kansas City on 22 December for the first leg of the 24-hour ferry flight to Fairbanks. Unfortunately the aircraft had been parked too long at La Grange and was not ready for the long flight. Shortly after departing Georgia N103 began to have problems. Condensation in the fuel manifold froze preventing fuel transfer, then the cockpit/cabin heater failed forcing Moss to divert to Tupelo Mississippi on 23 December. It was obvious that the problems were serious enough to force a returned to La Grange Georgia with an interim stop at Birmingham. Once again Frank Moss and N103, which was still affectionately known as "Plain Jane" were defeated.

Great Arctic Airways

"Plain Jane" remained at La Grange for three and a half months while maintenance items were cleared and Moss pursued new business possibilities. To raise the necessary cash to continue he sold half interest in the Carvair to Jay Moore and the two of them formed a new company named Great Arctic Airways. The Carvair was registered to the new company 07 June 1996. It departed La Grange on 08 April for the four-day trip to Fairbanks stopping at Kansas City, Billings, Spokane, and Prince George. It arrived at Fairbanks on 12 April 1996 to begin general cargo work as a FAR part 125 operator transporting cargo and vital supplies to the remote villages and construction sites in Alaska.

Final Flight of N103

On 28 June 1997 the crew completed unloading a consignment of building materials about 15:30 Alaska daylight time at Venetie Alaska. The flight was commanded by Frank Moss for

Great Arctic Airways and assisted by Jay Moore. It was being operated under the Hondu Carib Cargo certificate. At approximately 16:00 ADT N103 departed Venetie for Fairbanks on a positioning flight. The 14 CFR Part 91 flight was operating in visual meteorological conditions and a VFR flight plan was in effect.

Captain Moss stated in NTSB report ANC97LA093 that as the aircraft climbed out of Venetie the Number Two engine began to run rough and lose power. The engine was shut down and propeller feathered. It was noted the propeller stopped turning. During the process the fire warning light was illuminated on the console. The cargo handler visually verified the Number Two engine was ablaze. Moss activated the right bank of fire extinguishers. The fire warning light went out for about five seconds then came back on. The second bank or left fire extinguishers were activated with no affect on the fire warning light.

Captain Moss prepared to turn back for an emergency landing at Venetie when the fire became more visible around the Number Two engine. The engine separated from the wing and fell into a remote area starting a forest fire. The wing continued to burn and Moss stated it was apparent the aircraft would not make it back to Venetie. As the aircraft descended Moss was looking for a clearing to put it down. He pointed the burning Carvair toward a sand bar in the Chandalar River about two miles west-northwest of Venetie.[17]

He was able to make a successful belly landing through small brush and trees coming to rest partially on the sand bar. The three crew members jumped out and ran for safety as the left wing and fuselage forward of the wing was consumed by fire. The aircraft was destroyed and the fire was contained by and confined to the sand bar by the river. The separated engine started a small fire, which was extinguished by U.S. Department of Land Management SmokeJumpers. Because of the very remote sight of the incident the engine was not recovered and the aircraft was not salvaged. All three crewmembers were rescued and not injured.

NTSB report ANC97LA093 stated that the probable cause was because of a fire on Number Two engine of undetermined origin or reason. The report is very brief because of the lack

On 28 June 1997 N103 suffered an engine fire, forcing Frank Moss to make a crash landing on a gravel bar in the Chandalar River near Venetie, Alaska. The crew of three walked away, but the aircraft was destroyed (courtesy Warbelows Air Ventures).

of access to the extreme area and recoverable wreckage. Captain Moss was interviewed and unable to offer any additional data that could explain the incident. The total elapsed time from takeoff from Venetie to impact in the Chandalar River was approximately five minutes. Because of the low altitude there was little time to evaluate the situation. It was apparent the aircraft could not stay airborne and all effort was to get the aircraft on the ground before it became completely engulfed in flames.

Moss stated that he regretted the loss of N103. He had owned and operated it for many years and had actually lived in it during hard times. The aircraft was written off and Moss was instructed to remove the remains but because of the remote area nothing was salvaged.[18]

14

Carvair Eight

c/n 10448-8, C-54B-1DC, Aer Lingus

Carvair eight is one of three C-54B-1-DC airframes used for conversion. There are twelve older C-54/DC-4 selected for the program. The C-54 was built in Chicago and delivered to the USAAF 15 November 1944 with U.S. military serial number 42-72343. The airframe configuration is similar to the C-54A except the fuselage fuel tanks were relocated to the wing. This Skymaster was built as a staff and troop transport without the cargo hoist and a single cabin door. The max ATOG is 73,000 pounds and max fuel capacity of 3,740 U.S. gallons. This series of C-54 was equipped with either R-2000-3 or -7 engines. It was transferred from the USAAF to USAF on 18 September 1947 and remained in active military service until 1950. It began civilian service as N88819 with Twentieth Century Airlines and was immediately transferred to North American Airlines when the two carriers merged.

Twentieth Century/North American

North American was founded by Stanley Weiss in 1945 as Fireball Air Express and became the first airline to offer no frills cheap air travel to America. Weiss began his Fireball Air Express service as a freight carrier later changing the name to Standard Airlines when the CAB issued him a certificate as an irregular air carrier. In 1948 Standard Airlines and Viking Air Lines merged forming North American Airlines. The CAB set restrictions on large irregular carriers of eight round trips per month. Weiss in 1950 acquired Twentieth Century Airlines, Trans-National Airlines, Trans-American Airlines, and Hemisphere Air Transport. With four carriers and certificates he was able to operate more flights. This DC-4, which was originally purchased from the USAAF in 1950, came to North American through the acquisition of Twentieth Century. The low cost North American was so successful that by 1954 it had carried 194,000 passengers. This was more than the three smallest scheduled carriers combined. As a result the other carriers opposed Weiss through the CAB. Through court action in 1955 they were successful in forcing North American to sell the aircraft to Resort Airlines and cease all operations in 1957.

Resort Airlines/Slick Airways

N88819 is the fifth C-54/DC-4 to pass through Resort Airlines ownership on the way to becoming a Carvair. Resort operated N88819 until 02 July 1960 when it was leased to World Airlines for Logair military contract work. After one year with World it was returned to Resort and the following year leased to Slick Airways. Slick was formed in January 1946 by Earl Frates Slick and Colonel Sam Dunlap III. Slick was the son of wealthy Texas oil tycoon known as the "King of the Wildcatters." Slick attended Yale and joined the Army Air Corp in 1941. He was a

DC-4 N88819 in 1954 with North American titles. The single airline type cabin door verifies it was built as a staff transport for the military. Double cargo type doors were fitted when it was converted to Carvair standards (via Jennifer M. Gradidge).

B-24 pilot and met Colonel Dunlap while serving in the military. Dunlap had a vision of the airline industry and convinced Slick of the fortunes to be made. After Slick was awarded Scheduled Carrier Certificate 101 his nine C-46 aircraft began transcontinental freight service on 04 March 1946. The Carrier eventually expanded into a variety of larger aircraft including DC-4s and DC-6s. At one time Slick operated the largest fleet of DC-6s in the United States. Slick only owned two DC-4s that later became Carvairs four and eight, but his airline is woven into the Carvair story. In 1957 Slick executive and original board member Erban L. Drew left and formed Seven Seas Airlines. That carrier operated the Skymaster N5520V that became Carvair five. Slick abandoned scheduled freight service in 1958 when he got into financial difficulty and leased out two DC-4s to California Eastern Airways. Four of the Cal-Eastern DC-4s were later converted to Carvairs.

Earl Slick, company president Hank Huff, and senior Captain Bob Morgan financially regrouped and were able to re-establish the company. In 1959 Slick was granted a Logair military contract requiring seven DC-4s. He leased N65142 (Carvair four), which is one of five Resort C-54/DC-4s that became Carvairs. A $7.7 million Navy Quicktrans contract was awarded to Slick in June 1962. The aging Skymasters were used to supply six Naval Air Stations across the United States. The U.S. Military Logair, Quicktrans and MATS contracts were very lucrative. However, in 1964 the Department of Defense issued new guidelines that required Civil Carriers to operate turbine aircraft exclusively on all MATS contract charters. This is the primary reason the DC-4s of Slick, Resort, California Eastern, and others were surplused making them available to ATEL.

Slick upgraded to the CL-44 but the CAB suspended its scheduled operations on 27 August 1965 allowing only the existing contract work. Airlift International acquired the carrier 01 July 1966 and continued to operate the contracts until they expired.

Skymaster N88819 was returned to Resort in 1962. After only a few months service it was purchased by Aviation Traders on 23 October 1962 while still wearing the contract U.S. Navy Quicktrans titles. It had worn the Logair contract titles while leased by World Airways in 1960. Slick operated a number of Skymasters under both of these military contracts.

Carvair Conversion

Two days after purchase from Resort Airlines it was delivered to ATEL. It entered the line as Carvair eight, the second conversion for Aer Lingus International. On 18 February the tanks were drained and work began to remove the engines and outer wings. It arrived with the standard R-2000-7 engines fitted with the non-ram carburetor air intakes identified by the long scoop fairing the length of the top of the nacelle. The first Aer Lingus Carvair, EI-AMP was equipped with the short ram-air intakes. For Aer Lingus fleet uniformity the long fairings were removed and short carburetor intake scoops installed. Airframe eight is the first of only two Carvair conversions, both for Aer Lingus, converted from non ram long fairings to short ram-air systems.

The passenger cabin entry door on this DC-4 is quite different from other airframes used for conversion and the only airframe that arrived at ATEL with this unique single door feature. It was originally built for the military as a staff transport and not intended for cargo work. The majority of military C-54 aircraft were built with a double cabin entry door, which was retained after entering civilian service. The double unit was designed to allow for side loading of large cargo and is easily identified by round corners on the hinge side and square corners in the center where the two doors meet. Some carriers opted to bolt the rear half shut to reduce noise and draft. The single door is similar to the DC-6 passenger cabin door with all four corners rounded. The Carvair required the single door unit changed to doubles creating additional work for Aviation traders. Considerable sheet metal work was required plus the installation of door frame and threshold plate. A rear lavatory was also removed and the small window on the upper rear fuselage skinned over.

The special transporter delivered the new nose to Stansted on 18 December 1962. The Irish registration of EI-AMR was issued 05 February 1963. The fitting of the Carvair nose and tail and mounting of the engines was completed by 30 March. The dry weight was recorded as 41,063 pounds and it was roll-out on 11 April in full Aer Lingus livery after a conversion of just over five months. The first flight as a Carvair was made from Stansted on 18 April 1963. After 10-hours and 44-minutes of test flights over two days the C of A was issued. The same day it was flown from Stansted to Southend for minor detail work before delivery.

It is the second Carvair built with the 34-seat passenger cabin, and four car positions. To further capitalize on changing market conditions an additional configuration was devised. Like EI-AMP, it was designed to accept an additional 21 seats a total passenger cabin capacity of 55 with three car positions. The 12-seat cabin for the 34-passenger configuration required and additional emergency exit at the second to last round window on the starboard side. If operated in the proposed 55-seat configuration with the 21-seat forward cabin the over wing emergency exits were to be utilized for a total of four emergency exits on the starboard side. The Aer Lingus airframes were the only Carvairs with four starboard emergency exits. It is doubtful if the proposed 55-seat cabin was ever used since none of the Aer Lingus crews interviewed recalled flying in this configuration.

EI-AMR was delivered to Aer Lingus at Dublin 29 April 1963 at a purchase price of £175,000 ($490,175) including spares. The total contract package price was £350,000 ($980,350) for Carvair six and eight. It was named "St. Jarlath," which appeared on the port side and the Gaelic name "Iarfhlaith" appeared on the starboard side.

The second Aer Lingus Carvair EI-AMR on 25 April 1963, undergoing final checks at Southend before delivery. In nine months it would return to ATEL for re-routing of the control cables to raise the cargo hold ceiling and the installation of the Rolamat floor system (courtesy Guy Craven).

Aer Lingus

Crew training had begun with the delivery of EI-AMP to Aer Lingus. When EI-AMR arrived it was immediately added to the training program to prepare for the Irish car-ferry service beginning 08 May 1963. Car-ferry bookings were good and the number of cars transported exceeded original estimates but Aer Lingus officials saw the Carvair as marginal and only as a stop-gap to help Irish tourism until the arrival of car sea ferries.[1] Their lack of enthusiasm added to concern about what they perceived as excessive mechanical problems. It is now known that Aer Lingus did not encounter any more mechanical problems than other Carvair operators. The excessive engine failures they experienced were later determined to be from an outside source. It was originally believed that the flight crews had problems adapting to the characteristics of the R-2000 radial engines, which was a factor in ordering a third aircraft after the first season of operation. Officials at Aer Lingus were intent on exploiting the horse charter market and felt the third aircraft was needed to serve as backup to maintain a high usage schedule.

Car-ferry operations were cancelled during the 1963–64 winter due to seasonal traffic. The opportunity was used to return EI-AMR to ATEL at Southend for maintenance in December 1963. It was ferried to Stansted in February 1964 for installation of the "Rolamat" floor and raised ceiling modification of the cargo bay. Because of the lack of hangar space the work was performed outside in cold and snowy conditions. To utilize unused space of the hump in the cargo compartment behind the cockpit the ceiling was raised by re-routing the control cables. This gave the forward section of the cargo bay a cathedral appearance. The real reason to raise the ceiling was to transport horses and expand into bloodstock market, which was dominated by BKS Air Transport. Since the Carvair was reluctantly accepted under the direction of the Irish government, increase utilization was exploited wherever possible. The "Rolamat" floor system developed by ATEL made Aer Lingus a pioneer in "Quick Change" aircraft for changing cargo roles. This feature became quite popular on cargoliners and combi aircraft, which are identified as "QC" or "Quick Change" aircraft.

After modifications were completed at Stansted, EI-AMR returned to service for the 1964 summer car-ferry season. By April the third Carvair was on line and Aer Lingus planned phasing out the DC-3 fleet. The larger capacity Carvair was operated on overnight mail flights as well as car-ferry schedules. In June 1964 one week after the third Carvair, EI-ANJ, was delivered the fleet took over all mail service from the DC-3s. On 29 April EI-AMR had been damaged on the night mail flight from Manchester. The nose wheel collapsed on landing at Dublin crushing the gear doors. There was no structural failure and damage was considered minor. The aircraft was returned to service after repairs.

Returning to Dublin with Captain Dible in command on 27 June 1966 EI-AMR experienced wild fluctuations in RPM on Number Two engine. The crew feathered the prop and landed on three engines without further incident. On 08 August 1966 again commanded by Captain Dible EI-AMR departed Dublin at 03:15. Shortly after takeoff the Number Four engine went into an over speed condition and could not be brought under control. The prop was feathered and it promptly un-feathered with a loud noise. Captain Dible instructed the F/O to hold his finger on the feather button while he made an immediate turn and descent back to the field. Captain Dible told the Dublin tower, "We need terra firma pronto!" The prop ran away on touch down as all the oil had been pumped out. As the speed decayed the engine seized and was destroyed. The crew stated they considered themselves quite lucky they were able to keep the aircraft under control and Captain Dible was able to land with the damage confined to the engine.[2]

After an engine change and repairs EI-AMR returned to service. Under the command of Captain Jon Hutchinson it was the first Aer Lingus aircraft to land on the new Runway 09/27 at Liverpool on the evening of the official opening by HRH The Duke of Edinburgh. The longer runway allowed the evening freight flights from Dublin to roll to the end with minimum use of brakes. The extra length avoided the possibility of blowing tires when inbound at maximum landing weight.

The Carvair never really met Aer Lingus expectations and business began to deteriorate after 1964 with the improvement of drive on ferries across the Irish Sea. Aer Lingus purchased nine Viscounts from KLM in 1965 and cargo Viscounts were scheduled to be on line a year later allowing the Carvairs to continue into 1966 with reduced schedules. EI-AMR soldiered on but Aer Lingus Carvairs seem to have more than a fare share of problems. On 05 October 1966, again commanded by Captain Dible, EI-AMR experienced a complete hydraulic failure at Liverpool. A maintenance team was dispatched from Dublin for repairs, which resulted in a 21-hour delay. Captain Dible flew EI-AMR only once more on 28 October 1966. Three days later it was withdrawn from service.[3]

It was removed from service and officially stored at Dublin in October 1966 but remained in semi-active status while Aer Lingus attempted to find buyers. Aerovias Nacionales Transcontinental Ecuador (ANTENA) made serious inquires into the purchase of one or more Aer Lingus Carvairs. The carrier was negotiating rights to operate all cargo service between Ecuador and the United States (Miami). A very optimistic rumor circulated at Aer Lingus in the summer of 1967 that a buyer had been found in Ecuador. Carvair eight, EI-AMR, was removed from storage in September and returned to airworthy status. Retraction checks and test flights were made only to find the prospective sale collapsed. The aircraft was kept active in an effort to resurrect the sale.

In a rather bazaar twist EI-AMR was called back into service on 07 October 1967 to recover an Aer Lingus Viscount, EI-AKK that crashed at Bristol. The Viscounts had replaced the Carvair and was a contributing factor to them being removed from service. EI-AMR was dispatched to Bristol and transported the remains of the Viscount back to Dublin. It made the flight without incident and returned the next evening. It was again withdrawn from service and stored at Dublin with the other two Carvairs.

Eastern Provincial

After three more months in storage EI-AMR was purchased by Eastern Provincial Airways of Gander. The purchase was officially completed on 10 January 1968 in the amount of C$167,000 Canadian dollars for each of the Aer Lingus Carvairs. The price included spares and support equipment. Eastern Provincial needed high volume aircraft to support the Churchill Falls and smaller Twin Falls Hydroelectric projects in remote northeastern Canada.

Since EI-AMR had only been in storage for three months, it became the first of the three Aer Lingus aircraft to ferry to Southend arriving on 16 February 1968. It was scheduled for repainting and the installation of high capacity heaters along with nose door insulation for cold weather Canadian operations. Dual high capacity heaters were installed and the nose door intake was sealed and skinned over along with removal of the inside duct to cockpit. Dual air intakes were relocated about two-thirds back on the sides of the hump behind the cockpit. A hatch was installed in the cargo hold floor through the nose wheel bay for mail drops. A small ladder could be placed in the nose wheel well for crew access. In the event serviceable airfields had nothing to board the flight could make a pass NSI (No Stop Inbound) and drop small parcels or mail sacks.

The overhauled and modified Carvair was rolled out on 20 May 1968 without titles and the anti-glare paint below the windscreens. It flew for the first time with Canadian registration CF-EPV on 21 May. The UK C of A was issued on 22 May and the Canadian C of A was issued on 28 June 1968. The Irish registration EI-AMR was cancelled on 29 May. All the cold weather modifications were completed and the anti-glare paint applied to the nose by 23 May before

CF-EPV departed Southend on 24 May 1968 fitted with a solid nose wheel as opposed to the slotted spoke type. It is the only Eastern Provincial Carvair with EPA letters on the stripe. The titles were applied above the windows after it arrived at Gander (courtesy Transportraits/Richard Goring).

departure the next day. It was painted with a red stripe separating the white top from gray bottom and is the only Eastern Provincial Carvair with EPA titles over the stripe on each side of the nose. The Eastern Provincial Airways titles were not added above the windows before delivery but were eventually applied by Eastern Provincial in Canada. The next EPA Carvair CF-EPW was painted slightly different. Eastern Provincial officials made the decision that any livery changes could be made after delivery to Canada. It arrived Gander with 58,310:46 TT airframe hours.

Eastern Provincial operated the Carvair in eastern Canada in a combi-cargo role between Moncton, New Brunswick, Goose Bay Labrador, Twin Falls, Churchill Falls and Wabush until September 1973. Because of a drop in traffic and little need for heavy lift to the remote Hydro-electric and mining projects during the winter of 1969 it was withdrawn from service in October. It was also in need of maintenance because of operating more than a year into unimproved gravel fields. It remained parked at Gander through the summer until September 1970. The airline did not report any serious or higher than normal engine problems during the five years in use. This seems to reinforce the theory that Air Lingus crews were unable to adapt to the temperament of the R-2000 radial engine. However, EPA used a different vendor for engines, which also validates reports by Aer Lingus crews that engine failures were reduced when overhauls were contracted to Air France.[4]

CF-EPV suffered serious damage in an incident at Gander on 03 May 1972. The nose gear collapsed on landing at 19:28Z when Captain Mitton applied the brakes causing the lower fuselage to contact the runway. The aircraft had slowed considerably before the gear collapsed but damage was severe. The nosegear doors, surrounding skin, stringers and ribs were crushed resulting in structural damage. Engines two and three also suffered damaged when the props struck the ground. The aircraft previously experienced a nose wheel collapse while serving with Aer Lingus. This time the damage was more severe requiring major repairs to the skin and stringers. Fortunately at least one of the two spare nose sections that were unused at Aviation Traders was still intact in England. When the two unused noses were returned to Southend they were stored in the Military Freight Organization (MFO) hangar. This was the black T-2 hangar where Carvair one, G-ANYB was originally converted. In mid April 1971 they were moved to the scrapping area outside. At least one nose had not been reduced to scrap allowing EPA engineers to acquire the lower nose section. The common DC-4 parts needed were available in Canada to complete the engine rebuilds. After the repairs it was returned to Eastern Provincial Service.

As the Churchill Falls development project was maturing the aging Carvair was no longer needed and no longer cost effective. It was sold to R.C.S Graham on 23 July 1973 for $62,000. Robert Graham was a broker and shipping agent from Prince Rupert British Columbia. The Carvair only transited Graham pending sale to Norwegian Overseas Airline and a pending monthly lease by BAF from 21 September 1973 to March 1974, which was not taken up. It was ferried back to Southend from Gander for servicing at ATEL prior to potential sale to Norwegian Overseas. The hours were recorded as 61,861:52.

The Carvair generated considerable revenue for EPA having the highest charter rate of any aircraft in its fleet of two dollars per mile. Operating into remote northern Canada gravel airfields with oversize and sometimes overweight cargo put CF-EPV to the limits. The crews accepted the Carvair but were not overly fond of it. It was often called names like "Humpty Back Camel," "International Ballistic Blob," "Lumbering Lizzie," and most often "The Blob." In spite of the names it was the right aircraft at the right time for EPA.[5]

British Air Ferries/Norwegian Overseas/J. Jorgensen

Upon arrival at Southend in 1973, Carvair eight was scheduled to be overhauled and prepared for lease to Norwegian Overseas Airways. All work except painting in Norwegian livery

After work in Canada, CF-EPV was returned to Southend to be overhauled for Norwegian Overseas Airways. It never returned to service and was stored for five years with the engines and many parts removed (courtesy J.W. Bossenbroek).

was stated as completed but the mounting of new engines has never been verified. Norwegian Overseas specified the removal of the Pratt & Whitney R-2000 engines and re-fitting with Wright R-2600 engines. The fitting of Wright R-2600 engines to the DC-4 was developed by Charlotte Aircraft Corporation of Miami in the late 1950s and proved quite satisfactory giving considerably more power. The R-2600 engine is limited to a maximum 2,400 rpm versus 2,700 rpm maximum takeoff power and 2,550 maximum continuous power of the R-2000. This is because of the use of the same prop, the 13-foot Hamilton Standard, on both type engines. However, the R-2600 can produce 1,500 hp with maximum continuous power or 1,300 hp to 8000 feet, which is significantly better than the R-2000. The few aircraft that were fitted with the R-2600 were designated as the Super Skymasters DC-4ME2. Only one other Carvair (airframe 14) was considered for Wright R-2600 conversions, also for a Norwegian operator.

Payment of £16,075 ($45,026) was never received for services and ATEL filed suit against the owner for monies owed. The court ordered the aircraft impounded for non-payment. British Air Ferries acquired the aircraft from the courts and removed the R-2000 engines for spares with the airframe placed in storage for parts salvage at Southend.

It was not immediately stripped for spares and it remained in limbo for nearly five years. Since it was complete without engines, several Norwegian individuals and companies attempted to resurrect a purchase and return it to service. Norwegian Overseas Airways negotiated a second tentative purchase of Carvair eight and was assigned Norwegian registration LN-NAB on 12 June 1974. The registration was originally reserved for Carvair 14 and not taken up. The airline was attempting to acquire an aircraft for contract relief work in Asia but it remained at Southend in EPA colors through the summer. A second Norwegian registration of LN-MDA was issued on 09 September 1974, when the Canadian registration of CF-EPV was cancelled. A

The cockpit of CF-EPV has reverted back to the Aer Lingus identity of EI-AMR Larfhlaith after breakup and sits at Holton/Halesworth, Suffolk, England. The interior is relatively intact with seats, controls and gauges (courtesy Michael Zoeller).

letter of intent to purchase was signed the following day by J. Jorgensen of Oslo Norway. It was then leased to Norwegian Overseas Airways by Jorgensen on 17 October 1974 but remained at Southend. It never flew again and remained stored without engines at Southend. In 1974 it was clearly in poor condition with the engines and gear doors removed. The EPA lettering was painted out but the Canadian registration CF-EPV remained on the decaying aircraft until it was broken up for parts on 13 September 1978.

Many parts including the nose door were used on Carvair 12, G-AOFW. The cockpit was purchased by John Flanagan and displayed at Fritton until loaned to the USAAF 100th Bomb Group museum at Thorpe Abbots, Norfolk in 1982. The cockpit was moved several more times and came back to Suffolk in July 1997. It is currently at the 8th Air Force museum at Holton/Halesworth Suffolk England. As of May 2006 it is sitting outside in a grassy area and not maintained. Surprisingly most of the interior is still in tact with seats, gauges, and controls as they were when it last flew. The nose gear and a few other parts were saved and are now on display at the Thameside Aviation Museum at Tilbury Essex, England.

15

Carvair Nine

c/n 27249-9, C-54B-20-DO, British United Air Ferries

Carvair nine is the older of the two surviving aircraft in 2008. Only five newer model C-54 airframes were selected for conversion. Although quite active early on it was stored for 11 years at Griffin Georgia and used for spares support until Carvair five crashed in April 1997. After that tragic loss work began to restore N89FA to flying condition.

The C-54 was originally built at Santa Monica as a C-54B-20-DO and delivered to the USAAF on 11 January 1945 with military serial 44-9023. It is the only C-54 of this series converted to Carvair standards. It is equipped with integral fuel tanks in the outer wings increasing the total fuel capacity to 3,740 U.S. gallons and fitted with Pratt & Whitney R-2000-11 engines. The C-54 was built with the double type passenger/cargo-loading door and equipped with cargo loading hoist. On 15 November 1945 under Executive Order 9425 it was declared redundant and transferred to the Reconstruction Finance Corporation for disposal and assigned civilian registration NC88816 on 23 November. Western Airlines signed a letter of intent to purchase but the purchase was not completed because of financing difficulty.

The Skymaster was re-purchased by Douglas becoming the 10th conversion to DC-4 civilian transport standards. After conversion it was certified for a maximum of 54 passengers and empty weight recorded at 40,886 pounds. The overhaul included removing of the cargo hoist and the rear half of the double cargo doors. The hinges were removed and area skinned over. After conversion to DC-4 civilian standards Western Airlines acquired financing from United Airlines and purchased it for $90,000. It was assigned Western fleet number 209 and operated for 10 years before being sold to Guest Aerovias Mexico on 18 December 1956. The U.S. registration was cancelled on 19 December after receiving Mexican registration XA-MAA. Ten months later it was removed from Mexican service and leased to Aerovias Panama Airways. On 11 September 1957 it was assigned Panamanian registration HP-256 and in 1958 it was changed to HP-268.

After two and a half years with Aerovias Panama it was returned to Guest Aerovias Mexico on 21 March 1960. Instead of returning to Mexican service, Lloyd Aereo Bolivano negotiated a one-year lease for $96,000. At the same time the board of Guest Aerovias Mexico voted to surplus several DC-4s and DC-6s. The Skymaster was sold the next day to The Babb Company a division of New York based International Aircraft Sales. The Babb Company re-registered it in the U.S. as N9326R and within a week entered into a lease-purchase agreement with Lloyd Aereo Bolivano re-registering it in Bolivia as CP-682.

The Skymaster was based at Cochabamba for nearly two years before Lloyd opted out of the agreement. It was returned to The Babb Company and transferred in January 1962 to its parent company International Aircraft Sales where the registration N9326R was re-instated. Some records indicate it transited Remmert World Airways, which is possibly Remmert — Werner, a Distributor — Broker in Saint Louis. No other details were found. It remained parked

DC-4 N9326R as it appeared when purchased by Aviation Traders. Even though the plane was operated on lease in Mexico, Panama and Bolivia and operated in primitive conditions, it was in considerably better condition than most of the other airframes purchased for conversion (via Jennifer M Gradidge).

at Orly with U.S. registration until purchased on 24 October by Aviation Traders. British registration G-ASHZ was assigned on 06 May and remained on the aircraft for 16 years.

Carvair Conversion

The DC-4 was parked at Orly without titles for some time prior to being ferried as N9326R. Maintenance work was completed and final checks were run on 25 October 1962. It ferried on 29 October to Stansted where it cleared customes. G-ASHZ is the ninth Carvair with only six other ships built from later model airframes.

The nose assembly arrived by road from Southend on 27 January 1963. The double cargo doors had to be fitted since the rear half had been deleted and skinned over when converted from to C-54 to DC-4 civilian standards by Douglas for Western Airlines. The skinned rear door half had a window that was deleted when the doors were added during Carvair conversion.

Aviation Traders announced that Carvair nine would probably go to Interocean to fill the option for a third aircraft. Interocean was currently operating Carvairs four and five on the United Nations contract and considering additional aircraft. After a seven month conversion G-ASHZ was rolled out at Stansted on 01 June 1963. It became the fifth BUAF Carvair and no further mention was made of possible delivery to Interocean. The new BUAF Carvair made the first flight from Stansted to Southend on 08 June 1963. It was painted with British United titles in red over a white upper fuselage, light gray bottom separated by a dark band along the windows and on the tail. The registration G-ASHZ was in small letters on the lower rear fuse-lage.

Like other short range cross channel ships Carvair nine was designated an ATL-98-A with wing 250-pound lead bob-weights in the wingtips. This is explained in more detail in the engi-neering and development section. It was configured for 22-passengers and five cars but weighed in both 22 and 55 seating configurations and recorded as 40,385 pounds and 41,002 pounds respectively. In May 1964 a convertible "Quick Change" (QC) 33-seat cabin was installed for-ward of station X758 for use on the high density Ostend route with 55 seats.

Flight trials went as scheduled requiring only 5.50 hours of test to acquire the C of A, which was issued just hours before the first revenue flight to Rotterdam on 14 June 1963. The next day in honor of a famous Rotterdam bridge it was christened "Maasbrug" by the wife of Rotterdam's Burgermister Gerard van Walsum. As part of the celebration local dignitaries and BUAF officials were treated to a local flight to demonstrate the positive aspects of the Carvair. It returned to Southend that afternoon and was officially turned over to BUAF to begin sched-uled service. The following season G-ASHZ originated the deep penetration service from Lydd on 21 March 1964 and continued in scheduled service until 1967.

When Carvair nine first flew in the BUAF white/dark blue stripe color scheme the red letter-ing on the forward fuselage was "British United." A common fleet color scheme was adopted late in 1966 for all British United aircraft with "BUA" in black and a blue and sandstone stripe. G-ASHZ is the only ship that remained in old colors. British United Air Ferries (BUA) became British Air Ferries (BAF) on 01 October 1967. To reflect the new company name on 17 October the red lettering on the forward fuselage was changed from "British United" to "British Air Ferries." This interim lettering change made it the only Carvair to wear this one-of-a-kind livery. Since it was not due in for overhaul, it skipped the 1966 "BUA" winged bird logo (toppled mushroom) blue and sandstone livery. It was scheduled in for overhaul on 18 February 1968. The two-tone blue stripe livery had not been adopted and it was repainted in sandstone and blue stripe with "BAF" letters without the toppled mushroom. The livery is identified by the straight cut in the stripe behind the BAF letters. Only three months later BAF adopted the two-tone blue stripe livery.

British Air Ferries often participated in joint promotional campaigns with shippers. When the first shipment of Panter Cigars arrived from Holland BAF promoted it with great fanfare with a photo promotion of models in Leopard skin costumes unloading the aircraft.

The Channel Air Bridge and British United Carvairs were originally given "Bridge" names. G-ASHZ carried the name "Maasbrug" until 1970. By this time it was in the two-tone blue stripe colors and as part of a new advertising strategy it received a new character name of "Fat Annie" on 01 October 1970. British Air Ferries officials were looking for less formal names in an effort to stimulate business and change the airline image.

T. D. "Mike" Keegan purchased British Air Ferries in 1971 prompting another livery change for 1972. The two-tone stripe was changed to a wide solid blue stripe and the lower fuselage was painted navy blue. The aircraft retained the character name "Fat Annie" and was still wear-ing the two-tone blue stripe BAF livery in early 1973 indicating once again it was late receiv-ing new colors.

It remained in loyal service until June 1976 when it was placed up for sale but there were no offers. The car-ferry service had declined to only a fraction of the business of previous years. BAF had six Dart Heralds for passenger traffic displacing the Carvair for lack of demand. The decision was made to repaint it in the last BAF colors in an attempt to generate interest and find a new owner. The light blue stripe ran the length of the fuselage and up the vertical stabi-lizer with no titles.

With no active buyers coming forward G-ASHZ was offered for lease. Tentative plans were made to transfer it to another Keegan company, TransWorld Leasing. Société Anonyme de Con-struction (SOACO), a large French construction company with major projects in Libreville

Above: G-ASHZ Maasbrug departing Southend. It is the only Carvair to have British Air Ferries titles in red on the first BUAF color scheme. The other eight ships painted in this livery carried British United titles in red (courtesy Transportraits/Richard Goring). *Below*: The Panter Girls, Jill, Jennifer, Jo and Jayn, unload the first consignment of Panter Cigars from Holland. The aircraft is still painted in the blue and sandstone stripe, which is identified by a straight cut behind the BAF letters. The aircraft in the background has the italic cut in the stripe behind the BAF letters indicating it is two-tone blue (courtesy BAF).

During the winter of 1976 G-ASHZ, Fat Annie, was leased out for work in Gabon. All titles were removed to match Carvairs TR-LUP and TR-LWP that were previously purchased by SOACO (courtesy Richard Vandervord).

Gabon had purchased Carvair 10 in 1975. In March 1976 it purchased a CL-44 and in June acquired Carvair 13. SOACO was involved in development and expansion on a mammoth scale in Libreville and other cities throughout the country of Gabon. Additional lift was needed for construction materials and to support several oil drilling projects. A short term lease was negotiated by SOACO for G-ASHZ in November 1976 for a term of slightly less than three months. The activity and number of flights in Africa is not well documented but believed to be considerable because of the volume of building materials needed to support the projects. It was back at Southend in February 1977 for maintenance. SOACO only purchased two ships but the lease technically placed them in the small group of carriers and operators of three or more Carvairs.

Upon return from the SOACO lease the registration was transferred to TransWorld leasing on 24 February 1977. TransWorld leasing received financial backing from real estate magnate John Gaul and to a lesser extent Mike Keegan. TransWorld owned a number of aircraft leased to Transmeridian. It was not an aircraft user company but a title used by Keegan to shuffle his properties around under TWL lesseeship to other organizations. Little is documented of this transaction or where G-ASHZ was operated. It is possible that this was a book keeping transaction for tax purposes between two Keegan companies. Whatever the purpose G-ASHZ was returned to BAF in May and went in for periodic overhaul.

The car-ferry bookings continued to decline prompting the decision to convert G-ASHZ to an all cargo configuration. Twenty-eight months earlier Carvair seven, G-ASDC, was converted to all cargo and generating considerable revenue. In June 1977 "Fat Annie" became the second ship stripped to bare metal to reduce weight with the rear passenger cabin removed. The last row of seats against the rear bulkhead was left for extra crew and cargo handlers.

Unlike G-ASDC, which was first operated as a cargoliner in bare metal, G-ASHZ was rolled out on 06 July 1977 in the new cargo yellow and black sash and stripe tail livery. The name "Fat

All paint was stripped from G-ASHZ in June 1977 at Stansted. The passenger cabin was removed and it became the second of the two aircraft converted to all cargo with yellow and black sash livery (courtesy Kenneth Tilley collection).

Annie" was not applied to the nose after it was converted to all cargo. Only two Carvairs were painted in the yellow and black sash BAF "CARGO" livery. Later a variation of the color scheme became the Falcon Airways livery by changing the yellow to blue.

During a cargo operation in September 1977 the nose door was ripped of in heavy winds at Belfast. It was replaced with a spare from one of the earlier ships that had been broken up. In July 1978, one year after cargo conversion, G-ASHZ was leased for two months to the International Red Cross for disaster relief work in the Middle and Far East. Norwegian Carvair 14 had been involved in Red Cross relief work in Southeast Asia in 1975 and was still grounded and deteriorating at Bangkok. G-ASHZ was returned to BAF on 07 September 1978 reverting back to cargo service. Because of declining cargo bookings it was withdrawn from service and again offered for sale. With no interested buyers the engines were removed and it was stored at Southend on 24 December 1978.

Falcon Airways

The Texas based cargo carrier Falcon Airways operated a mixed fleet of cargoliners including DC-4s and was shopping for a suitable high volume aircraft for fleet expansion. Falcon already operated several DC-4s making the Carvair an easy fit to its fleet. At the suggestion of Jim Blumenthal, another DC-4 operator, Falcon president Lewis May reviewed the surplus BAF Carvairs for transporting oil field equipment.[1]

Falcon Airways vice president and operating manager Jim Griswold along with Ruth May wife of Falcon Airways owner Lewis May negotiated the purchase. They inspected the Carvairs and submitted a letter of intent to purchase on 11 April 1979.[2] (Purchase of Falcon Carvairs covered in history of ship five) As the terms of the sale were being finalized U.S. registration N83FA was originally reserved by Falcon for Carvair nine. However, the registration was not taken up and eventually was assigned to Carvair five when purchased by Ruth May. Carvair five was

With an uncertain future, G-ASHZ was stored without engines at Southend in December 1978 until purchased by Falcon Airways in April 1979 (courtesy Richard Vandervord).

leased to her husbands company Falcon Airways. The purchase of ship nine was finalized on 01 May 1979 for $166,667 and signed by Mike Keegan. Carvair seven was purchased at the same time for the same amount. The British registration of G-ASHZ was cancelled on 14 May 1979 and the following day it received U.S. registration of N89FA.

It was repainted at Southend late in April while the terms were being worked out. The livery consisted of changing the yellow to blue in the sash color scheme, deleting BAF and adding the Falcon Airways lettering above the windows. With the confusion of the registration N83FA lettering was painted on the aircraft in error. After some embarrassment and sorting out of the proper registration N83FA was removed and the correct registration N89FA added. The U.K. certificate of airworthiness for export was issued on 02 June 1979.

Carvair nine, registered N89FA, departed Southend for Dallas on 03 June 1979. The 35-hour four-day trip took the same route as N80FA via Lisbon, Azores, and Newfoundland. The United States C of A was issued on 11 June 1979 with total airframe time recorded as 58,698.7 hours. After arriving Dallas the wing "Bob-weights" were re-installed by Falcon on 17 July restoring it to ATL-98A status. The nose door was modified for a weather radar unit by adding the di-electric patch and gill liners were installed.

N89FA was active with Falcon for less than a year when the cargo carrier fell on hard times and ceased operations. It was repossessed by The Mercantile National Bank of Dallas on 21 March 1980. Falcon maintenance was somewhat lacking because of the financial problems in the last days of operation adding to the deteriorating state while it sat idle. The aircraft registration was transferred to the Mercantile Dallas Corporation for disposal on 07 April 1980.

Nasco Leasing/Calm Air

Nasco Leasing Corporation purchased the N89FA on 10 October 1980. Nasco listed a mailing address of Incline Village Nevada. Little is known about Nasco or if contracts were negotiated. It appears to have been poorly financed and lacking any firm lease agreements for the aircraft. After only three months it was repossessed for the second time by the Mercantile National Bank of Dallas on 07 January 1981. The registration was transferred back to Mercantile Dallas Corporation for disposal on 19 February. The condition and whereabouts are not clear during this period but it was moved to Tucson Arizona and placed in storage in June 1981. Several unsuccessful attempts were made to sell it during 1981 and early 1982 however, because of the age and condition the offers were low and not accepted. The only negotiation with promise was in 1982 when Calm Air Canada of Manitoba seriously reviewed the possibility of purchasing the two ex–Falcon Carvairs N89FA and N80FA. (See ship seven for more on Calm Air.) The sale negotiations did not progress pass an informal inquiry. N89FA remained at Tucson until 01 December 1982 when rescued by James Blumenthal.

James Blumenthal

Jim Blumenthal had been considering a front or rear loading aircraft to add to his DC-4 operation. He purchased the two ex–Falcon Carvairs N89FA and N80FA from Mercantile National Bank on 01 December 1982. He first saw the pair in 1978 while flying N67019, a DC-4 chemical sprayer used to break down the oil spill from the Amoco Cadiz disaster. At that time he realized the Carvair with a "high bulk — low weight" capability had potential in supporting oil field operations. Since Falcon Airways contracted to the oil industry and was expanding, he suggested their possible use to Lewis May. Now, four years later in a twist of circumstances, the fate of two of the Falcon ships was passed to Blumenthal.[3]

After purchasing Carvair nine from the Texas bank Blumenthal soon found that it had not been maintained and corrosion was a problem. He stated that most Carvair operators after the original owners were poorly funded and abused the aircraft and N89FA was no exception. Blumenthal's meticulous attention to detail is obvious in the appearance of his DC-4, which is extremely well maintained. It is also evident in his well-maintained C-123N. His philosophy is very simple, "The safety and performance can only be expected if the aircraft is well maintained." He installed new engines on N89FA with the short exhaust and after considerable effort and cost the corrosion was brought under controlled. It was repainted in white with yellow stripe to match his DC-4 making an impressive pair of cargoliners. The fact that Blumenthal's DC-4 was still flying in 2006 in the same color scheme as the Carvair is a testament to his maintenance standards. He stated that everything he touched on N89FA cost him money and it never produced any revenue during the time he owned it.

Prior to purchasing the Carvair Blumenthal actively pursued the purchase of a surplus C-124 Globemaster II but the FAA had no intention of certifying it for civilian use. Consequently they were not made available. With the C-124 out of reach his second choice for a high bulk front loading aircraft was the Carvair. The fact that N89FA still exist today can probably be credited to his effort to put it in top operating condition. His desire for quality ultimately became so expensive that he reluctantly made the decision to sell and today someone else realizes his efforts.[4]

After three years and nine months of continuous headaches and restoration N89FA was sold to NCBA Incorporated on 14 September 1986. NCBA is actually a small contract operator headed by Steve Kehmeier. He listed two addresses for his operation, a street address in Denison Texas and a post office box in Eckert Colorado. None of the individuals contacted commented on his operation. His stationary letterhead consisted of hand drawn sketches of a Carvair and a DC-4 with the company name. He took possession in August and in a few months it was advertised for sale. Kehmeier stated in a letter dated 23 November 1986 to Bob McSwiggan, managing director of Custom Air Service, that the FAR 121 (approved for over 7500 pounds of cargo) aircraft with Cardex had only 59,098 TT hours. He also stated that Number One and three engines were new with only 4 hours each.[5] McSwiggan later stated that the engines were not what they were represented to be, with one of them failing at 15 hours.[6]

Custom Air Service

Bob McSwiggan purchased the aircraft on 04 March 1987 for $175,205. It was flown from Tucson to Griffin Georgia as a planned back-up for N83FA which he had purchased in 1985 from Ruth May. It was operated under the Hondu Carib certificate owned by Frank Moss, who eventually owned the third ex–Falcon Carvair based at LaGrange Georgia. There was not enough business to support McSwiggan operating both aircraft and when he discovered the engines to be problems N89FA was used for spares to support N83FA. McSwiggan believed he was getting two serviceable new engines on N89FA that have a replacement cost of nearly $30,000 each but all of them failed in short order. By all accounts the aircraft was in exceptional condition when it left Blumenthal. The airframe was well maintained but McSwiggan encountered engine problems with the first engine failure immediately after taking possession from Kehmeier.[7]

It is difficult to determine how many hours previous owners logged on the aircraft. Kehmeier stated that it had 59,098 when he sold it to McSwiggan. The logbook showed 59,102 on 20 May 1998 just after completing the FAA certification flight to return to service. That is only four hours in twelve years. The aircraft had flown nearly four hours on the morning it ferried from Tucson to Griffin Georgia. It is possible that it was not flown during the four months between the date of Kehmeier's letter and the purchase by McSwiggan. The C of A issued by the FAA in 1979 when it arrived in the U.S. list the hours as 58,698. This would indicate that Falcon Airways, Blumenthal, and Kehmeier flew it for exactly 400 hours before McSwiggan purchased it.

The excellent condition of N89FA when sold by Jim Blumenthal suggests that Custom Air Service could have put it in active service. McSwiggan believed that he was getting a good aircraft and was considering using N83FA for spares support. After evaluating N89FA the plan was reversed. Carvair N83FA had 10,000 less hours at slightly more than 49,000. N83FA also had the enlarged forward bulkhead, which McSwiggan was not aware of when purchased. It allowed more clearance for oversize cargo than N89FA. Consequently N89FA was withdrawn from service at Griffin and remained inactive for many years. After the loss of N83FA at Griffin in April 1997 work began to bring N89FA back to flying status.

It underwent extensive overhaul under the direction of propliner maintenance engineer George Dyess, who maintained McSwiggan's DC-3 fleet for years. In addition to a complete check of the airframe and all systems it received four overhauled engines and props. The cockpit was completely refurbished and gauges re-conditioned. George Dyess also performed the maintenance on Carvair N83FA for Custom Air Service for years before it was lost at Griffin. He felt that the spirit of that aircraft should be carried on in N89FA. Dyess went through the wreckage of N83FA and found the knob from the nose door actuator pump. He installed it on the nose door pump handle of N89FA where it remains today.[8]

N89FA was test flown out of Griffin 06 April 1998 landing at Tara field in Hampton Georgia. The avionics were installed by Custom Air Service and certified on 08 July 1998. It returned to service after years of storage and was in the Bahamas late in 1998. After operating a number of successful charters it suffered a nose wheel collapse in March 2000 at Opa Locka Florida and was grounded pending repairs. Custom Air Service holds the largest supply of Carvair spares that exist but had sold the last remaining set of nose doors to HawkAir in June 1997. McSwiggan

Above: Custom Air Service returned N89FA to service at Tara Field, Georgia, in May 1998 after ten years of storage. It was still wearing the same colors applied by Jim Blumenthal in 1983, validating his attention to detail and high quality maintenance standards (author's photograph). *Below*: The cockpit of N89FA after the aircraft was restored to service for McSwiggan at Custom Air Service. The control surface lock pin lanyard is extended from the ceiling reel in the captain's field of vision. This aircraft was brought back into service to replace N83FA that crashed due to the pin not being removed because the lanyard was tucked behind the captain's seat on takeoff (author's photograph).

contacted master fabricator Hugh Smith who made a new set of nose gear doors in order to put the aircraft back in service.[9]

Avignon Inc.

On 09 February 2001 N89FA was purchased by Ohio attorney Deems Clifton for $275,000 and registered to Avignon Incorporated of Cincinnati Ohio. Clifton who signed the Bill of Sale as President along with two partners formed an ad hoc charter business catering to the automotive and oil industry. He applied for transfer of the registration on 13 February. It was relocated to Texas from Georgia in order to be closer to oil industry of Houston and Oklahoma area. The aircraft reverted back to the 1967 British Air Ferries name "Fat Annie" and wears 1940s World War II style nose art although the DC-4 airframe was built in 1945 and the 1963 Carvair conversion never saw military service.

Carvair nine took up the ad hoc charter service that N83FA flew before it was lost in a tragic accident. The winter snow and ice prevent overland shipments of auto parts from vendors to automobile manufacturing plants. In order to receive approval to operate in icing conditions and generate more cargo business N89FA was flown to Columbus Indiana in November 2001 by Frank Moss for the fitting of de-icing boots. Moss has flown Carvairs for over 20 years and lost his own Carvair to fire at Venetie Alaska on 28 June 1997. He also served as Captain on Carvair 20, C-GAAH, for HawkAir while operating at Bronson Creek Canada. In mid 2002 N89FA was returned to its base at Sherman-Grayson County Airport at Sherman Texas.

Some time after 2001 another attorney, Alfred K. Nippert, Jr., became president of Avignon. FAA records indicate Deems Clifton had paid Nippert $139,000 for part ownership in the aircraft.

Gator Global

The company name changed again on 01 November 2002 to Gator Global of Cincinnati. Once again the FAA recorded a Bill of Sale for N89FA. It indicates that Gator Global purchased the aircraft from Avignon for $603,535 on 06 August 2002. Alfred Nippert signed the document as President of Gator Global. It appears that another partner was added on 09 January 2003 when another Bill of Sale for $330,000 was executed and signed by Dick Zerbe as Director of Operations. The inner workings of the two companies is not clear during this period.

The aircraft has remained idle and only flew one charter in over a year. Frank Moss was called in fly the charter of transporting an oilrig from Houston to Ciudad del Carmen Mexico on the Yacatan peninsula. Before leaving Houston the Number Three engine required work. The load crew had difficulty loading the drilling equipment and the operation had to be taken over by Moss and the flight crew. After unloading and departing Mexico the Number Three engine failed and was shut down arriving in Miami on three engines.[10]

In August 2005 after being idle for sometime N89FA arrived at Rantoul Illinois for a Skydiving event. The Carvair made several flights with 80 Skydivers and a crew of five for a total of 85 on board. It was set up with 40 seat belts down each side of the cargo compartment for the jumpers who sat on the floor. The aircraft was flown to 10,300 feet where the Number Two engine was throttled back and the flaps extended to reduce prop-blast. Two passes were made at 110 mph dropping 40 jumpers on each pass. Even with the extra precautions the jumpers were tossed around in the propwash then tumbled several hundred feet after exiting. It is believed this is the only time that Carvair nine ever lifted 85 passengers and crew. Some Carvairs were used on relief and refugee missions where the number of passengers was not recorded but this is without question the most uplifted in the United States.

16

Carvair Ten

c/n 10382-10 C-54A — 15-DC British United Air Ferries/Alisud

Carvair 10 was originally built as a C-54A-15-DC at Chicago and delivered to the USAAF on 24 August 1944 with military serial 42-72277. With the war coming to an end the aircraft was declared redundant and transferred to the Reconstruction Finance Corporation for civilian sale. It was purchased by Braathens SAFE on 13 March 1947 and registered in Norway as LN-HAU. Braathens was formed by four Norwegian shipping lines naming Captain J. Rad as Chief pilot. The DC-4 was used to efficiently ferry relief crews from Norway to ships throughout the world.

It was converted to DC-4 passenger cabin and christened "Norse Trader," later "Delhi Air Trader," flying mostly to the Middle East and Asia with occasional immigrant flights to Johannesburg via Khartoum. The crew flights to the Middle and Far East continued until 1949 when an eight-day turn around flight was initiated from Oslo to Hong Kong. Later that same year Braathens became Braathens South America and Far East Transport operating additional service between Oslo and Caracas. The service did not last and the DC-4 was sold to Seaboard & Western Airlines on 17 October 1950, registered in the U.S. as N1221V, and re-named "Oslo-Airtrader."

New York based Seaboard & Western operated five DC-4s at the time on worldwide cargo charters and the U.S. military Trans-Pacific supply line. The 40-hour route began at Travis Air Force Base in California stopping at Hickam Air Force Base Honolulu, Wake Island, and Iwo Jima before arriving in Tokyo. It is the first of two DC-4s acquired by Seaboard & Western that eventually became ATL-98s. The same pair of Seaboard & Western DC-4s, N1221V and N54373, were two of the last three Skymasters operated by the carrier. Eventually all three were sold to Interocean, the second of the original Carvair operators.

Seaboard converted N1221V back to cargo configuration in order to support an expanding freight operation. The carrier continued to operate N1221V until 20 January 1953 when it was dry leased to New York based Trans Caribbean Airways (TCA). The carrier operated service between New York/Miami and Cuba, San Juan, and South America and maintained bases at Miami and Teterboro. Trans Caribbean applied to the CAB in 1952 for Trans-Atlantic freight certificate. N1221V was returned to Seaboard in 1955 after slightly more than two years but TCA continued leasing DC-4s from Seaboard & Western for many years afterward.

Seaboard & Western secured a MATS contract in 1953 operating routes between the United States and Japan. The Skymaster flew this route as well as to Europe becoming quite worn and tired. On 16 September 1957 while landing at Brindisi Italy en route to London from Delhi, it ran off the runway collapsing the nose gear. The aircraft had just landed to refuel and was taxiing when a jeep crossed in front of the aircraft forcing the pilot to swerve to avoid a collision. The aircraft sustained considerable superficial damage but no major structural damage to the airframe. After repairs it was put back into service and continued with Seaboard until 01 September 1959 when it was sold to the Lewman Corporation.

Seaboard and Western N1221V, the Oslo-Airtrader, was converted to Carvair ten for British United and is the tenth oldest airframe used for conversion. It was operated for several years on the 40-hour turn-around from California to Tokyo before work in Europe and Africa with Interocean (via Jennifer M. Gradidge).

The Lewman Corporation was only a transition company that re-sold it to Interocean in March of 1961. It was registered as LX-BBP and operated under contract to the United Nations peacekeeping forces in the Belgian Congo. Luxembourg based Interocean shared the same management team of Intercontinental Airlines based in New York. (See chapter on Interocean) The fate of the Skymaster took an unexpected twist when Interocean via Intercontinental purchased two Carvairs to transport military equipment for the United Nations in the Congo. Rather than wait for an Interocean DC-4 to be converted the decision was made for this aircraft to become part of a swap purchase agreement for Carvairs four and five. They were on the production line when LX-BBP was transferred to ATEL at Southend on 11 November 1962.

Aviation Traders was expecting Interocean DC-4 LX-LMK c/n 7469 but unexpectedly the aircraft that arrived was LX-BBP which was damaged in another incident on 27 October 1962. The details are vague and the extent of damage not recorded. It appears that after being repaired to airworthy condition Interocean officials decided to substitute this C-54 for LX-LMK and retain the better Skymaster. The terms of the contract allowed the substitution of another aircraft and ATEL accepted LX-BBP for LX-LMK. It was reasoned that it was to be rebuilt and poor repairs were not a problem. This later proved costly since damage was more severe than represented resulting in additional man-hours for retro work to complete the conversion.

Consequently an Interocean C-54 LX-BBP became a BUAF Carvair and a BUAF Skymaster became an Interocean Carvair. The Interocean DC-4 LX-LMK, for which LX-BBP was substituted, never became a candidate for Carvair conversion. Two years later in March 1964 the CAB refused to renew the certificate for U.S. based associate company Intercontinental leaving Interocean to stand-alone. Under financial pressure LX-LMK was sold to ATL in April 1964 two years after it was expected. Later that same year ATL re-sold it to Trans Australia Airlines. This was the second time LX-LMK missed being converted to Carvair standards.

Conversion/BUAF "Channel Bridge"

LX-BBP is the second of the two Interocean trade-ins and the 10th oldest C-54/DC-4 airframe used for Carvair conversion. It was transferred to Stansted on 12 November 1962 for disassembly. The engines were removed and returned to Interocean while the eight fuel tanks were drained, and purged. In an effort to standardize the Carvair fuel system it was converted to six fuel tanks. After being broken down and the nose removed it was ready to be moved to the conversion line on 11 December 1962. Four months after LX-BBP arrived at ATEL it was still outside. The stripped fuselage with nose and empennage removed sat in the snow with temporary work shelters over the wings. It was not moved inside until the nose arrived from Southend on 24 March 1963.

The damaged original reinforced cargo floor was replaced with a new laminated floor and cargo tie down system for securing automobiles. Like other BAF ships G-ASKG was designated an ATL-98-A with the 250-pound wing "Bob-Weight" modification ATL 98/5150 for cross-channel service. Additional explanation can be found in the engineering and development section.

After eight months in conversion the Carvair was rolled out on 27 July 1963 and new dry weight recorded as 40,585 pounds. On 29 July the first flight was made from Stansted to Southend with Captain Don Cartlidge in command. The new Carvair was fully painted at Stansted prior to the first flight. The C of A was issued 06 August 1963 after a total of only 4.50 hours of check flights over nine days. It was delivered to BUAF at Southend on the same day, christened "Channel Bridge" and scheduled to begin service.

The former Interocean LX-BBP fuselage used for Carvair ten sat outside at Stansted in March 1963 for four months after being stripped waiting for hangar space (courtesy Ian Callier).

A very rare glimpse at G-ASKG on 06 August 1963 with the name Channel Bridge over Alisud titles. The bridge name was removed the next day (courtesy Aviation Traders Ltd.).

Alisud — Compagnia Aerea Meridional of Naples

Alisud a very small privately owned Naples based company is closely associated with Alitalia. As the ramp-handling agents at Naples it became Italy's first car-ferry air service and first commercial airline to fly the tenth Carvair. Two days after delivery to BUAF the Italian carrier wet leased the new Carvair. The lease was all-inclusive providing aircrews, engineers, traffic-staff and two "Hylo" cargo loaders from British United. Alisud entered into a short-term lease agreement in an effort to establish vehicle ferry service between Naples and Palermo, Sicily. The four-month minimum lease agreement negotiated by BUAF sales manager Dick Worthington also contained an option to purchase.

On 07 August, one day after delivery to BUAF, the titles were painted out and the green Alisud logo added. The BUAF paint scheme was used with only the change in the titles on the forward fuselage. The British United titles were removed and replaced with the "Alisud" title over a winged dolphin painted in green. The ship name of "Channel Bridge" remained on the aircraft for a short time after the Alisud titles were added but was removed from the aircraft before it went into Alisud service.

G-ASKG departed Southend for Naples on 08 August at 10:40 BST. After the loaders were in place, the first revenue flight operated between Naples and Palermo on 15 August 1963. Known as the Tirrenian Air Bridge, the 450 mile flight across the Straits of Messina from Naples to Palermo was at first successful. The airline announced a planned network linking Naples and Rome to Sicily and Sardinia. North Africa service was also planned and the fleet scheduled to be increased to three Carvairs. The fare for an automobile was £13 ($36.41) compared to £11 ($30.81) by sea. Even though the flying time of one hour 10 minutes was a vast improvement over the nine hours by sea, the Alisud Carvair operation was extremely short. Company officials believed that the time factor alone would promote the business.

The majority of the flights for the first four months were operated with BUAF Captain Don Cartlidge in command under the wet lease agreement. Cartlidge began turning the operation over to other BUAF crews in the fifth month. The service continued only one more month before being withdrawn. The contract was actually extended and G-ASKG returned from Palermo to ATEL at Southend for maintenance on 25 February 1964 but did not return to Alisud. After only six months and 10 days from the first Alisud flight the service was suspended. The Carvair retained the British registration of G-ASKG while on lease to Alisud.

It appears the service was destined for failure from the outset for several reasons. Alisud did not promote the new service and there was very little fanfare for the first flight. In addition the carrier was not able to obtain permanent traffic rights causing service to be suspended in February 1964. Interocean operated some supplemental service for Alisud in the same market with the two Carvairs just returned from the United Nations Congo contract.

British United Air Ferries

After returning to Southend in February, G-ASKG remained out of service a month for maintenance. It became the seventh Carvair added to the BUAF fleet. It was scheduled to be the sixth, however the lease to Alisud one-day after delivery allowed Carvair 13 to enter BUAF service first. The Alisud titles were painted out and it returned to BUAF service without titles

BUAF became British Air Ferries (BAF) on 04 October 1967, which called for a livery change. G-ASKG touches down in the grass retaining the blue and sandstone stripe. With the change from BUA titles, the stylized bird logo has been painted over by extending the stripe, which is obvious in the straight cut behind the BAF lettering (courtesy Transportraits/Richard Goring).

although the Alisud option to purchase did not expire until 25 April. G-ASKG flew from Southend to Rotterdam without titles the evening of 25 March commanded by Captain Dudley Scorgie. It continued on the Rotterdam route and by the summer of 1964 the red British United titles were re-applied. The name "Channel Bridge" that it was given prior to Alisud lease was not re-applied below the cockpit windows until early 1965. It was rotated into other BUAF car-ferry destination of Basle, Geneva, Ostend, and Liège until December 1965.

Air Ferry

G-ASKG along with a sister ship G-APNH became involved in world politics in November 1965. The U.D.I. declaration by Rhodesia (now Zimbabwe) declaring independence from England closed the surface route to Lusaka Zambia. BUAF served East and South Africa and was still operating its "Africargo" DC-6 service out of Gatwick making it too politically sensitive to operate anti–Rhodesian aid flights in British United colors. The solution was quite simple. The Carvair was transferred to Air Ferry, which had recently been acquired by the Air Holdings Group. Since the DC-4s of Air Ferry were already contracted to be utilized in the African oil lift, the Carvairs would be transferred to the same fleet. The titles were changed to Air Ferry and they were flown by British United crews.

It departed Southend 28 December 1965 stopping at Malta en route to Dar-es-Salaam. This is the second time Carvairs were dispatched to Africa for supply and relief work because of political turmoil. Interocean Carvairs were used in the Congo in 1962–63 where they proved themselves under remote primitive conditions. This Carvair was converted from Interocean DC-4 LX-BBP, which had been operated in the Congo for the United Nations. It was traded to ATEL as part of the Interocean deal for two Carvairs.

The African mission for G-ASKG only lasted from 28 December 1965 to 11 March 1966. During this 10-week period it was scheduled to fly daily from Dar-es-Salaam to Lusaka and back. It transported 26 oil drums containing a total of 1,200 gallons of badly needed fuel for the military operation. (The Rhodesian mission is covered in more detail with Carvair 11.) Once again it proved Carvair ruggedness and durability returning to Southend in March 1966 for service and maintenance. All British registered aircraft were required to return for maintenance every 300 hours. It had been quite abused in Africa while operating in a primitive environment during war conditions. After undergoing maintenance to bring it back up to standards it was returned to service with BUAF in March 1966 operating through the summer and fall. In November it was repainted in the newly adopted blue and sandstone stripe livery with stylized bird logo and black BUA titles. The "Channel Bridge" name was maintained below the cockpit windows.

G-ASKG returned to scheduled service by winter in the new livery but bookings were down and British United financial losses were mounting. The 1966–67 winter season saw major cut backs in service. On 25 February 1967 G-ASKG flew the last Southend to Basle flight. Carvair flights to other destination continued with reduced frequency. BUAF became British Air Ferries (BAF) on 04 October 1967 and once again the fleet was scheduled for a livery change. The new livery dropped the stylized bird logo, retained the blue and sandstone stripe and changed BUA titles to BAF. It was repainted at Southend on 06 January 1968 retaining the "Channel Bridge" name below the cockpit windows. It continued in service with BAF until March 1969 when it was transferred to French carrier Compagnie Cie Air Transport (CAT).

Compagnie [Cie] Air Transport (CAT)

Compagnie Air Transport, known as Cie Air Transport or CAT, is a small French airline that began operating Bristols in a joint agreement with Silver City prior to the Channel Air Bridge merger. Joint service continued with British United from Le Touquet and Calais to Lydd and also Nimes to Corsica and Palma with leased BUA Bristol Freighters. CAT initially purchased the two Interocean Carvairs in 1965 in an effort to expand its route system and update the fleet. One of those Carvairs was lost at Karachi in 1967 and replaced with Carvair 15 allowing the carrier to maintain the two Carvair service. In 1969 CAT increased service expanding to Southend prompting the purchase of a third Carvair, G-ASKG. It departed Southend on 03 March 1969 still painted in BAF colors.

The overhaul and repainting was completed at Le Touquet. It was re-named "Commandant Charles," which was not immediately applied to the aircraft. "Commandant Charles" is actually L.J. Ottensooner, who was Chairman of CAT. He had been de Gaulle's personal pilot during the war and was given the code name of "Commandant Charles." He had also been a member of the board at Silver City Airways beginning in 1961, a post he acquired in a business arrangement when CAT purchased three Silver City Bristols.

The new livery was no more than a single blue stripe and two small lines on the vertical fin with Carvair titles at the top. Cie Air Transport titles were applied over the stripe on the forward fuselage. It received French registration F-BRPT but because of CAT financial problems BAF officials were unsure if the transaction would be successfully completed. It was returned to Southend where it remained until 17 March. The financing was thought to be resolved but the aircraft was only with CAT for two months when the on going financial problems forced ownership to revert to BAF for a second time. It remained in limbo until 08 June when it was withdrawn from service and parked at Lydd Kent wearing the CAT livery.

Transports Aeriens Reunis (TAR)

BAF reviewed a tentative lease with Transports Aeriens Reunis (TAR), a French charter airline, for F-BRPT. At the same time officials at TAR were negotiating a separate lease purchase agreement with CAT for the other two ships. The French registration of F-BRPT was transferred to TAR on 27 June 1969 although it still carried the CAT livery. Transports Aeriens Reunis (TAR) moved F-BRPT back to France and parked it at Nimes pending the completion of the

French carrier TAR purchased Carvairs 5, 10, and 15 from CAT. Only F-BRPT received TAR titles while on lease to British Air Ferries (courtesy Transportraits/Richard Goring).

The name Big Joe was added in 1970. The TAR titles were replaced with BAF, creating another one-of-a-kind British livery. In 1973 BAF repurchased the aircraft and it was repainted in the final BAF dark blue livery (courtesy Richard Vandervord).

purchase agreement. In March 1970 the TAR purchase of F-BRPT along with the other two CAT Carvairs was finalized. The titles were changed to TAR with the CAT colors. The stripe remained blue with T.A.R. in large block letters on the side of the forward fuselage. It is the only ex–CAT Carvair to receive a TAR livery.

Some limited charter flights were operated by TAR before a rather unique arrangement was negotiated to lease two of the ex–CAT Carvairs back to British Air Ferries. It was named "Big Joe" as soon as it went into British Air Ferries service but the French registration of F-BRPT was retained. It operated on the reinstated Southend — Basle route that had been cancelled because of lack of traffic in February 1967. At this time BAF operations considered this to be the best Carvair in the fleet. The crews preferred it over other ships. It would go anywhere any time and never experienced major problems.[1]

The TAR title remained on each side of the forward fuselage of F-BRPT and the titles "ON CHARTER TO BAF" were added above the passenger door. The second TAR Carvair purchased from CAT, F-BMHV retained the CAT dark blue band with small BAF letters above it and the third, F-BOSU, was used for spares and remained in CAT livery until broken up. F-BRPT, went through a number of livery changes in a short time. The TAR titles were removed in September 1972 and it was operated with no titles through December. In January 1973 it received the BAF letters on a white square over the stripe on the forward fuselage giving it a one-of-a-kind livery that no other British Carvair received.

British Air Ferries

The name "Commandant Charles" was replaced with character name "Big Joe" when BAF leased if from TAR in March 1970 and remained on the nose when the lease was renewed in

April 1972. British Air Ferries re-purchased Carvair 10 from TAR in March 1973 and it was returned to Southend for overhaul and repainting. The British registration G-ASKG, which it held as a DC-4 and Carvair prior going to France, was restored on 26 April 1973.

Car-ferry demand had declined by 1973 prompting the change to three-car 55-seat configuration. G-ASKG made its debut in the last BAF livery of white upper fuselage, light blue stripe and navy blue lower fuselage on 08 April 1973. It remained in this livery until early 1975 when it was removed from service because of the decline in car-ferry demand.

SOACO/Société Anonyme de Construction

Société Anonyme de Construction, a French based company with major projects in Libreville Gabon, purchased G-ASKG from British Air Ferries on 23 February 1975. The Gabon registration TR-LUP was assigned on 21 February and the British registration G-ASKG cancelled on 25 February. The aircraft remained in BAF color scheme with titles removed. SOACO, as the construction company was known, had large-scale projects in the capital and many other cites throughout the country. Aircraft were needed to transport building materials from suppliers in Marseilles. The cargoes included plumbing fixtures, tile, windows, lumber, beams, furniture, and cement. The need for cargo lift was so great that SOACO purchased a CL-44 in March 1976, a second Carvair (Ship 13) in June, and leased a third Carvair (Ship 9) in November to support the projects in Gabon. SOACO contracted to Air Africa in March 1976 to overhaul the aircraft. TR-LUP was flown to Brazzaville Congo for maintenance and repainting. The new paint scheme was nothing more than a white upper fuselage and bare metal bottom. The name "Big Joe" was retained under the cockpit windows.

After overhaul by Air Africa it was returned to service transporting construction materials and workers. Most of the missions were concentrated in western Africa and were no more than two hours in duration. During the time with SOACO it logged on average 100 hours per month. TR-LUP was withdrawn from service in 1979 and abandoned at Brazzaville. The Empire of Central Africa government confiscated it in March 1979 for monies owed. It remained stored at Maya Maya airport Brazzaville Congo through 1980 and did not fly again until purchased by Aero Service in 1982.

Société Aero Service (Afrigo)

Aero Service is the oldest operator in the Congo (Zaire). It was formed in 1967 at Agostino Neto Airport in Pointe Noire Congo to support the Total Elf oil drilling projects and is still operating some 40 years later as a schedule carrier and air taxi service known as Afrigo. The carrier is based in The Congo and services the bordering countries. In early November 1981 the Carvair was purchased in order to bid on lucrative oil company contracts of oversize cargo.

It was registered in the Congo as TN-ADX and was airworthy by June 1982 flying a busy schedule making 24 domestic flights from home base at Brazzaville to Djambala during July. Through the end of 1982 it was active operating multiple charters transporting oil-drilling equipment to and from Pointe Noire on the Atlantic coast of the Congo. In December 1982 it flew a segment to Yaoundé, the capital of Cameroon. The Carvair was proving itself once again as a rugged workhorse continuing the heavy schedule into 1983 with 15 domestic flights to Pointe Noire in January. By the spring of 1983 the projects were near completion and the Carvair was once again idle and withdrawn from service on 16 March. Aero Service continued to periodically run up the engines and maintain the aircraft with optimism. On 03 February 1984 after 11 months in storage it was returned to service, flying twenty segments during the year of a greater distance to Luanda Angola, Libreville Gabon, and Bangui Central African Republic.

Some of the flights were oil equipment charters to Cabinda Angola, which is on the Atlantic coast south of Pointe Noire.

The Aero Service Carvair operation was the most successful in 1985 with 37 charters. The majority of the segments were two and half-hour flights to the northern remote area of Impfondo. This kind of operation is a testament to the ruggedness of the Carvair. Another Carvair was being used for spares but it is a noteworthy accomplishment to operate in such remote areas of the world where you are your own support.

During 1986 TN-ADX only flew 10 revenue flights, which were split between oil exploration support to the domestic destination of Pointe Noire and general cargo to Kinshasa. The Kinshasa (formerly Leopoldville) flights are hardly worth operating since it is situated across the Congo River from Brazzaville. It is a 10-minute direct flight but because of political dictates and procedure flying between the two countries the flight actually takes about 30 minutes.

As rugged and proven as the Carvair is, operating in remote Africa stresses even the best-maintained aircraft. During 1986 TN-ADX only flew two non-revenue maintenance flights totaling two hours. It remained idle until July 1987 operating only 14 flights during the second half of the year. Nine of the flights were revenue generating to Luanda Angola and Impfondo Congo and the other five were maintenance check flights. The aging Carvair had three engines replaced by the end of the year and was becoming very costly to keep airworthy. Because of many mechanical problems it remained idle and was sporadically maintained until July 1989. After a maintenance check flight out of Brazzaville it was again declared airworthy and flew 28 oil exploration support flights to Pointe Noire between July and October 1989. After seven years of commercial charter flights for Aero Service it was in need of a major overhaul. TN-ADX flew into Brazzaville from Pointe Noire on 07 October 1989 and was parked in front of the Aero Service offices as an impromptu aviation monument. It was generally thought that Carvair 10 had seen its last working days and would sit silently deteriorating in the sun at Maya-Maya airport at Brazzaville.[2]

EclAir (Entreprise Consolidé pour des Livraisons par Air)

The founder and driving force behind EclAir and the restoration of Carvair ten is Mike Snow, who for many years wanted to start his own air service. Formerly with SAS he acquired a private license and later a commercial rating. He claims to have flown for Interocean Airways in Africa on famine relief work missions where he gained experience. It is not clear what missions and relief contract this was or amount of flying he logged. This second DC-4 operator named Interocean was formed in 1981 and flew relief work in Mozambique. It is often confused with the Luxembourg based Interocean that flew the UN (ONUC) Congo mission in 1963.

While living in South Africa Mike Snow gained knowledge of the tobacco distribution industry in Zaire and was well known to the management of Tabazaire distributing. There are 26 distribution warehouses for cigarettes and tobacco products in Zaire that distribute 250 tons per month. It is very difficult to transport these products overland because of corrupt ground cargo companies and 50 percent currency inflation creating an obvious need for a high volume aircraft with low operating cost. The primary distributor in the area is Tabazaire based at Lubumbashi (Elizabethville) at the extreme southern tip of Zaire.

Snow saw a photograph of the derelict Carvair that was taken by Belgian aviation enthusiast Michel Anciaux. It was obvious to him that this was the perfect aircraft for the Zaire tobacco industry. He went to Brazzaville in 1992 to meet the owner Raymond Greisbaum and inspected the Carvair. At the time it was one of the last six Carvairs still in existence. To his amazement it was in relatively good condition and still wearing the British Air Ferries name "Big Joe." It had not flown since 1989 and the paint had cracked and peeled because of the intense African sun and humidity. Raymond Greisbaum of Aero Service ran-up the engines periodically and performed maintenance to keep the aircraft airworthy. While Snow was in Brazzaville the engines were run-up and the aircraft systems checked out. The results were good enough that he returned to Johannesburg to gather and convince investors that the aircraft was worth the $85,000 purchase price. In addition it had considerable spares and ground support equipment, which would be needed for the new cargo operation.

The political situation in Kinshasa Zaire, which is only a 10-minute flight away, was rapidly deteriorating but Snow was determined to follow through. He put the investors together consisting of Tabazaire as the primary with himself and crew as junior partners. He agreed to manage the Tabazaire transport operation on a fixed per ton rate. The aircraft was required to fly 19 days per month amounting to 110 hours. The balance of the time each month would be required for maintenance. Snow was also allowed to operate ad-hoc charters. The 19-day, 110-hour requirement would only transport approximately half of the 250-ton requirement. Plans were made to purchase a second aircraft to operate under EclAir to transport the balance.

Tabazaire installed 270,000-liter fuel tanks at Luano airport in Lubumbashi. It also arranged transport of the fuel and 4000 liters of oil from Durban South Africa. In January 1993, six months after Snow inspected the Carvair, he returned to Brazzaville with engineer George de Mar. It had been seven weeks since the last run-up and the aircraft needed considerable maintenance and system checks to become airworthy. All work had to be completed before it could be ferried back to South Africa for overhaul. The final terms of the sale and transfer to the new company of EclAir also had to be finalized.

The majority of the peeling paint would need to be stripped and the skin treated with an acid solution before repainting. The Carvair had not been fully stripped of paint since owned by British Air Ferries. The dark blue paint on the lower half of the fuselage had been removed after serving with SOACO. The BAF block letters were clearly visible as the layers of paint were

TN-ADX stored at Brazzaville, Congo, in January 1993. The engines had been run up periodically to keep it active. The name Big Joe is faintly visible under the cockpit window (courtesy VIP/Nicky Scherrer).

peeled away. The painting of the new livery would wait until the aircraft ferried for overhaul at Lanseria, which is northwest of Johannesburg. It was not practical to paint the aircraft at Brazzaville since some modifications had been planned.

It was immediately discovered on the January visit that Number Four engine was not running smoothly and backfired when powered up. The cylinder compression checks did not reveal any bad cylinders. After considerable checks and inspection the carburetor was found to be the problem. Fortunately the parts needed were on hand in the spares that came with the aircraft. Finally after days of work in the African sun the systems were considered satisfactory leaving only the retraction check to complete the deal.

After a successful retraction check the Carvair was put through many high speed taxi runs at Brazzaville. After an exhausting three months it was ready to fly again. During the time Mike Snow was working on the aircraft at Brazzaville he was consulting with Fields Aircraft Engineering at Rand Airport, which is two and one half miles south of Johannesburg near Germiston. The plans were to fly the aircraft to the overhaul facility at Lanseria, which is north of Johannesburg.

During a visit to Lanseria while in a restaurant bar Mike Snow observed a Frenchman named Pierre Cote sitting at the bar reading an aviation magazine. They began a conversation about aircraft and by strange coincidence he discovered that Cote had logged more than 200 hours in Carvairs. He flew for Nice based Secmafer the French contract cargo carrier that operated Carvair 17, F-BVEF, under the name SFAir. Even more strange, the Carvair Pierre Cote flew in France had been an Interocean DC-4 in its previous life and flown by Interocean under United Nations contract in the Congo. It was the third DC-4 Interocean sold to ATEL for conversion. Mike Snow's earlier experience with Interocean in Africa combined with Pierre Cote's SFAir Carvair time and considerable propliner experience in Africa gave them enough in common that Cote soon became a member of the EclAir team.

While the maintenance was on going at Brazzaville, even with Tabazaire's influence, it took a month and a half to wade through the bureaucracy processing the papers in Zaire. Finally on 17 March 1993 the Carvair received the Zaire registration 9Q-CTI. After three and a half years on the ground, the first test flight lifted off from Brazzaville on 19 March. The aircraft handled well but had a number of write-ups and returned to Brazzaville to correct the problems. The second test flight late in the afternoon of the same day was satisfactory and the decision was made to depart the next day, Saturday 20 March 1993, to South Africa.

The planned early morning departure from Brazzaville was delayed because of thunderstorms. The Carvair departed Brazzaville at 11:15 A.M. for the five and a half-hour flight to the first layover at Lubumbashi. With the exception of some erratic gauges the flight was without any major incident until the approach to Lubumbashi. It was dark when the field came in range and the Carvair was cleared to land. The anti-collision light was working but the landing lights did not work when switched on. Mike Snow continued the decent and made a smooth landing with the runway lights as his guide. Lubumbashi is the headquarters of Tabazaire and planned base of operation for the Carvair after overhaul in South Africa.

After over-nighting 9Q-CTI was back in the air for the next leg to Bulawayo Zimbabwe. The three-hour flight took the Carvair across the center of Zambia, which is the narrowest part of the country near Lusaka. Twenty-eight years earlier this same Carvair was in Zambia transporting fuel from Dar-es-Salaam to Lusaka. The Carvair entered Zimbabwe airspace at the most northern tip and crossed the country from end to end. After a short time on the ground at Bulawayo and completion of the proper paperwork 9Q-CTI was off again for Lanseria. It crossed into South Africa at the eastern tip of Botswana shortly after leaving Bulawayo with the final two hours passing without incident. The crew could not resist making a low pass over the field to proudly announce their arrival at Lanseria then smoothly touched down on 21 March.

The overhaul work at Lanseria began the following day. The instruments were removed and sent to Fields Aircraft for testing and overhaul. All the spares along with the magnetos, generators and fuel pumps were also overhauled. The fuel lines were found to be brittle and easily broken, needing replacement. Weather radar was added with a small radome since the original nose door was never replaced or updated with the di-electric patch. A Global Positioning System along with new radios was also installed. The rear passenger compartment was completely restored and fitted with 15 vintage passenger seats, a chemical toilet lavatory, and a video screen on the forward bulkhead for the Tabazaire staff and guest. The flight deck seats were reupholstered in leather and a new engineers seat was ordered from the United States to complete the restoration.

All Carvairs have several small round windows, which follow the contour of the fuselage descending from the flight deck. This ship has five windows, two on the port side and three starboard. Two of the windows on the starboard side are in the compartment behind the flight deck and the third is high up in the forward cargo compartment. All of these were sealed at Lanseria and one larger round window installed on each side of the upper deck crew compartment.

After the new porthole windows were installed, the upper half was painted white with three cheat lines in shades of brown and orange. A logo emblem was designed for the vertical stabilizer that resembles African tribal marks and is said to symbolize the word "EclAir." The new paint scheme was finished with the letters "EclAir" on the forward fuselage.

The overhaul at Lanseria was completed in five and a half weeks. The first test flight on 29 April 1993 was a short hop from Lanseria northwest of Johannesburg to Rand Airport where the aircraft was scheduled for a weight check by Fields Aircraft Engineering. Fields had previously overhauled the gauges, pumps and generators. The aircraft performed well mechanically

Carvair ten receives final work at Lanceria in April 1993 after a complete overhaul to begin African service with EclAir under Zaire registration 9Q-CTI (courtesy VIP/Nicky Scherrer).

Carvair ten was unnecessarily destroyed at Kinshasa in 1995, just over two years after it was returned to service across the river at Brazzaville (courtesy Cassiman via M.A.).

on the flight cruising five knots faster with the new paint but the radios were a problem that required two more days of testing. A second test flight on 01 May was completed at Rand with all systems performing well as 9Q-CTI returned to Lanseria.

The Carvair loaded with spares including a spare R-2000 engine departed Lanseria on runway 24 the morning of 04 May for the long flight to the new home base at Lubumbashi. With only six Carvairs surviving at the time this was a moment of achievement for Mike Snow and the crew of EclAir. The enthusiasm would not last as the political condition in Zaire continued to deteriorate. No one could imagine that in two years 9Q-CTI would be grounded at Kinshasa Zaire.

Transair Congo purchased the Carvair in April 1995 for spares support. It was parked at Kinshasa without engines and avionics on 31 July. Because of political instability it was unnecessarily destroyed 05 September 1995 at Kinshasa across the Congo River from Brazzaville where Mike Snow originally located it. It also began service there with Aero Service in 1981.

17

Carvair Eleven

c/n 18333-11, C-54B-5-DO, British United Air Ferries

Carvair 11 was originally built at Santa Monica as C-54B-5-DO and delivered to the USAAF on 22 June 1944 receiving military serial 43-17133. It is the oldest of the four aircraft of this series used for conversion. Only four older DC-4/C-54s airframes were used in the Carvair program. It was declared redundant by the Air Force only six months after acceptance and returned to Douglas Aircraft to become the 34th conversion to DC-4 civilian standards.

After the upgrade was complete it was purchased by Delta Air Lines on 27 April 1946 and assigned civilian registration N37477. Delta operated it in regular schedule service for eight years before it was sold off to North American Airlines in 1954. Under the direction of Stanley Weiss North American became the first low cost carrier in America and was successful in the market. In order to circumvent the CAB restriction of eight round trips per month for irregular carriers North American acquired four other carriers in 1950. By 1954 the airline had become embroiled in litigation because of success and the CAB reviewed its certificate as an irregular carrier. North American managed to stay operational by appealing the suits and postponing the obvious. In 1955 the CAB revoked the North American operating certificate.

The DC-4 belonged to North American but operated for more than two years under Trans-American Airlines. After a long court battle that ended in 1957 over the North American company name identity it was officially transferred to Trans-American. The transfer was dictated when North American Aircraft Company won the litigation against North American Airlines regarding similar names and North American was forced to cease operations. A year later N37477 was transferred to Twentieth Century Airlines, another carrier that North American had absorbed in 1950. It remained with Twentieth Century for less than a year having transited three of the airlines in the Stanley Weiss group. Four of the DC-4s used for Carvair conversion were owned and operated by one or more of airlines under the North American umbrella.

With North American out of business the DC-4 was tentatively sold to Lufttransport-Untermenhen (LTU) on 13 May 1958 and registered in Germany as D-ANET. The sale was never completed and a second sale was negotiated on 17 June 1958 with Independent Air Transport, a British operator based at Hurn. The DC-4 was registered in England as G-APNH. This would become the registration the aircraft would carry after Carvair conversion and balance of its operational life until lost in a landing mishap in France.

The Skymaster ferried to Hurn arriving 19 June 1958 to begin service with Independent Air Travel. Independent was founded in 1956 by Captain Marian Kozubski and became one of Britain's major charter airlines. Kozubski was a Polish war hero and veteran of the Berlin Airlift. He was considered one of the most dashing and controversial operators in British aviation.[1] Two months after G-APNH arrived his success changed with the crash of an Independent Vickers Viking G-AIJE. The Viking crashed into houses at Southall causing fatalities on the ground as well as the flight crew. It brought into question the operational standards of the carrier, which ultimately spelled its demise. The investigation discovered inferior maintenance, unskilled and fatigued crews, overweight aircraft and bad weather as contributing to the crash. The Inquiry Commissioner stated that the captain should have never taken off clearly knowing it was overweight.

Captain Kozubski resigned on 16 March 1959, the day the inquiry began on the operation practices of Independent. He took with him a number of other Independent employees and formed a new carrier known as Falcon Air (Not to be confused with Falcon Airways of Texas). Captain Frank Lloyd became Managing Director of Independent along with Captain William Bright as Chief Pilot and Norman Ebbutt as maintenance director. The publicity caused the loss of contracts leaving the G-APNH as the only active aircraft in the Independent fleet during the 1958–59 winter.

It returned to Hurn on 02 February and operated charters to West Berlin on 13 February and Dublin on 19 February. It was ferried to Prestwick and operated a freight charter to Malta on 20 February under the command of Captain Frank Lloyd. On 16 February 1959, the day the inquiry began on the Southall crash, G-APNH operated Hurn to Frankfurt. The negative publicity was slowly destroying Independent. The officials of the airline met that day and formed a plan to save the carrier by operating as Blue Air. The DC-4 received Blue Air titles almost immediately. Most of the flights were inclusive tour charters with some ad-hoc work spread far and wide. Captain Bright commanded G-APNH on a sports charter from the U.S. to Gander and Blackpool. The charters became less frequent placing the carrier in serious financial condition. Captain Banfield flew the DC-4 on 13 July from Hurn to Beirut and Captain Cole flew the return to Newcastle and a ferry to Hurn.

Bookings were so few that Blue Air ceased all contract charter operations by October. The

Skymaster G-APNH was flown in Independent markings in 1959 prior to the company being re-named BlueAir. A year later it was acquired by Air Charter and eventually merged into British United to become Carvair eleven (via Jennifer M. Gradidge).

DC-4 continued ad-hoc charters and operated Madrid to Manchester on 12 October then ferried to Blackbushe where it was grounded and Blue Air ceased all operations. Only 10 months after the arrival of G-APNH with Independent it was idle and Blue Air was bankrupt. The name change and reorganization had done little to help.

G-APNH only remained at Blackbushe for a short time before it was transferred and tentatively sold to Netherlands based carrier EurAvia late in 1959. It was painted with EurAvia titles and the application for registration PH-EUR requested. The registration was granted in November but it retained British registration G-APNH. The sale was never finalized and the registration was never applied to the aircraft however, it did operate a charter for EurAvia to Bahrain on 30 November. It returned to Blackbushe then ferried to Rotterdam on 04 December 1959 for at least two more charters. Through default it would ultimately become the property of Air Charter at Southend.

It is interesting to note that Ted Langton later used the name EurAvia for his tour group operations with Constellations. Given his close association with Independent Air Tours, it is possible this was his attempt to use the carrier's assets to form a new tour operation using the previously registered EurAvia name.

It remained registered to Blue Air as G-APNH after the company collapse on 15 October and continued to be held under the company assets until 09 November 1959. It was ferried from Rotterdam back to Southend on 08 February 1960 but remained idle because of debt liens until 08 March. Records indicate Astraeus Limited acquired it for a period of 15 days ending 23 March 1960. It had been parked on the ramp at Southend for sometime awaiting settlement of monies owed to Laker's Air Charter for service. Laker acquired it through default action on 25 March and had it towed inside for maintenance. The tail fin was painted white, Air Charter titles added and British registration maintained. After transfer to Air Charter, it operated Southend — Ostend inclusive-tour work. In July 1960 Air Charter was merged into British United Airways and the fate of the DC-4 was now determined as it flew for British United until 1963.

Conversion/British United Air Ferries (Pont de Crouch) Menai Bridge

G-APNH was transferred to ATEL for conversion on 01 February 1963. Carvair production orders were slow and there were many idle workers at Southend. The nose sections and Carvair conversion components were fabricated at Southend but with few orders parts were not being produced. The decision was made to convert Carvair 11 at Southend to provide work for craftsman who would otherwise be furloughed. Most of the multi-qualified Southend craftsmen gained considerable experience on the mock-up and building ship one, G-ANYB. It was also reasoned that when a Carvair order came in the delivery time could be reduced with a conversion already in progress. Laker had implemented a practice at ATEL years earlier to transfer and even bus manpower where it was needed. During any down time on other projects engineers could work on G-APNH.

Work began on 01 December 1962 but was suspended after a few days and the DC-4 spent most of the winter in storage. Disassembly resumed on 28 February 1963 becoming the second longest conversion eventually taking nearly two years. It is also the second of only three ships built at Southend. It was originally intended for a fast conversion since ATEL officials still believed demand would be high and anticipated forthcoming orders. There had been many new inquiries about possible conversions and it was believed to be just a matter of time before it would be sold. Unfortunately the inquiries were not followed up by signed contracts and it reverted to an on/off conversion being built between other projects of higher priority. The nose arrived at the hangar 02 May 1963. The slow pace gave engineers the opportunity to further modify it with innovative features in hopes of attracting new orders.

As ATEL engineers considered many ideas, G-APNH was selected for upgrade to an

G-APNH is the second of three Carvairs built at Southend. It was stripped and in the stands in April 1963 with the reinforcing and strengthening of empennage in progress. Work continued at a slow pace, with the conversion taking two years to complete (courtesy Transportraits/Richard Goring).

extended long-range model. This required considerable more engineering and modifications to meet the long-range standards. It became part of a plan to develop a specialized version to obtain British Air Ministry contracts to transport rockets and other oversize hardware to the Woomera range in Australia. Carvairs were standardized with six fuel tanks but G-APNH arrived with eight tanks prompting the decision to retain them for long range work. All other conversions except AVIACO ship 16 were built with six tanks even though the DC-4s arrived in various fuel tank configurations. The extra tanks would extended the range for worldwide operations necessitating the installation of more sophisticated radio and navigation equipment.

Access to the rear of the aircraft would be restricted because of positioning of oversize cargo. The long-range version called for the installation of a lavatory, galley, two crew seats plus bunks or two sleeperette seats on the flight deck. These items were all part of a planned design that would be necessary for extended flights. A small galley was installed on the rear wall of the flight deck with hot pot for coffee and a hot ring similar to the two Interocean ships. Drawings were made for a crew lavatory in the far back left corner of the cockpit hump. The positioning of the crew lavatory or chemical toilet was later moved to the nose door similar to the Interocean ships without passenger cabins. G-APNH is the only BAF Carvair scheduled to have these options.

The special options were installed both for testing and demonstration. An ATEL Design Office construction report notes a "pre-freighter" weight of 40,256 pounds. It was designed and planned as a long-range freighter but the large passenger cabin windows were installed leaving the option of changing the configuration if an order was received. It had many special options but was not fitted with the "Rolamat" floor system, which was considered too heavy and would reduce the range. It also decreased the height of the cargo hold by four inches reducing loading clearance.

Aviation Traders announced in February 1964 that it expected Carvair 11 to fly by June 1964. The fly date was updated again in August 1964 when it was announced that G-APNH would fly by 08 December 1964 for operation on a new government freighting contract to Adelaide. To understand why the completion date continued to change an explanation of Freddie Laker's use of manpower has to be reviewed. Laker had very strict rules of not paying any overtime. He expected the staff at Air Charter and Aviation Traders to work extra when needed and take off when not needed, which worked very well and most everyone was happy. The DC-4 G-APNH was made redundant when the Air Charter freight contracts were handed over to the DC-6 aircraft in the fleet. Laker had G-APNH brought to Southend for conversion to utilize spare man-hours in the event orders came in. Actually work continued at an erratic pace and long periods passed with no progress at all.

One morning in the spring of 1964 after worked was postponed many times engineers arrived to find a new name applied below the cockpit window. In full sign writing the title of "Pont de Crouch" appeared below the cockpit window having been applied overnight. The engineer humor made reference to a small muddy river just north of Southend. It was insinuated that it would never receive a "Bridge" name and fly across the Channel and it would be more fitting to name it after a nearby river. It carried the name on the green metal protective coating for about a year until removed for final painting.[2]

On 07 April 1964 G-APNH was rolled out of the hangar substantially complete. It eventually made its debut in the first British United livery with dark stripe and British United titles in red. It was officially named "Menai Bridge" on 30 December 1964 and flew for the first time on 04 January 1965 from Southend. Only 5 hours and 25 minutes of flying was required for

In April 1964 G-APNH was rolled out of the hangar structurally finished. The faded trace of the name Pont du Crouch is visible under the cockpit window. It remained in limbo and did not fly until January 1965 (courtesy Transportraits/Richard Goring).

certification. The C of A was issued on 09 January 1965 before being delivered to British United the same day. G-APNH is the last of four Carvairs to have only one registration during its entire operational life. It took two years to complete, which included four months of cumulative time when no work was performed. It was engineered with many special modifications but they were not successfully utilized and it was only operational for six years and 2 months.

The long-range Carvair was initially chartered to the British government to validate the world wide potential. British United was very interested in obtaining British military contracts to transport rockets to Woomera Australia for testing. This work had been accomplished in the past with Bristols that had to be modified by removing the rear bulkhead. This was a grueling eight to twelve day trip with the short range Bristol. Aviation Traders officials were convinced that the Carvair would serve well as a replacement for the Bristol. It had the capacity and range that was viewed with optimism for future sales.

G-APNH departed Lyneham on 11 January 1965 westbound over the North Atlantic route with a rocket as cargo destined for Woomera. The aircraft aborted the mission at San Francisco on 15 January because of the rocket developing a crack. Officials stated changes of temperature, pressure and vibration caused the failure forcing a return to England. An extensive air-conditioning system had been installed on the aircraft to stabilize the rocket temperature, however it was not enough. The incident was met with great disappointment. After returning G-APNH remained on charter to the British government supplying units at Mediterranean airfields and Cyprus. It was reported in June to having been "adapted" (installation of passenger cabin) for BUAF and it was put into regular service during the summer of 1965.

Two years after work began G-APNH was completed at Southend and received the certificate of airworthiness. The British United livery was applied to nine British Carvairs (1, 2, 7, 9, 10, 11, 12, 13, and 15 (courtesy Transportraits/Richard Goring).

World Demonstration Tour

BUAF leased G-APNH back to Aviation Traders in October 1965 for a world demonstration tour to generate interest and promote sales. The Carvair departed Southend on 16 October for demonstration in New Zealand and Australia. It left England with BUAF chief pilot Dudley Scorgie in command and carried a cargo of computers for delivery in New Zealand. The tour followed the delivery of Carvair 19 to Ansett a month earlier. The demonstrations of the long-range Carvair included Ansett and Trans Australian Airlines (TAA) and two-month operational trials in New Zealand with Straits Air Freight Express (SAFE).

It was leased to SAFE, an Air Holdings Company, for evaluation and operational trials between the north and south islands. The titles were not changed and trials were conducted in British United colors. It was perceived to be a perfect fit since SAFE was an associate company and a potential market existed. Aviation Traders sales compared an operation across the Cook Strait to being similar to the car-ferry operation across the English Channel.

Straits Air Freight Express was established in November 1950 by Airwork subsidiary Tradair to secure a New Zealand railway contract. It is an Air Holdings company in the group of associated companies that includes Aviation Traders and British United. Officials believed these demonstrations would prove that it would be the suitable aircraft to replace the aging Bristols, which had been in service with SAFE for 13 years. The Bristols were used primarily to ferry palletized cargo 70 miles across the Cook Strait from Blenheim to Paraparaumu Airport near Wellington. In 1952 the carrier was granted approval to transport cars and passengers. With the success of the Carvair in England it was considered a good fit for the market since the stage length was relatively the same as across the English Channel plus late in 1965 service to the Chatham Islands was being reviewed.

There are about 750 permanent residents in the group of ten Chatham Islands some 480 nautical miles from Wellington and 536 miles from Christchurch. The Bristol was evaluated by SAFE for cargo service to the islands but it could not carry a full load of freight and mail because of fuel loads, which are dictated by the winds. Aviation Traders marketing saw the distance as similar to the "Deep Penetration" routes established by the Carvair from England to mainland Europe. After two months of service SAFE determined that the Carvair was too large for domestic use, not practical for the short haul routes with small fields and could not be turned in 25 minutes as the Bristol. The carrier continued flying the Bristol Freighter and Wayfarer combi.

In the early years SAFE acquired an Aer Lingus Bristol through Airwork and a number of them from the Pakistan Air Force. After further consideration in 1966 SAFE purchased three-second hand Bristols from ATL instead of placing a Carvair order. Two of the Pakistan Air Force Bristols were converted in 1967 for the Chatham Island service, which was initiated in January 1968. Straits Air Freight Express remained one of the primary Bristol operators in the world until 1988 upgrading to the AW 650 Argosy in later years.

Thirty years later fate added another strange twist. HawkAir of Canada eventually purchased one of the former Pakistan Bristols that SAFE operated to the Chatham Islands. When the Bristol was lost in a landing accident at Bronson Creek HawkAir replaced it with former Ansett Carvair 20, which had been operated in New Zealand by Nationwide.

G-APNH returned to Southend from the demonstration tour on 01 December 1965 without generating any immediate orders in Australia or New Zealand. The promotion was not successful but the Carvair did make an impression on several local operators. Ansett ordered the last Carvair built in 1968. After its three ships were surplused in 1972–73 several attempts were made by New Zealand operators to purchase and operate them. Nationwide operated two ex–Ansett aircraft for a short time and Bass Straits Freight Service attempted to start operations in Australia.

Air Ferry

After returning to Southend G-APNH was not idle for long. Two Carvairs, G-APNH along with a sister ship G-ASKG, soon became involved in world politics. The Rhodesian U.D.I. was declared while G-APNH was in Australia. Under the leadership of Ian Smith Rhodesia (now Zimbabwe) declared independence from England on 11 November 1965. The British settlement was established there in the 1890s and by 1965 there were 280,000 whites in the country. Smith did not subscribe to the British policy of "One man, One vote." He created a new government internally and defied London by declaring independence.

The British began organizing world sanctions to put pressure on Smiths government. The U.D.I. declaration by Rhodesia closed the surface route to Lusaka Zambia cutting off the supply route from the Indian Ocean. Initially the RAF Transport Command operated Britannias of the 99 and 511 Squadrons to airlift supplies and petroleum for the operation based in Zambia. It became apparent the airlift would last for some time. Civilian carriers Air Ferry, Caledonian, Lloyd International, and Transglobe were contracted to operate the transport service. Air Ferry supplied two DC-4s in support of the RAF Transport Command while the other carriers operated a mix of DC-4s, -6s and Britannias. One of the Britannias operated by Transglobe was a passenger aircraft fitted only with passenger doors making loading quite difficult. It soon became obvious that special aircraft were needed to transport oversize equipment.

The rebels in Rhodesia under Smith were quite resourceful and were obviously getting help from outside the country. Gabon turned a blind eye to the fact that black Africa was supporting white Africa against black opposition. South Africa was effective in covering any connection to the Rhodesian operations. British United operated scheduled service throughout East and South Africa. Because it was politically sensitive the British United operations were protected with a simple solution. Both BUAF Carvairs G-APNH and G-ASKG were leased to Air Ferry in mid December 1965 and the titles were changed. By using Air Ferry, another Air Holdings company purchased in 1964, they could be flown by BUAF crews in a wet lease arrangement.

Air Ferry Limited was actually formed in 1961 by Hugh and Audrey Kennard after they left the Britavia group. In April 1962 the carrier established a permanent base at Manston Aerodrome, Kent. The executive staff consisted of Captain Ken Sheppardson (General Manager), Captain D. Brooker (chief pilot), R. Illsley (chief engineer), and C. Carrol (commercial manager). Air Ferry began service with two Viking 3Bs and added two C-54As.

As a C-54 operator the leasing of the Carvair would not raise any questions. Air Ferry was originally formed to provide additional capacity for inclusive tour firms and to operate ad hoc passenger and cargo charters. Air Ferry applied in 1964 for vehicle ferry service between Manston and Le Touquet and Belfast — Le Touquet to Ostend. The service was never operated. Additional request were made for Manston to Dijon and Glasgow to Le Touquet, which were also denied. Air Holdings purchased Air Ferry in October 1964 for operation under BUA. Plans were made to operate the car-ferry services from Manston to Le Touquet, Calais, Ostend, and Rotterdam. The carrier was also operating a C-54A Skymaster on loan from British United Air Ferries (BUA) when the two Carvairs were transferred in.

The BBC required a large radio antenna tower to broadcast un-jumbled government sponsored anti–UDI broadcast to the Rhodesian people. The British United titles were quickly painted out and Air Ferry titles applied. On 02 December 1965 G-APNH ferried to Lyneham where the ill-fated Woomera rocket mission began a year earlier. The radio tower was loaded overnight and it departed the next day for Lusaka Zambia. After a successful delivery of the tower to Africa it returned to Southend for maintenance. The Carvairs began transporting the support equipment to Zambia for the other aircraft involved. It departed again on 30 December 1965 staging through Malta and Khartoum loaded with spares and equipment destined for

Dar-es-Salaam arriving on 03 January. G-APNH was the obvious choice for this mission since it was developed for long-range work. It previously operated flights for the British Air Ministry and ATEL officials welcomed another opportunity to demonstrate the ability of the long-range Carvair to the government and other carriers.

Dar-es-Salaam is a port city on the Indian Ocean in eastern Tanzania. The name means "Haven of Peace." The round trip flight to N'dola Zambia is eight hours and 45 minutes. If the mission was to Lusaka, which is due south of N'dola, another hour was added to the return. While in Africa G-APNH along with G-ASKG were scheduled on oil missions to transport twenty-six 45 gallon drums of fuel weighing 300 pounds each from Dar-es-Salam Tanzania to Lusaka and N'dola Zambia. The 26 drums totaling 7800 pounds is less than half of the Carvairs payload capacity but the near 1000 mile flight required the Carvair to carry enough fuel for the daily round trip. G-APNH was ideal for this since it was equipped with long range fuel tanks.

The other Carvair G-ASKG arrived back in England on 11 March. G-APNH remained another three months and was the last of the pair of Air Ferry titled Carvairs to return to the U.K. on 02 June. When it returned to Southend it was wearing a bit of "nose art" acquired in Zambia. Someone painted the head of a giraffe on the blue band above the Carvair title on the nose. The artwork had the name "Twiga," which is Swahili for Giraffe, below it. It remained with Air Ferry on the Africa U.D.I. mission until 02 June 1966 when it was returned back to British United Air Ferries.

During the African mission the two Carvairs not only transported drums of gasoline and oil but other cargo including oil drilling equipment, military personnel with equipment and spare parts. Rhodesia remained a rogue state until 1979 when the guerrilla forces overthrew Smith's government. In 1980 it became the independent state of Zimbabwe under Robert Mugabe.

The long-range Carvair was transferred to Air Ferry in December 1965 to deliver a radio tower to Zambia during the Rhodesian UDI. It remained in Africa for transporting fuel and materials for the military operation. It returned to Southend in 1966 with a small Giraffe head and the name Twiga on the stripe below the cockpit window (courtesy Transportraits/Richard Goring).

After the return from Africa Air Ferry again attempted to initiate Carvair service from Manston to Calais, Le Touquet, and Ostend. British authorities denied the service on the grounds that the now improved roll-on sea ferries were already servicing the routes. Air Ferry pleaded their case based on past history of their operation of Bristol Freighters out of Lydd using contract Silver City Airways crews. Despite considerable effort British authorities were not convinced by Air Ferry and the aircraft were returned to BUAF in June.

British United Air Ferries/British Air Ferries

After the Air Ferry Rhodesian operation G-APNH was returned to ATEL at Southend for maintenance and converted back to six fuel tanks and scheduled for repainting of British United titles. Sales of the long-range option were not successful and the aircraft was returned to regular service with BUA where the extra tanks were no longer needed.

In November 1966 the new BUA colors of blue and sandstone stripe were applied. The livery consisted of the BUA titles and a forward facing stylized bird logo after the BUA letters on the forward fuselage. The bird logo was incorrectly applied backwards on the port side of G-APNH making it the only Carvair with the identity of the bird flying backwards. The "Bird" was soon given the name "Toppled mushroom" with the comment the slipstream blew it over.

The British government chartered G-APNH for clandestine work in January 1967. The crew was required to sign official documents of secrecy and as far as anyone was concerned the flights never occurred. Ten pallets of gold bullion were loaded accompanied by eight armed military guards in plain clothing carrying automatic weapons and side arms. The bars with South Africa markings were stacked three high on the pallets. The cargo was flown to Dusseldorf while the guards rode silently beside. Upon arrival the gold was unloaded in the cold dark hours and G-APNH returned to Southend.[3]

It was wet leased to Shell on 12 April 1967 and departed Southend staging through Marseilles and Rome en route to Oman on an eight-day trip. Shell Oil maintained drilling operations in the

Menai Bridge at Berlin in August 1967 shows off the infamous backward bird logo that was never corrected. It remained on the aircraft until October 1967 when BUA became BAF and the titles were changed (courtesy Ralf Manteufel).

Trucial states near Muscat Oman. The Trucial states were created by a maritime truce created in 1853 between the Omani sheiks and Great Britain. The Carvair was needed to transport oil drilling equipment and crews in the group of sheikdoms along the southern coast of the Persian Gulf. The area, which is rich in oil, lacked developed roadways. The Carvair proved to be well suited for such work with the ability to lift oversize cargo such as oil drilling equipment in and out of short airfields. It performed well extending the mission of eight days to two months before returning to Southend. Over the years a number of Carvairs were used to transport oil drilling equipment in the Middle East, Africa, United States, and South America.

Carvairs were often called on to transport new or damaged aircraft. On 12 June 1967 G-APNH transported three Djinn helicopters F-BHOI, F-BIEV, and F-BMLO to Southend for potato crop spraying. The French built Sud Aviation Djinn is the worlds first production tip-jet helicopter utilizing compressed air jets to drive the rotor at the tips. On 30 August the G-APNH once again transported Sud Djinn helicopters F-BMLO, F-BIEV, F-BMLH, F-BIPY, and F-BIFP back to France.

BUA became BAF (British Air Ferries) on 04 October 1967 prompting an interim color scheme that was adopted on 25 October. The titles were changed from British United to British Air Ferries and the BUA lettering to BAF. The stripes remained blue and sandstone and were extended over the now infamous backward bird (toppled mushroom) eliminating it.

On 07 February 1968 G-APNH, under the command of Captain Mike Owen was called into recovery service. It was dispatched to Benghazi Libya to recover a Belgian registered Riley Dove conversion, OO-BPL, that was damaged. Captain Popp commanded the Dove on a test flight when it made a wheels up landing in the desert. The Dove was loaded into the Carvair and transported back to Luton where it was rebuilt.[4] The next recovery flight occurred in January 1969 when G-APNH collected the remains of BAC-111 G-ASJJ from Milan. The aircraft commanded by ex–Carvair Captain Martin suffered an engine failure. The wrong engine was shut down causing the crash. The remains of the BAC-111 were returned to Gatwick. Two months later in March the G-APNH retrieved two Pawnees ST-ADM and G-ASVR plus spares, a Piper Cub, and EP 9 from Sudan for transport back to ADS Aerial at Southend.

The Carvair front loading ability and capacity has proven its value in past recovery work. Carvair 11 was selected for these recovery missions because it was still considered best for long-range cargo work by BAF. The fuel capacity of G-APNH had been reduced from 8 tanks to 6, but it continued to function well because of the other features that made it ideal for such work. This was proven near the end of its operational life when G-APNH transported a live whale from RAF Binbrook to an aquarium at Nice. The Carvair was designed for a single purpose of car-ferry operations but continued to prove itself whatever the mission and whenever called upon.

British Air Ferries (Crash at Le Touquet)

On 18 March 1971 while attempting to land in a severe crosswind at Le Touquet G-APNH was damaged beyond repair. The aircraft was on a scheduled flight from Southend with a crew of seven and 11 passengers on board. A procedure approach was made on Runway 14 in the marginal conditions. The main gear contacted the runway with the aircraft in a nose high position continuing for a short distance. Because of the nose's high attitude the nose wheel contacted the pavement with considerable force. It collapsed, causing the aircraft to skid approximately 1550 meters from the threshold. It remained on the runway during the skid, causing severe damaged to the forward fuselage and engines two and three when the props contacted the ground. The forward fuselage was structurally damaged as the nose gear and mounting were driven upward into the floor of the cargo compartment. The lower skin, stringers and nose gear doors were reduced to scrap. Aviation Traders engineers were dispatched to evaluate the airframe. At first a plan was considered to transport the damaged aircraft back to England for a complete rebuild by replacing the nose with one of the unused assemblies still stored at Southend. Officials also considered transporting a nose to Le Touquet but adequate facilities were not available so the idea was soon dropped.

The British registration was cancelled 02 June 1971. In July no decision had been made regarding the disposition. The aircraft remained in limbo until ATEL engineers re-evaluated the extent of damage. A review determined that it would not be possible to repair it in France. Furthermore the car-ferry era was in decline and the cost to disassemble, transport to Southend and repair would exceed the value of the soon to be obsolete Carvair. The engines, control surfaces, instruments, and other parts were removed and transported back to Southend for spares and the remaining fuselage was broken up at Le Touquet.

This was not a fitting end to a very unique and one-of-a-kind long-range Carvair. It was never painted in the last BAF livery of dark blue bottom, white top and light blue stripe. It was dismantled the year before the new colors were adopted. The name "Menai Bridge" was the only name G-APNH carried, since it was destroyed before character names were assigned.

18

Carvair Twelve

c/n 12-10351, C-54A-15-DC, Iberia — AVIACO

Carvair 12 was originally built in Chicago as a C-54A-15-DC for the USAAF. It was delivered 15 July 1944 with military serial 42-72246. On 17 June 1946 it was declared redundant and transferred to the Reconstruction Finance Corporation for disposal. The C-54 was returned to Douglas Aircraft Company for upgrade to DC-4 civilian standards becoming the 25th conversion to commercial transport.

Pan American World Airways/Twentieth Century Airlines

The DC-4 began civilian service 15 March 1947 with Pan American World Airways. It was assigned U.S. registration NC88919 and in Pan Am tradition named "Clipper Panama" then later re-named "Clipper Talisman." The DC-4 was only in service with Pan Am for seven months when it was damaged in a landing incident in Alaska. After repairs it was sold to Twentieth Century Airlines in early 1948 and operated for two years. In 1950 Twentieth Century became one of the four airlines acquired by North American Airlines under Stanley Weiss. Shortly after the consolidation of companies, N88919 was acquired by Alitalia.

Alitalia

Italian carrier Alitalia purchased the DC-4 on 21 April 1950, registering it as I-DALV and naming it "Città di Napoli." In 1954 after only four years with Alitalia it was purchased by California Eastern Airways and returned to the United States. Cal-Eastern received U.S. registration N1346V and operated it until 09 December 1955. The carrier had suspended service in 1953 because of financial problems but after reorganizing in 1954 under the direction of Samuel Soloman the carrier was granted a five-year operating certificate. It was successful in obtaining MATS trooping contracts in the Pacific from California to Tokyo. California Eastern and North American Airlines jointly formed a leasing company to supply aircraft to other carriers. With the successful leasing operations, the carrier began selling off the older DC-4s in order to purchase better equipment. (See Carvair two for more on California Eastern.)

Freddie Laker's Air Charter purchased the DC-4 for British trooping and cargo flights. It was ferried Rome — Hannover — Stansted on 11 December 1955 for overhaul, named "Jason," and assigned British registration G-AOFW. The British United Airways group inherited it on 01 July 1960 with the Air Charter acquisition. It remained in service until early 1963, eventually becoming the second of three Air Charter DC-4s converted to Carvair standards. It flew as a Carvair under the same British registration it carried as a DC-4. Even after serving on lease to AVIACO after conversion the British registration G-AOFW was restored and remained on the aircraft until retired and broken up at Southend. On 23 April it flew from Southend to Stansted for the last time as a DC-4 and was transferred to ATEL for disassembly and conversion.

Conversion

G-AOFW arrived configured with eight fuel tanks, which were reduced to six for the Carvair. After de-fueling, the outer wings, engines and tail were removed and the DC-4 nose was sawn off. A trestling device was attached rear of the fuselage and it was towed into the hangar and placed on stands. The Carvair nose was transported by road from to Stansted on 09 June 1963 for installation on the waiting airframe. It was completed without the passenger cabin since it was not determined if it would go to BUAF, Alisud or AVIACO. The new empty weight was recorded at 39,965 pounds as the lightest aircraft to date with only Carvair 15 eventually weighing less.[1]

After nine and a half months in conversion G-AOFW flew on 11 February 1964 from Stansted painted white with bare metal bottom. The C of A was issued 01 April 1964 requiring only four hours and 55 minutes of testing. It was ready to begin commercial service but was delayed nearly two months in anticipation of an order from Italian carrier Alisud that never materialized. The carrier previously leased Carvair 10 in August 1963 with an option to purchase an additional ship. Alisud did not receive route authority allowing the option to expire 31 March 1964.

DC-4 G-AOFW before conversion at Southend with Carvair two, G-ARSD, parked beyond, giving an excellent comparison of the extensive modification. The position of the nose wheel gives perspective to the additional length of the fuselage (courtesy Guy Craven).

The completed nose unit at Southend waiting for loading on the transporter for Sunday morning delivery to Stansted, where it will be grafted to the fuselage of G-AOFW (courtesy Ian Callier).

Iberia/AVIACO

Spanish carrier AVIACO placed an order with Aviation Traders in February 1964 for conversion of two of its DC-4s and took an option to purchase Carvair 12 if Alisud failed to exercise its option. AVIACO traded in four Bristol Freighters. Aviation Traders also negotiated a separate sale of conversion 12 to BUAF if Alisud did not exercise its option to purchase. A condition of that agreement allowed ATEL to delay delivery and AVIACO to lease G-AOFW pending conversion of the two Iberia DC-4s. The AVIACO colors were applied at Stansted in ATELs north hangar before it ferried to Southend on 26 March 1965.

Iberia (AVIACO) delivered DC-4 EC-AEP to ATEL on 15 February 1964 for conversion to Carvair 16. In the interim ATEL leased G-AOFW to Iberia to begin the AVIACO service. The carrier operated domestic freight and passenger service in the Balearic Islands and planned expansion into the car-ferry market with the Carvairs.

The Spanish C of A was issued on 15 April 1964. AVIACO crews completed training at Stansted on 17 April, one day before delivery to Barcelona. All training was conducted in Carvair 12 wearing Spanish transport markings EC-WVD. It was delivered to Spain on 18 April with the transport registration then re-registered as EC-AVD for operations. AVIACO service began late in April 1964 operating from Barcelona and Valencia to Palma. The carrier planned to operate three times per week from Valencia to Palma after the second Carvair arrived. All service was operated in the 22-seat five-car configuration. Carvair 16, EC-AXI, was delivered in June 1964 giving AVIACO a two-ship operation until the end of the season. Leased Carvair 12 remained with AVIACO until 16 November when it was returned to Stansted for maintenance. Together the two Carvairs transported 3,300 vehicles and considerable freight during the five-month period.

AVIACO studies indicated three Carvairs would be required to service the market. Company officials were convinced there was considerable potential for growth in the market and began negotiations to purchase EC-AVD (G-AOFW) at the end of the lease and a second Iberia DC-4 was scheduled to be converted. The optimism was not to last, bookings declined rapidly after the initial summer season and freight contracts did not materialize. The leased Carvair EC-AVD in AVIACO titles was parked at Stansted while the option to purchase was put on hold. In anticipation of the AVIACO lease expiring and the purchase agreement with BUAF, ATEL removed it from storage on 14 January 1965 to return to service. On 25 January it received the British C of A but the AVIACO lease of one year was still in effect until March 1965. The aircraft remained parked and did not revert back to the British registration of G-AOFW until 08 March when the lease expired and it was purchased by British United.

When leased Carvair EC-AVD was returned to Stansted in November 1964 AVIACO future bookings were down and anticipated cargo contracts had not materialized. AVIACO discussed cancellation of the second conversion of DC-4 EC-AEO. AVIACO officials were reminded of the contract obligation noting deposits had been paid along with the Bristol trade-ins. Future bookings were down but there was guarded optimism that new business could be generated and conversion of a second ship went forward.

British United Air Ferries/British Air Ferries

After BUAF purchased Carvair 12 it reverted back to British registration G-AOFW. It arrived at Southend 26 March 1965 after being brought out of storage. It was overhauled and repainted in the original British United livery returning to scheduled service on 03 April. It is the first of only two BUAF Carvairs that never received a formal "Bridge" name. It wore the British United livery for the next two years operating many trooping flights from Southend to Cyprus. It was involved in a mishap at Dublin on 18 September 1965 when it overran the runway crashing through a fence and coming to rest with the main gear in a drainage culvert. The gear remained extended and damage was minimal.

BUAF announced the cancellation of the seven long haul routes on 23 January 1967 because of lack of traffic. As a result of the pull back G-AOFW was once again redundant. Instead of being repainted in the new BUA blue and sandstone colors it was ferried to Lydd Kent 07 February 1967 and parked pending a decision on BUAF market needs. It was officially withdrawn from service and placed in open storage on 05 March 1967. The first Carvair, G-ANYB was removed from service and parked beside it, never flying again.

G-AOFW remained in open storage for two years until brought back to flying status in 1969 after the sale of G-ASKG to Cie Air Transport (CAT) of France. During the period G-AOFW was in storage there were two livery changes. It skipped the first 1966 BUA sandstone/blue "toppled mushroom" colors. Eight months after being stored the company became British

Carvair twelve sustained minimal damage at Dublin on 18 September 1965 when it left the runway and crashed through a fence, coming to rest in a drainage culvert. Some dents are visible below the hinge of the nose door and rear quarter of the gear door (courtesy Peter Marson collection).

G-AOFW skipped two livery changes while stored and returned to service in 1969 in the two-tone blue stripe scheme. Seven Carvairs (7, 9, 10, 11, 12, 13, and 17) wore these colors (courtesy Richard Vandervord).

Air Ferries (BAF) and there was an interim livery change of titles and deletion of the Bird logo. Neither color schemes were applied to G-AOFW. It ferried from Lydd to Southend in April 1969 for overhaul and repainting in the latest two-tone blue stripe BAF colors. When it returned to service on 29 April the BAF fleet totaled six Carvairs.

Carvairs have transported many unique and sometimes bizarre cargo assignments. Sometime in 1969 G-AOFW became involved in world politics by transporting a consignment of Marconi TV equipment via Marseilles to Benghazi Libya. The equipment was destined to Muammar Qadhafi and his revolutionary television station. The BAF crew was confined to the aircraft overnight under tight security and armed guards. They were forced to spend a most uncomfortable night sleeping in the seats of the passenger cabin. Neither the terms of the charter or who booked it were disclosed.[2] The types of charters operated cover the spectrum. Missions were as exciting as a clandestine delivery of a radio equipment to third world countries or mundane as transporting reels of cloth from a factory in Yorkshire to Jersey (EGJJ) on the Channel Islands. G-AOFW operated a number of these bulk cargo charters while retaining the rear passenger cabin in the event it was needed for car-ferry service.

T. D. "Mike" Keegan purchased British United Air Ferries on 01 October 1971, which prompted a change in livery and a new promotional campaign beginning in 1972. During the 1970s BAF became involved in formula motor car racing via the chairman's son, Rupert Keegan. BAF sponsored the Formula Ford named "BAF Special Royale."

British Air Ferries was faced with the high cost of overhaul and corrosion repairs on G-AOFW in 1971. Withdrawal from service and scrapping was considered as an option. A decision was made as part of the racing sponsorship to use G-AOFW to transport the car and equipment around Europe. The aircraft did not receive the full livery change immediately and has the distinction of having a one of a kind color scheme. The two-tone blue stripe was removed by painting it over leaving the upper half of the fuselage white with a light gray bottom. The BAF lettering and British Air Ferries titles were left in tact. The name of "Big John," which was

The Star Truckin 75 cartoon logo was added to the last BAF livery in 1975 when BTM Productions wet leased G-AOFW to transport touring rock groups around Europe (courtesy Paul J. Hooper).

adopted at an earlier time was also retained during the Keegan racing program. This is the only name it ever carried since it never received the traditional "Bridge" name.

The repainting in white is most likely an outcome of the corrosion control efforts before Keegan used it for the formula car race season. The aircraft was in the two-tone blue stripe scheme in 1969–70–71. The last BAF colors of white upper, dark blue lower fuselage, and light blue stripe began appearing on other ships in 1972. The exact date of the white livery is not known. Photos taken of Rupert Keegan were used during his racing career and possibly were taken as it began. He was born 28 February 1955 making him only 17 during the 1972 Formula I racing season. In 1974 he was a Formula Ford competitor. He drove the Formula 3 in 1975 and 76. In 1977–78, 1980, and 1982 he was a Formula I Grand driver. Late in 1972 G-AOFW was painted in the blue stripe-dark blue bottom scheme. This would indicate the solid white paint scheme was in the spring of 1972. Whatever the date the odd white livery was on the aircraft for a very short time.

Aviation Traders and BAF were no longer under the same corporate group after the Keegan purchase. Keegan established his own aircraft maintenance company at Southend known as Hawke Aircraft Parts. In an effort to utilize the facility and idle manpower fiberglass racecar components were manufactured. It is not clear if Keegan's son got into racing because of the parts business or the manufacturing of parts was because of his racing.

The livery of G-AOFW was further modified in August 1975 when it was wet leased to a music promotion company. The name "Big John" was retained and a large cartoon like character was painted on each side of the fuselage behind the cockpit windows. Pop music promotion company BTM leased the Carvair to transport primarily the rock group "Status Quo" on their

On 02 May 1977 the Certificate of Airworthiness for G-AOFW expired. The engines were removed at Southend and it was allowed to deterorriate. Early in 1979 a wind storm ripped off the nose door, which was replaced with a spare salvaged from Eastern Provincial's CF-EPV, Carvair Eight, when it was broken up. The conflicting paint schemes called attention to the poor condition of the airframe (courtesy Ian Haskell).

fall concert tour around Europe. BAF provided crews and maintenance in the wet lease agreement. The large cartoon like tour logo BTM presents "Star Truckin 75" was applied to the sides of the fuselage. The rock groups found the Carvair ideal for concert tours. The musicians could relax in the rear cabin while the stage equipment and instruments were transported in the forward cargo bay. This was not the first time Carvairs were chartered for concert tours. The Who, Small Faces, Shirley Bassey, Alvin Stardust, and even the Vienna Philharmonic Orchestra chartered Carvairs for music engagements. Because of lack of music group bookings for G-AOFW the BTM contract was cancelled and G-AOFW was temporarily withdrawn from use in January 1976.

SFAir/SECMAFER

Secmafer a French contract carrier formed a secondary company that operated under the name of SFAir. The parent company entered into an agreement to purchase G-AOFW from BAF in June 1976. The carrier previously purchased Carvair 17, G-AXAI, in January. After reviewing the aircraft and the cost to convert it to cargo, SECMAFER considered it for spares to support the other aircraft rather than bare the cost of overhaul for cargo. The transaction was never completed and the aircraft remained with British Air Ferries.

British Air Ferries

BAF began the transition to Dart Heralds acquired from Eastern Provincial in 1975. In 1976 only four Carvairs (-ASDC, ASHZ, AOFW, ASKN) were active. Carvair seven, -ASDC had been converted to all cargo a year earlier and BAF considered keeping only one in car-ferry service with the other three in cargo. The transition to Heralds on all cross-channel passenger flights was completed by 1977. The Carvair was no longer profitable in any role other than cargo. G-AOFW made the last car-ferry flight in January 1977. It was parked at Lydd temporarily pending a decision of disposition. In February it was withdrawn from service at Southend allowing the C of A to expire on 02 May. The engines were removed and it remained next to the BAF building in poor condition. In early 1979, while in storage, G-AOFW was damaged by wind in a severe storm when the nose door, which was not secured, blew open. The force was so great the nose door broke at the hinges and separated from the aircraft.

The engines were removed, but G-AOFW was still considered stored and not yet slated for scrap. The nose door was replaced in May 1979 with the spare salvaged from Eastern Provincial Carvair CF-EPV. It was purchased in 1973 and brought back to Southend where it was stored and eventually scrapped in September 1978. Although G-AOFW was in poor condition and damaged it was still viewed as available to restore to flying status sometime in the future.

In June 1976 BAF placed large ads in trade publications picturing G-AOFW in flight. The advertisement stated that the fleet of well-maintained Carvairs was for sale or lease. The aircraft were capable of 7-ton payloads or 65 passengers and available with spares. Support, aircrew training, and contract maintenance was also available. Secmafer had purchased G-AXAI in January and made an offer for G-AOFW in June. The transaction progressed to final stages before Secmafer cancelled. There were no other serious inquiries and it soon became obvious that G-AOFW would never return to service. In May 1978 the airframe was advertised without engines and other components and un-flown for several years in disrepair.[3] With no response, it was painted and engines were installed to give the appearance of a complete aircraft.

In 1979 Ruth May, who purchased Carvair five, was shown G-AOFW "Big John" when she was negotiating the Falcon Airways purchase of Carvairs seven and nine. BAF officials believed they could convince Falcon, who indicated an interest in three aircraft, to purchase G-AOFW along with -ASDC and -ASHZ. However, F-BYCL, stored in France was in better condition. She related the story of the BAF efforts during an interview on the night of 03 April 1997 unaware that Carvair five, which she had personally owned, crashed at Griffin Georgia.

G-AOFW was placed on temporary static display at Southend in October 1981. Plans were formed at that time to turn it into a permanent static memorial as a tribute to the dedication of BAF and ATEL employees. In January 1982, as a tribute to the community and the many people who worked for Aviation Traders and British Air Ferries, the repainted Carvair was displayed in front of BAF engineering. It was officially transferred to Panavia Air Cargo, a subsidiary of British Air Ferries but remained non-airworthy while on display. It continued to deteriorate until December 1983 when the decision was made to dispose of it. The last remaining BAF Carvair, "Big John" was moved from display in front of BAF engineering to a less conspicuous area to be scrapped.

Former British Air Ferries Captain Laurie Rowe attempted to purchase it in 1982 on behalf of the Southend Historic Aircraft Society. He believed the Carvair was a significant part of aviation history and should be preserved. Captain Rowe agreed to maintain the aircraft and attempted to negotiate with the Southend airport authorities for parking space to display it on a permanent basis. The authorities were not in agreement and rejected Captain Rowe's £2000 offer. It was sold to a scrap dealer for £1600 in December 1983 and in the same week hacked to the ground in just two hours with a backhoe. The engines were not removed prior to the destruction. Much of the interior, which was still in tact was ripped out and piled on the ground.

Captain Jack "Laurie" Rowe who flew Carvairs for eight years along with high time senior

G-AOFW was placed on static display at Southend for several years. Sadly, it was broken up in December 1983. Several attempts, including one by BAF Captain Jack Laurie Rowe, were made to preserve it, but all failed (courtesy Keith Burton).

Carvair Captain John Baard in who's honor it was named "Big John" solemnly watched. Rowe said, "It was a museum piece. I think it is absolutely disgraceful. It was a piece of the regions history."[4] Witnesses stated that Captain Baard never spoke during the destructive event.[5] The British registration of G-AOFW was cancelled 30 May 1985. Some parts were donated to the Thameside Aviation Museum at Tilbury, Essex. Captain Rowe managed to save one of the propellers and displayed it in his garden for many years. The cockpit side windows and the lower cargo compartment door are in the possession of avid enthusiast and aviation writer Paul Doyle of Hertfordshire.

19

Carvair Thirteen

c/n 3058-13, C-54-DO, British United Air Ferries

Carvair 13 was converted from a C-54-DO, which is the oldest airframe selected by Aviation Traders. It was built at Santa Monica as the 9th C-54 airframe produced by Douglas and one of the original 26-passenger troop transport versions. This early model Skymaster was not equipped to transport cargo because of four long-range fuel tanks mounted in the fuselage giving it a total fuel capacity of 3620 U.S. gallons. The fuselage-mounted tanks reduced the cargo capacity and limited the use to staff or troop transport with a max payload weight of 6,000 pounds. It was fitted with Pratt & Whitney R-2000-3 engines and is the only C-54-DO used for Carvair conversion. The C-54 was delivered to the USAAF 08 February 1943 with military serial 41-37272. In 1946, after three years with the Air Force, it was declared redundant and transferred to the Civil Aeronautics Administration (CAA) receiving U.S. civil registration N79000.

Eastern Airlines/Aero Leases/California Eastern Airways

The CAA operated the aircraft until 11 April 1951 when it was purchased by Eastern Airlines and named "721." Eastern flew it until May 1955 when it was sold to Aero Leases. Stanley Weiss of North American Airlines along Jorge Carnicero of California Eastern Airways formed Aero Leases as a subsidiary of their airlines. Weiss, who founded North American and acquired Twentieth Century, Trans-American, Trans-National, and Hemisphere Air Transport, needed more DC-4s for his rapidly expanding operation. Carnicero's California Eastern had been a DC-4 operator since its inception in 1946. It was reorganized in 1949 and continued to have numerous financial and certificate problems. Weiss became embroiled in a dispute with the CAB with his irregular carrier certificate and jointly formed Aero Leases. He also became involved with Air World Leases operated by Andre de St Phalle. Air World and Aero Leases were active in a number of joint dealings.

The lease agreements and change in corporate strategy made the primary business the leasing of aircraft to other operators. (See Carvair two for more on California Eastern) Stanley Weiss purchased the entire DC-4 fleet of Eastern Airlines with spares in 1955. When the delivery of new replacement aircraft to Eastern was delayed after the DC-4s were sold, Weiss formed Aero Leases with the assistance of Carnicero to lease the DC-4s back to Eastern Airlines. He also returned the spares at the original purchase price. The one Skymaster that did not go back to Eastern was N79000, which was leased to Grimley Engineering. This is the fifth C-54/DC-4 that passed through a Stanley Weiss company that was later converted to Carvair standards. It is of note that Eddie Rickenbacker, CEO of Eastern Airlines at this time, was so impressed with the integrity of Stanley Weiss that he recommended him as his replacement when he retired.[1]

Grimley Engineering/Northwest/Transocean/World

After a short lease of two months to Grimley Engineering, N79000 returned to Aero Leases in June 1955. It was purchased on paper by associate company California Eastern and immediately resold to Northwest Airlines. After three years with Northwest it was re-purchased by Aero Leases. After heavy maintenance it was leased on 28 February 1958 to Transocean Airlines, which was beginning to have financial difficulties. The Skymaster was operated for one flight then returned to California Eastern. Four of the C-54/DC-4s that became Carvairs passed through Transocean. (This is detailed on Carvair 17) In 1959 the DC-4 was purchased by World Airways but after less than a year with World it was re-sold to California Airmotive in 1960.

Global Airways/President Airlines/California Airmotive/Continentale Deutsche Leftreederei

Global Airways leased the DC-4 from California Airmotive for only two months. It was returned to Cal-Airmotive and leased on 14 October 1960 to President Airlines until early 1961. In May California Airmotive sold it to Continentale Deutsche Leftreederei and it was re-registered in Germany as D-ADAM. A year earlier the same airline purchased another DC-4, which later became Carvair 14. The DC-4 was operated for an unspecified period of time before being grounded at Hamburg on bad debt claims. When the Carvair was introduced in 1961 Continentale Deutsche Leftreederei held talks with ATEL to review the conversion of several of its DC-4s. The deal was never formalized and no contracts were signed. After Continentale Deutsche collapsed and went into financial default, Kunderkreditbank Essen liquidated the

A rare view of the DC-4 at Prestwick in Global titles only worn for two months. It was registered N79000 from 1946 to 1961 until purchased by Aviation Traders in 1963. It is the oldest airframe converted to Carvair standards (via Jennifer M. Gradidge).

assets. Aviation Traders successfully bid at auction on two of the same Skymasters that Continentale Deutsche planned to convert two years earlier.

Carvair Conversion

Aviation Traders purchased D-ADAM at auction specifically for Carvair conversion. It was transferred to ATEL on 22 July 1963 and assigned registration G-ASKN before the ferry flight back to Stansted. It would carry the same registration after Carvair conversion. DC-4 airframes were purposely purchased in poor condition because the cost was much lower. It was reasoned not to be an issue since the airframes would be completely disassembled and rebuilt. It appears that the extremely poor condition of D-ADAM was either not disclosed or overlooked. Whatever the reason, when the aircraft was inspected at Hamburg it was determined not to be airworthy and not capable of a flight to Stansted. The condition of the aircraft was so bad because of severe corrosion on the wing spars the possibility existed that it would have to be scrapped.

Besides being the ninth C-54 Douglas airframe manufactured, it had been poorly maintained by a financially marginal carrier prior to default and parked in the open at Hamburg. During the nearly two years the aircraft sat idle in the damp air it was not maintained. The age and condition made it so questionable that ATEL stress technicians were dispatched to Hamburg to determine if it could be made airworthy.

G-ASKN was delivered in the first British United livery and eventually wore all five BU, BUA and BAF car-ferry color schemes. It is also the first production conversion built with raised control runs and ceilings and the only British Carvair so equipped (courtesy Aviation Traders Ltd.).

The C-54 is engineered with a wing that is technically known as three-spar fail-safe. Even with this engineering feature many reinforcing plates had to be fitted to the wing to make it airworthy. It was ferried without incident to Stansted on 25 July 1963 to clear customs and two days later it was inspected by engineers. De-fueling and disassembly began the next afternoon and the nose was transported to Stansted the next day. The noses were not usually delivered the day after the disassembly begins but the conversion had been planned while waiting to take possession in Hamburg and completion of maintenance to make it airworthy. Considering the amount of stress work that had to be done by ATEL engineers, it is quite an accomplishment that it was completed in record time. It is surprising that there were 13 other conversions that took longer.

G-ASKN is the first assembly line ship and only British Carvair built with the raised ceiling in cargo bay behind cockpit by default. The two earlier Aer Lingus ships had been returned to ATEL to be retrofitted with raised ceilings. Aer Lingus Carvair EI-AMP was severely damaged on landing at Stansted. The first assessment suggested the repairs plus the re-routing of control runs and raised ceiling would be lengthy. G-ASKN was earmarked to replace it and fitted with the Collins radio and elevated controls. The Aer Lingus ship was repaired in time and G-ASKN was delayed until 26 March before going into service.

One interesting feature of note on this ship is the carburetor intake scoops. All other ships had either the non-ram type with the long fairings that tapered back along the nacelle or short ram-air type. The portion above the cowl flaps on G-ASKN is the same as the non-ram long fairing yet the rear portion tapers off quite abruptly only half the length of the long fairings. There is little explanation for this unique feature other than this was the ninth C-54 built and the only conversion from the C-45-DO series. Many components on these early airframes were still being refined when produced. Early C-54s had a series of different intakes that were running model changes making these intakes appear to be a hybrid variety.

The new Carvair conversion was rolled out at Stansted on 19 December 1963, seven and a half months after arriving in questionable condition. The new empty weight was recorded as 40,542 pounds. The old DC-4 that was destined for scrap was given a second life as a Carvair. It flew for the first time from Stansted to Southend on 08 February 1964. At Southend it was fitted with the three-car 55-passenger interior and re-weighed and recorded as 41,067 pounds. This configuration was designed for BUAF Southend-Ostend service, which was licensed for supplementary inclusive tour passengers. The configuration could easily be converted to the original 5-car 22-passenger cabin upon demand. All subsequent BUAF conversions were fitted with the dual standard interior feature. BUAF ships nine and 10 were later re-configured like G-ASKN for 55-passenger 3-cars for planned service on the high density Ostend route. This configuration was first designed for Aer Lingus Carvairs six and eight but the carrier opted for the 34-seat versions. The BAF forward cabin was fitted with 34 additional seats, which combined with the original 22-seat cabin for a total of 55-seats.

Carvair 13 made its debut in the first British United dark stripe livery without the "British United" titles in red or "Carvair" in black. The C of A was issued 20 February 1964 after six hours and 52 minutes of test flights. It was purchased by BUAF on the same day and remained at Southend without titles. They were added the next week before it was delivered on 02 March 1964. In the "Bridge" name tradition it was christened "Pont D' Avignon" and began scheduled service to Ostend on 26 March,

British United Air Ferries (BUAF)

Carvair 13 is the sixth new ship delivered to BUAF just days ahead of Carvair 10, which was scheduled for British United but diverted to Alisud on short term lease. It operated for nearly three years in the original British United colors. The new BUA sandstone and blue color scheme was introduced in October 1966 when British United adopted a common fleet livery and G-ASKN was repainted on 14 December 1966.

CB Flying Showcases (Colin Beale)

G-ASKN continued to operate in the BUA colors until August 1967 when it took on a unique role as a flying exhibition hall. The titles were changed above the windows from British United to CB Flying Showcases. In place of the BUA letters and "toppled mushroom" logo the title "SHOWCASE" was applied in block letters for the short term lease.

Colin Beale, a London based marketing executive conceived the idea that the Carvair with a 665 square feet of floor space would be ideal for flying mobile exhibitions. Beale did extensive studies of American sales and promotion and wanted to duplicate global sales presentations using aircraft with the concept that had been pioneered by IBM and Alcoa. He reasoned that the cost of displays for corporate advertisers in each city would be reduced my setting them up inside the aircraft and flying to each presentation. The novel approach to market British goods abroad would attract potential clients for the particular product that was being presented. Electrical and electronic products were the primary featured items. Up to 15 executives could be transported in the Carvair and an agreement was even negotiated with Avis rent-a-car to provide city-center to airport transportation.

Beale wet leased G-ASKN from BAF and it was modified for the project. Besides changing the British Air Ferries titles to "CB Flying Showcases," the seats were removed from the rear cabin and it was converted into a sales office complete with a conference table. The cargo area was fitted with fluorescent strip lighting powered by a portable 115-volt generator. The cabin was set up with a communications center and miniature film show theater. A Marquee style tent was designed in the various colors of companies being represented to fit around the nose of the aircraft with entrance steps. If chartered by a single company the "SHOWCASE" lettering on the forward fuselage could be changed to the product or company name. The cost for an eight-day (seven exhibitions) European tour was set at £14,000 ($39,215). A 15-day North American tour was set at £29,000 ($81,229).[2]

Landing at Southend in 1968, G-ASKN is the last of five British Carvairs (7, 9, 10, 11, and 13) that wore the BAF blue and sandstone colors identified by the straight cut in the stripe behind the F (courtesy Transportraits/Richard Goring).

The "Flying Showcase" ferried to Luton on 14 August and the next day sixty top companies gathered at the airport for a demonstration of the program and the additional services BUA could offer. The concept was not successful even though the press reviewed several sales programs. A British Brewery considered one such North American promotional program. The plan was to fly to selected major airports in America and promote the product by passing out samples. The campaign was presented to the press at Luton Airport but after consideration promoters realized that it would be impossible to transport enough beer on the Carvair for a long campaign. The program was dropped and only a small number of products were ever presented in the Carvair Showcase. The exhibition hall program was over in a few months, the lease cancelled and G-ASKN returned to BUA.

In October shortly after returning to BAF and removal of Colin Beale titles BUA was in transition to BAF. It was not practical to pull the aircraft from service since it was going to Lydd at the end of the month for a "check four." The decision was made to leave the aircraft in the BUA blue — sandstone colors and on an overnight at Southend the two-tone stripe was extended by painting over the "toppled mushroom" logo and the BUA titles were changed to BAF.

British Air Ferries (BAF)

On many occasions the Carvairs were used to recover damaged aircraft and transport new aircraft. G-ASKN was dispatched to Le Touquet on 27 October 1967 to recover a Cessna 172 G-ATLN that was blown over and damaged by wind while in France. After arrival at Southend

In 1969 G-ASKN was converted from the 55-passenger back to the 22-passenger five-car configuration. It was often used in advertising campaigns to promote bookings in a dwindling market (courtesy BAF).

the 172 was transported by road to Rogers Aviation for the rebuild. G-ASKN then ferried to Lydd on 30 October for a "check four" overhaul. It emerged in November still wearing the tidied up blue and sandstone colors with BAF titles. It remained in this interim BAF livery until December 1968 when it returned to Southend for checks and full repainting, becoming the first ship to receive the new BAF colors with two-tone blue stripe. In 1969 it was in the 22-pax 5-car configuration, which was demonstrated in advertising photos.

Because of the unique cargo carrying capacity of the Carvair, G-ASKN was chartered during July and August of 1969 by the British government for a clandestine operation. The crews were required to sign "Official Secret Act" documents and the aircraft registration was blanked out on each flight. Large crates marked as machine parts were transported but the crew was not told the exact purpose of the mission. The actual contents of the cargo has never been disclosed. At least five of these flights were made at night to an unspecified dirt airstrip in Algeria. The runway was lighted with barrels burning some type of fuel. Very quiet government agents accompanied the cargo and heavily armed group met the flight, quickly unloaded and disappeared into the night. The BAF crews were not allowed off the aircraft. All of the missions were without incident except the final flight, which diverted to Lyon France on the return because of weather.[3]

On 07 December 1970 G-ASKN took a serious bird strike causing major damage to the nose door. The door was replaced with an unpainted bare metal unit and flown for a week before the new nose was painted. It is most likely the replacement nose door came from one of the spare nose assembles that had been built for ships 22 and 23, which were never completed. In November 1971 G-ASKN was returned to ATEL for maintenance and storage. It remained idle until 18 December 1972 when it was removed from storage and the process began to bring it back to flying status.

By 1973 BAF was the only car-ferry airline in the world. Under the direction of Mike Keegan, BAF planned to offer additional service on the long haul route to Basle and announced, subject to government approval, additional service to Bordeaux France. Service to points in Germany was also suggested to the press. Keegan sought to improve the image of the aging

Departing Jersey in the last BAF color scheme, G-ASKN had just returned to service in 1973 after being grounded because of a shortage of engines. During the Keegan era, seven Carvairs (5, 7, 9, 10, 12, 13, and 17) were painted in the last BAF livery (courtesy Dave Heaney).

Carvair by refurbishing the interiors and repainting them in the new white and dark blue color scheme. The seats were changed to first class type and the cabins were convertible for seating between 17, 40, and 65 seats which was achieved with a moveable bulkhead.

After being repainted in the new livery and re-furbishing of the cabin G-ASKN received the new name "Big Bill." It had been in storage since 1971 and was brought back at considerable expense to begin service in the spring of 1973. After a series of test flights it was ready for active service on 27 May. On 27 June -ASKN was grounded because of a shortage of engines and Mike Keegan made the following statement in a memo to pilots and engineers.

"I do not have to tell you, of all people, of the desperately serious position we are now in as a result of engine failures, this is aggravated by Scottish Aviation being on strike, and we have now had to ground an aircraft (KN), on which we have just spent a small fortune, to enable us to keep our services reliable."[4]

A similar letter was issued to flight crews on 29 June 1973 and a Pratt & Whitney representative was brought in to re-train crews on handling the big radials. In due time Scottish Aviation settled their labor dispute. G-ASKN returned to service where it remained in both car-ferry and cargo service until 03 June 1976. The car-ferry era had long ended because of lack of traffic causing it to be withdrawn from service and stored at Southend as the remaining ships were being put up for sale.

Société Anonyme de Construction (SOACO)

After considerable success with Carvair 10, TR-LUP, Société Anonyme de Construction (SOACO) required more lift to support construction projects in central Africa. SOACO approached BAF on several previous occasions with the prospect of purchasing an additional Carvair. The Carvair was long past its prime but BAF was not actively looking to sell since all serviceable ships were being utilized. At the same time SOACO was pressuring BAF the two Ansett Carvairs were being offered for sale with spares in New Zealand. Apparently SOACO was not aware of the two ships being available or the price was not attractive. SOACO was in the process of purchasing a CL-44 in March 1976 and G-ASKN was not withdrawn from service and available until June.

In early June an agreement was reached with SOACO. The British Air Ferries titles above the windows were painted over and the sale was completed on 09 June. It has been reported in error that SOACO titles were added but in reality the light blue stripe was painted the full length of the aircraft and up the vertical stabilizer with no titles. G-ASKN departed Southend on 10 June 1976 for Libreville Gabon. Shortly after arrival the British registration was cancelled and it received the Gabon registration TR-LWP. It appears that SOACO was rather short sighted in their need for an additional lift or it was purchased for spares support. The few records that are available are not complete and the intention of SOACO officials is not clear. However, TR-LWP was operated for only five months and made its last flight in December 1976 from Libreville Gabon to Menongue Angola then a 3.5 hour flight to Brazzaville.[5] In early 1977 TR-LWP was withdrawn from service and impounded at Brazzaville because of monies owed for servicing. The engines were run up periodically and it remained at Brazzaville until confiscated along with SOACO Carvair TR-LUP in 1979 by the Empire of Central African government.

AERO SERVICE (AFRIGO)

Mr. Raymond Griesbaum purchased the two SOACO Carvairs from the Central African Government and gave the following information on Carvair thirteen. The aircraft was registered to SOACO in Gabon 09 July 1976 as TR-LWP. It made its last three and a half-hour flight on 08 December 1976 from Libreville Gabon to Brazzaville. The aircraft was not flown again. The engines were run up periodically to keep it airworthy. The last engine run up was on 20 June 1978 and lasted only 20 minutes. The total time on the aircraft was recorded at 60983.47 hours when it was withdrawn from service.

Aero Service purchased it in 1982 but with no specific flights scheduled and no contracts pending it was used for spares support. In February 1982 it still appeared in good condition and as late as mid 1984 it was still intact less engines and control surfaces. In the final days it could be seen sitting on the tail with the rudder missing at Brazzaville. It was scrapped in May 1986. Even in the final days in Africa it remained painted in the last BAF colors of white upper fuselage, light blue stripe and dark blue bottom. There was no lettering except for the BAF character name "Big Bill" which was still visible on the nose.

TR-LWP was not scrapped because of deterioration but lack of spare parts and pressure from the airport authority to remove it. The Number Two engine s/n 108869 was removed 06 June 1978 for use on Carvair 10. The Number Three engine s/n 108991 was removed on 05 August 1978 again for Carvair 10. The airframe was found to be in exceptional condition with little corrosion when broken up. It is unfortunate that another Carvair was destroyed because of economics and not for wear. Had it not been so expensive to make airworthy and so many parts used to support Carvair 10 it possibly could have seen more service. Reports persist that the nose door still remains in the storage area at Aero Service in Brazzaville. The airframe carried the TR-LWP registration when broken up in 1986.[6]

It is of interest that during the time Aero Service owned Carvair 13 (1982–86) and Carvair

Carvair 13, registered TR-LWP in its final days, deteriorating at Brazzaville, Congo. It was grounded and eventually broken up, not because of condition but lack of spare parts (photograph by P. Kirsdren/G. Cassiman Archive).

10 (1981–93) it also owned the only Swing-Tail DC-4. The aircraft was the potential Carvair competitor and less expensive alternative in the 1960s. In 1976 Aero Service purchased the Swing-Tail DC-4 that was developed by SABENA at the request of Air Congo. It remained with the carrier until sold in 1984. (This aircraft is detailed in Design, Engineering and Development under potential buyers.)

20

Carvair Fourteen

c/n 10458-14, C-54B-1DC, Aer Lingus

Carvair 14 was originally built in Chicago as a C-54B-1-DC c/n 10458. It was delivered to the USAAF on 27 November 1944 with military serial 42-72423. A total of 120 C-54B-DC aircraft were built and are basically the same as the C-54A except the two fuselage tanks were relocated in the outer wing. They were equipped with a reinforced floor and some were fitted with the swing out cargo hoist. The hoist was not installed on consecutive C-54s making it difficult to determine which airframes were so equipped. A total of four C-54B-1-DCs series were converted to Carvairs. This series had a max weight of 73,000 pounds and max fuel capacity of 3,740 gallons and was fitted with either the P & W R-2000-3 or -7 engines.

The Air Force declared the Skymaster redundant after the war selling it back to Douglas Aircraft on 18 January 1946. It became the seventeenth C-54 converted to DC-4 commercial standards before being re-sold to Western Airlines on 14 September 1946. It received U.S. registration NC88721 and remained in service with Western for more than nine years.

SOBELAIR purchased it from Western Airlines on 12 January 1956 with 27,467 hours on the airframe. It was re-register on 15 February as OO-SBO then ferried to Brussels on 06 March and named "Lomani." SABENA leased it between 25 March and early July 1956. SOBELAIR operated it until 25 November when it was leased to Union Aeromaritime de Transport (UAT) and registered in France as F-BHVR. After return to SOBELAIR on 15 March 1957 it was leased a second time to SABENA. The French registration was cancelled and it reverted back to the previous Belgian registration of OO-SBO. SABENA operated the leased aircraft for two more years before returning it to SOBELAIR on 31 October 1959. SOBELAIR put it back in service for another year before selling it to Continentale Deutsche Luftreederei, which was the last carrier to operate it as a DC-4.

Continentale Deutsche Luftreederei applied and received registration D-ABIB with intent to purchase but it was not taken up because of financing. The second attempt to purchase was completed on 08 March 1960 with the DC-4 recorded as having a total of 33,992 hours. A month later on 18 April it received German registration D-ANEK. Continentale Deutsche operated the DC-4 for an unspecified number of flights if at all before going into financial default. The DC-4 was grounded at Hamburg where it remained for two years on bad debt claims. Aviation Traders acquired it on 09 July 1963 in a default liquidation auction sale. It is possible the registration of D-ANEK was never applied to the aircraft. The Belgian registration of OO-SBO was clearly visible during disassembly at Stansted along with the freshly applied British registration of G-ASKD but there was no evidence of any German registration.

When the Carvair was introduced in 1961 Continentale Deutsche Luftreederei held preliminary talks with ATEL to review the possibility of converting several of its DC-4s. The deal was never formalized possibly because of the poor financial condition of the carrier. When the Continentale Deutsche defaulted, German bank Marcard & Company took possession for

The DC-4 was operated by Western Airlines over nine years before SOBELAIR purchased it in 1956. It was sold to Continentale Deutsche in 1960 and grounded at Hamburg within a year before being acquired by Aviation Traders (courtesy G. Cassiman Archive).

liquidation. Aviation Traders acquired it along with another DC-4 for the express purpose of Carvair conversion.

Carvair Conversion

Aviation Traders purchased D-ANEK 09 July 1963 and after being made airworthy it was registered as G-ASKD for a planned ferry to ATEL at Stansted for inspection and to begin conversion. Airframes were purchased in a questionable state and this DC-4 like the other from Continentale Deutsche was in extremely poor condition. When the aircraft was inspected at Hamburg the airframe was found to have considerable corrosion. It had not been maintained by financially strapped Continentale Deutsche prior to default and allowed to deteriorate even further during the two years in the open at Hamburg. It was not airworthy prompting the dispatch of ATEL technicians to Hamburg to determine if the two newly acquired DC-4s could be made flyable. It was determined not to be as bad as the other DC-4 D-ADAM, but did require considerable work before safely ferrying direct to Stansted. Aviation Traders obtained registration G-ASKD for the DC-4 once it was made airworthy.

Conversion/Aer Lingus

After clearing customs at Stansted, G-ASKD entered the line on 31 July 1963 for conversion fourteen. Carvair thirteen was built with the raised control runs in anticipation of an Aer

The stripped airframe at Stansted before being moved to the hangar. The faded SOBELAIR registration OO-SBO is visible in the stripe above the British registration G-ASKD (courtesy Ian Callier).

Lingus order but went to BUAF. When the carrier placed a third order in November 1963 a line number was not allocated. Carvair 14 was already in progress when Aer Lingus made the decision to exercise the third aircraft option. The official announcement that G-ASKD would be used to fill the order came in January 1964 and it was disclosed this would be the third and final Irish Carvair. The airline originally ordered two Carvairs in 1962 with an option for a third and reviewed the possibility of a fourth. After limited success during the 1963 summer season and fears of mechanical reliability the option for a third Carvair was reluctantly exercised late in 1963.

Aviation Traders originally purchased the DC-4 to convert for British United. There were thirteen older airframes used for conversion but this later model was in extremely poor condition. The DC-4 had considerable wing corrosion and other deterioration because of standing out in high moisture so long at Hamburg. It was inspected again by ATEL engineers after arriving and there was still some question of whether it should be used for spares or conversion. Upon arrival at Stansted, the fuel tanks were drained and outer wings, and engines were removed. The trestling device was attached to the rear and the balance of disassembly completed outside before being towed inside and placed on stands.

The R-2000 engines on this DC-4 like the second Aer Lingus conversion were fitted with the non-ram air carburetor intakes identified by the long fairing across the nacelle. Aer Lingus rigid standards for fleet uniformity required these fairings to be changed to the ram-air short air scoop. This attention to details made the three Aer Lingus Carvairs the closest to being uniform.

The nose assembly arrived by road from Southend on 22 September 1963. Aer Lingus did not formerly purchase EI-ANJ until after the first flight as a Carvair, but conversion was completed with additional specific options specified by the carrier. It was configured with the 12-seat forward cabin for total of 34-passengers and four-cars. As with the previous two Aer Lingus conversion, an additional emergency exit was installed at the next to last round window on the starboard side.

Unlike the two previous Aer Lingus aircraft that were retrofitted after the first season, it was built with the "Rolamat" floor system for palletized cargo or car-ferry work and the raised ceiling. It is the first assembly line Carvair with "Rolamat" floor and the second with the raised ceiling in the forward section of the cargo compartment. The aircraft control runs and heater ducts were relocated from behind the cockpit in order to raise the ceiling. This was done to obtain a higher ceiling clearance for bloodstock charters. The horse pallets were always loaded in the 2nd and 3rd positions to take advantage of this option. It is suggested that the high ceiling had some calming affect on the high strung horses. Aer Lingus proposed the original concept in planning sessions with ATEL for the express purpose of expanding into the bloodstock ferry market formerly dominated by BKS Air Transport. Even with these modifications Aer Lingus was never able to achieve the success sought with horse charters.

The new empty weight was recorded as 41,313 pounds prior to the first flight on 17 April 1964. It required only four hours 22 minutes of test flights to receive the C of A issued on 22 April. Aer Lingus took delivery of EI-ANJ on 24 April and it was set to enter limited service on 17 June. The name "St. Seanan" was applied to the left side of the cockpit and the Gaelic equivalent "Seanan" was applied to the right side. The delivery of a third Carvair enabled Aer Lingus to retire the last two cargo DC-3s.

Shortly after delivery of EI-ANJ young First Officer Michael O'Callaghan did his flight training at Dublin. His logbook shows 6 hours for the conversion highlighted by blowing all the main gear tires on an exercise that was carried out at the maximum landing weight. The aircraft came to a halt at the intersection of runways 06/35 virtually closing Dublin Airport while the wheels were changed.[1] He survived the subsequent lecture and went on to a career of nearly 40 years, becoming the last active Aer Lingus Captain that had flown the Carvair.

The addition of the third Carvair increased frequency in some Irish markets with the extra capacity. EI-ANJ took on more ad hoc charter roles. In 1965 EI-ANJ transported a replica Pfalz bi-plane from London Heathrow for a movie becoming the first recorded account of an Irish Carvair being used to support motion picture production. It was the first Aer Lingus Carvair to land in Scotland transporting a consignment of racehorses to Edinburgh/Turnhouse on 16 February 1965 utilizing the option for which it was built.

In another incident EI-ANJ was called in to rescue an Aer Lingus BAC-111 that ingested a bird on landing at Barcelona. The spare engine was flown down to Spain without incident. Upon arrival on a Spanish holiday it was realized that the local staff was not enthusiastic with assistance. The lifting equipment for the BAC 111 engine failed increasing the downtime. The return flight commanded by Captain Donal Hannigan, who was known to be quite competent on the Carvair, was routed over the Pyrenees where the minimum en route altitude was 12,000 feet. EI-ANJ struggled to reach this altitude and finally ran out of boost to climb at 11,000 feet. The only way forward was to complete the crossing visually and when clear descend to 8000 feet, which is the normal cruising altitude for the longer segments operated by Aer Lingus Carvairs. The flight was completed without further incident taking a total block time of 6.5 hours to return home. Captain Hannigan was very popular with his peers and went on to 707s and rose to Chief Instructor on the B747 to complete his career. He often noted the similarity of the Carvair and 747 profile. Captain Hannigan passed away in his retirement in 2002.[2]

On 13 May 1966 Aer Lingus Captain Joe Dible experienced his third in-flight emergency in three weeks on a Carvair. The previous two incidents were on EI-AMP, Carvair six. On a routine flight to London the Number Two VHF caught fire. The damaged avionics combined with marginal weather at London prompted Captain Dible to divert to Manchester. When questioned he calmly stated, "The idea of arriving in the London zone with the possibility of no radios was not appealing."[3] No further explanation was required.

The sea-ferry competition with improved service and comfort forced Aer Lingus to drop the car-ferry service from Dublin to Bristol/Liverpool/Cork and Cork to Bristol by 1965 but freight service and bloodstock charters continued. The need for the Carvair continued to decline prompting the removal of EI-ANJ from service in September 1966 and placing it in storage at

Dublin on 31 October. After only two years of service the third Aer Lingus Carvair was declared redundant. It remained in storage for two more years until purchased by Eastern Provincial Airways on 10 January 1968.

Eastern Provincial

On 15 February 1968 EI-ANJ became the second of the three Aer Lingus Carvairs ferried to Southend for repainting and installation of high capacity heaters, nose door insulation and nose wheel bay hatch. The Canadian registration of CF-EPW was assigned on the same day it arrived. The Eastern Provincial livery was slightly different from the previous Carvair with the omission of the EPA letters on each side of the nose. After overhaul was completed CF-EPW flew for the first time on 04 June 1968. On 05 June it received the UK C of A for export. The refurbished and modified Carvair departed Southend on 06 June, twelve days after the first for EPA. It returned to Dublin to pickup spares and support equipment en route to Gander. The Irish registration of EI-ANJ was cancelled on 29 May 1968. It arrived at Gander with 43,793:10 TT airframe hours and received the Canadian C of A on 04 July.

Eastern Provincial withdrew CF-EPW from service between October 1969 and September 1970 because of a drop in traffic and suspension of heavy lift to the remote Hydroelectric and mining projects during the winter. The withdrawal carried into summer as the aircraft received maintenance due to operating into unimproved airfields that were particularly hard on the airframe. Quite by chance Aer Lingus Captain Michael O'Callaghan, who had graduated to the 707, discovered the two remaining EPA Carvairs in a hangar at Gander in May 1970. He reported

EI-ANJ arrived at Southend in February 1968 for overhaul and cold weather modifications for Eastern Provincial. Note the solid non-spoke hub on the nose wheel (courtesy Transportraits/Richard Goring).

The crew used an entry ladder through the nose wheel bay mail drop hatch, which is visible on CF-EPW at Gander in 1970. The EPA Carvairs were the only three aircraft fitted with an access door in the belly for cold weather operations (courtesy Rick Gaudet).

them to appear in very good condition.[4] The airline never reported any serious maintenance problems with CF-EPW during its five years of operation. The Carvair was the highest revenue producing aircraft in the Eastern Provincial fleet with a rate of two dollars per charter mile. In November 1972 the hours were recorded at 47,964:02.

By 1973 the project at Churchill Falls was well along and there was little need for the Carvairs. CF-EPW was well worn, but the better of the two surviving EPA aircraft. It was parked at Gander with titles removed and put up for sale. The advertised price is not known, but it sold for $62,000.

R.C.S. Graham a Canadian broker and shipping agent based at Prince Rupert British Columbia became the interim owner of CF-EPW on 17 April 1973. A contract was signed to lease the Carvair to Norwegian Overseas Airways and sub-lease to British Air Ferries. The Norwegian Overseas crews were already standing by at Gander to begin flight training. During the crew training at Gander on 20 April the Carvair was damaged in a nose wheel collapse incident. The Norwegian training crew landed to change pilots. The pilot not being familiar with the aircraft was making a 180-degree turn off the active to the taxiway and return to the take-off point. After turning a few degrees the steering did not respond and the engineer was dispatched to inspect the nose gear. He reported that the nose gear oleo strut was fully extended. Although the scissors links were properly aligned they were rendered useless by the full extension of the strut. The scissors angle was diminished to a point where they were almost parallel to the strut. The steering mechanism has no leverage in this position, hence little or no steering. This situation is normally caused when the aircraft is loaded with too much weight to the rear creating a tail-heavy situation.

The pilot attempted to remedy the problem by steering with the engines. It was thought that this would aid the steering mechanism and castor the nose wheel. When this did not work the pilot applied the brakes to compress the nose strut with the weight of the aircraft. The crew agreed this should reduce the angle of the scissors and restore steering. This was not the case; the oleo strut fractured and collapsed dropping the nose on the pavement. Both crews except the pilot in command fearing fire exited the aircraft through the left over wing emergency exits. By the time they were on the ground the pilot had completed the shut down and no fire was found.

The aircraft sustained damage to the nose strut, gear doors, skin and light structural damage to the underside. The Number Two and Three props and engines were damaged by contacting the pavement. The damage was very similar but not as serious as the other EPA Carvair CF-EPV suffered a year earlier at Gander. After repairs it was scheduled to ferry from Gander to Southend on 10 June 1973. The lease with option to purchase by Norwegian Overseas Airways was scheduled to expire on 03 August 1973.

British Air Ferries

T. D. "Mike" Keegan owner of Trans-Meridian Air Cargo purchased British Air Ferries in October 1971. The two companies were merged maintaining separate identities. By strange coincidence Keegan, who was one of the founders of BKS Transport, was the primary bloodstock charter operator that Aer Lingus evaluated in an effort to compete in the market. The attempt to move into this market prompted Aer Lingus to order this Carvair with the raised ceiling for bloodstock charters in 1963.

The car-ferry business had been on the demise for some time. BAF had been leasing additional Carvairs from French carrier TAR since 1972. In an ambitious attempt in 1973 to revive the car-ferry market Keegan expanded the BAF schedule, creating an additional need for Carvairs. The Norwegian Carvair was leased by BAF from Norwegian Overseas Airways in June 1973 under rather strange circumstances. Canadian broker R.C.S. Graham technically still owned the Carvair, which ferried to Southend. The aircraft was leased with option to purchase by Norwegian Overseas during crew training and remained in the EPA colors with no titles while in Canada. It also retained the Canadian registration of CF-EPW. It returned to Southend from Gander that same month for repainting in the Norwegian livery.

Thor Tjontviet, who was poorly financed, contracted to ATEL for overhaul in order to operate pending charter contracts. No money was advanced for services and it was sitting outside the ATEL hangar awaiting payment before work could continue. Since Keegan had negotiated a lease and needed the aircraft, in a rather daring move he dispatched engineers Brian Mees and Neddie Underhill to go across to ATEL and snatch the aircraft. Underhill hooked up the tow bar while Mees jumped up the nose hatch, climbed into the cockpit, and released the brakes. The aircraft was taken to BAF engineering before anyone at ATEL missed it. The details are a mystery but it is known that Thor Tjontveit gave Keegan a considerable sum of cash and agreed to lease conditions with BAF in exchange for the work that had been completed. The amount owed to Aviation Traders was not disclosed, however Keegan worked out a deal and settled with ATEL on monies owed. Thor Tjontveit was forced to use what little funds he had to pay R.C.S. Graham on the lease-purchase agreement. BAF Engineering rather than ATEL finished the work to bring the aircraft back up to standards.[5] The British registration G-ASKD was assigned on 06 August 1973 but it was completed in BAF colors with Norwegian registration LN-NAA.

The DC-4 was originally purchased from Continentale Deutsch in 1963 with the express purpose of converting it for British United Air Ferries. In an effort to supply Carvairs to more carriers it was used to fill the Aer Lingus option and went on to Eastern Provincial. It had gone

The original DC-4, G-ASKD, was slated to become a BAF Carvair. Nearly ten years later it actually came to BAF with Norwegian registration LN-NAA. It received a one-of-a-kind light blue stripe livery and was named Viking North Sea Express (courtesy Tony Rogers).

full circle and after ten years it was flying for BAF. It was not painted in the current BAF dark blue color scheme but rolled out on 22 August 1973 in a variation of the previous BAF livery with light gray bottom, wide light blue stripe with white top and dark BAF titles. The name "Viking North Sea Express" was applied below the cockpit windows.

The Norwegian Overseas lease with option to purchase from R.C.S. Graham expired on 03 August 1973. British Air Ferries signed a sub-lease agreement with Norwegian Overseas on 06 August 1973. The agreement was pending the renewal of the Norwegian Overseas lease-purchase with R.C.S. Graham, which was signed 02 September 1973. The details of this convoluted transaction with Thor Tjontveit are not clear and possibly will never be known. The status of the aircraft was in limbo, however it was already in the possession of BAF to pay off the bill that was owed to ATEL and BAF Engineering. The CAA gave temporary C of A approval because it did not meet British standards. It was originally delivered to the Irish and modified for Canadian service. The British registration G-ASKD, which it was assigned ten years earlier when built, was applied for BAF service. The revival of the car-ferry operations was not a success and the decision was made in October 1974 to reduce the Carvair fleet again. When the BAF lease for G-ASKD expired on 11 November 1974, the British registration was cancelled and it was returned to Norwegian Overseas Airways.

Norwegian Overseas Airways/Rorosfly Cargo

Norwegian Overseas originally negotiated to purchase Carvair eight and the other EPA Carvair in June 1974. The Norwegian registration LN-NAB was reserved but never taken up or purchase completed. Four months later on 17 October after a second failed attempt to purchase Carvair eight, Norwegian Overseas exercised the lease-purchase for Carvair 14 from R.C.S Graham. It was registered in Norway as LN-NAA. Norwegian Overseas had been leasing it since April 1973 from Graham and sub-leasing it to BAF since August 1973.

LN-NAA was transferred on 02 December 1974 to Norwegian Overseas associate company Rorosfly Cargo Norway. The registration was transferred on 19 December to Thor Tjontveit of Oslo Norway. The exact date the registration was applied to the aircraft is not known. It did wear the Norwegian registration LN-NAA in BAF livery before the titles were changed to Rorosfly Cargo Norway.

Eventually the convoluted ownership leasing arrangements of LN-NAA ended and the BAF titles were changed to RFN, Rorosfly Cargo Norway (courtesy Ola Furulund).

LN-NAA was stored in Athens for a period beginning late in 1974 and continuing into January 1975. The aircraft movements are not well documented but there are reports of it being stored at both Bahrain and Southend. Once again financial difficulties were a factor. BAF flight engineer Brian Mees was on the recovery team dispatched to Athens to make it airworthy in 1975. It was test flown on 24 January and the ferry team returned it to Southend from Athens. Mees doubted very much that it was ever stored at Bahrain.[6] Rorosfly took possession on 28 January then sub-leased it to Rode Kors. It remained at Southend for servicing until 01 February 1975 still painted in the one-of-a-kind BAF livery.

The titles were changed from BAF to RFN and Rorosfly Cargo Norway in the same style as British Air Ferries. This is another example of a modified BAF livery for the new owner. It retained the name "Viking North Sea Express." Rorosfly took possession for the second time and ferried it from Southend to Oslo on 02 February 1975. The carrier operated limited cargo service between Oslo and Kristiansund on the southern coast of Norway. The service lasted for little more than a month and it was idle again until a lease was negotiated with the International Red Cross for refugee relief service in Southeast Asia.

Red Cross/Croix Rouge Internationale

On 01 April 1975, four months after Norwegian Overseas Airways transferred the Carvair to Rorosfly Cargo a lease agreement with the International Red Cross for disaster and famine relief work in Southeast Asia. The Carvair was repainted solid white with a Red Cross on the forward fuselage and vertical fin and Croix Rouge Internationale titles (International Red Cross). It was intended for relief work in war torn Southeast Asia from a base of operations in Thailand.

With the fall of Phnom Penh Cambodia to the Khmer Rouge the refugee crisis intensified in the region. The Red Cross and other world charity organizations contracted to any cargo carrier that could deliver relief supplies along the Thailand border with Laos and Cambodia (Kampuchea). LN-NAA was reported to have first arrived at Bangkok on 04 April 1975. The date has not been verified but is almost certain to have been a planned date. It was at Basle Switzerland on 27 April and reported en route to Bangkok. Basle had been a regular destination for British Air Ferries during the peak of the Carvair era but the presence there in the Croix Rouge livery was not ordinary and noted.

It arrived in Bahrain from Basle on 07 May 1975 with an in-flight failure of Number Three engine. The crew was Norwegian except BAF engineer Dick Rawlinson, who had been assigned by BAF to assist Thor Tjontveit in getting it to Bangkok. The Norwegian crew left the aircraft at Bahrain leaving Rawlinson and BAF engineer Brian Mees to change the engine. Mees arrived that same day on G-AREK delivering the Pauling International managers Rolls-Royce for servicing. The two engineers spent the next week changing the Number Three engine completing the work on 13 May. Rawlinson expressed his concern about getting the Carvair to Bangkok when he discovered one of the Norwegian crew had taken the ADF radio to his hotel to adjust it. The Norwegian crew told him that the final destination was Saigon. Tjontveit purchased a commercial ticket for BAF Engineer Mees to fly back to Abu Dhabi after assisting in the engine change. Mees arrived only to learn that G-AREK on lease to Paulings was at Thamarit with engine problems. A replacement engine was flown out to Abu Dhabi by commercial transport. Anxious to generate revenue Tjontviet volunteered the Norwegian Carvair in Red Cross livery to fill in for G-AREK while Pauling waited for the engine to arrive. Upon arrival it was loaded on LN-NAA and flown down to Thamarit and mounted on Carvair five.[7]

Norwegian owned LN-NAA left Basle painted in Croix Rouge Internationale titles. It was grounded at Bangkok for monies owed and eventually abandoned, never flying again. It was confiscated and left to deteriorate until finally broken up about 1988 (courtesy VIP/Nicky Scherrer).

The final destination and mission for LN-NAA were never clear. It appears the aging Carvair operated few if any relief missions under the Red Cross. Some reports list LN-NAA withdrawn from use at Bahrain in 1975. It was at Abu Dhabi on 27 May 1975 and continued to Bangkok sometime thereafter. BAF engineer Dick Rawlinson was highly qualified on the Carvair and R-2000 engines. He became so concerned about the engineering practices and operating procedures of the Norwegian crew that he left the Carvair at Bangkok and returned to Southend.[8]

LN-NAA is reported to have operated several relief flights to Cambodia. It was grounded at Bangkok in June 1975 because of monies owed to several companies for service and repair. The Norwegian crew abandoned it there and the Red Cross lease expired on 09 October 1975. In November 1975 Bangkok authorities confiscated it and declared it not airworthy then listed it as stored on 31 December 1975.

In an effort to recover losses and dispose of the derelict Carvair Norwegian Overseas Airways ran very detail advertisements offering a Carvair for sale in May 1976. The Carvair in the ads is not identified, however it is most certainly LN-NAA. The only other Carvair Norwegian Overseas ever owned was CF-EPV and it was at Southend minus engines and control surfaces waiting to be scrapped. Bangkok authorities grounded LN-NAA in November 1975 and declared it not airworthy. It was complete on 12 October 1976 with little indication of any major problems. Robert C.S. Graham requested authority to ferry Bangkok, Thailand to Terrace, BC Canada on 03 January 1978 in an attempt to recover some of his losses in dealing with the Norwegians. It was verified to be still intact on 13 November 1979. Debts owed for servicing, maintenance and storage were possibly a greater contributor to the demise of the aircraft than condition. Monies owed plus the cost of storage, expense to make it airworthy and ferry cost back to Canada exceeded the value of the aircraft. It was allowed to deteriorate at Bangkok and remained grounded on a remote parking area of the airport until 01 June 1981 when the Norwegian registration LN-NAA was cancelled.

After 1981 the accounts of the status of the aircraft are questionable. It was reported broken up in 1985 but not confirmed. During a nine month period in 1987/88, the FAA worked with the Don Muang Airport Flight Police Air Wing and Thai Navy in breaking up 10 C-47s, one Caribou and a Carvair. The R1830-90/92/94 and R2000-7M2 engines were recovered for the FAA repair shop at Malta. The exact date LN-NAA was scrapped was not confirmed.

21

Carvair Fifteen

c/n 27311-15, C-54E-5-DO, British United Air Ferries

Carvair 15 is the first of two C-54E-5-DO aircraft selected for conversion. It was built at Santa Monica with a redesigned fuel system vastly improved over the previous series. Trans-Pacific flights were limited on the volume of bulk cargo that could be transported because of fuselage fuel tanks. To overcome the problem the fuselage fuel tanks were deleted and collapsible fabric bag fuel tanks were designed and fitted in the inner wings increasing the range with a 3,520 U.S. gallon capacity. The C-54E-5-DO was designed with a quick-change interior that could be configured for a 32,500-pound cargo load or a troop transport with 50 canvas seats or 44 airline seats for staff service. It had a maximum takeoff gross of 73,000 pounds and was fitted with P & W R-2000-11 engines.

This Skymaster was delivered to the USAAF 05 April 1945 with military serial 44-9085. With the war ending it was declared redundant and transferred to the Reconstruction Finance Corporation. It was purchased by Douglas Aircraft and returned to the Santa Monica where it became the second C-54 converted to DC-4 commercial standards receiving civilian registration NC88886.

Pan American World Airways purchased it on 14 November 1945 placing it in civilian service as "Clipper Mandarin." Pan American utilized the increased range operating it worldwide for 16 years. During those years it was also named "Clipper Frankfurt," "Clipper Pegasus," and "Clipper Hannover."

It was sold to Beirut based Trans Mediterranean Airways on 16 January 1961, converted for cargo and registered in Lebanon as OD-ADW. The carrier was formed in 1953 as a cargo operator and initially operated from Beirut to the oil fields in the Arabian Gulf. In 1959 the carrier received certification for regular scheduled routes and at the same time received government freight contracts. In 1967 it was sold to ATEL when TMA began phasing out older transports and leasing CL-44s. TMA evaluated the Carvair as a possible fleet choice in 1962 and followed up with preliminary talks of converting its DC-4s. TMA was more interested in DC-6 Carvair conversions and opted to wait until they were offered but they were never produced. If Trans Mediterranean had ordered the DC-4 Carvair, this aircraft would have been converted earlier since TMA acquired it in 1961 just prior to talks with Aviation Traders.

Conversion/BUAF/BAF

Aviation Traders purchased OD-ADW from TMA in November 1963. It departed Beirut 01 November via Basle arriving Stansted 08 November. The airframe is one of the latest models C-54/DC-4s used in the program with only three newer airframes used for conversion. With no firm orders Carvair 15 was not allocated to a customer and converted on speculation using surplus manpower and taking two years and four months before it actually flew. The nose

After military service this C-54 became the second airframe converted by Douglas to DC-4 standards. It was purchased by Pan American Airways and operated worldwide until 1961 (via Jennifer M. Gradidge).

A rare look at G-ATRV, in the British United livery it wore for only one year. It was stored without engines for 18 months after completion and is the only British Carvair never named (courtesy Ad-Jan Altevogt).

assembly arrived from Southend on 10 November 1963. The fitting to the stripped airframe was completed by 20 December. Work progressed slowly until August 1964 when it was reported near completion and would be ready to fly on 04 October. In October ATEL moved production from Hangar One to the former Britannia overhaul Hangar Four. OD-ADW was transferred to Hangar Four on 07 November for completion. Without a firm order it was stored at Stansted, structurally complete, unpainted, and without engines, until April 1965 when it was placed up for sale. It received British registration G-ATRV on 07 March 1966 after being stored 18 months. It was painted in the original British United dark blue stripe livery. As another example of no two ships being the same this aircraft had the dark trim on the top of the fin painted with a straight edge rather than following the contour of the fin like previous conversions.

It did not receive a formal "Bridge" name as previous British United Carvairs. Nine of the original eleven Channel Air Bridge, BUA, BAF Carvairs received "Bridge" names in the tradition established by Freddie Laker. Of the final two, G-ATRV was never named and G-AXAI only received the character name of "Fat Albert" after BAF changed marketing strategy.

Two and a half years after entering the line at Stansted, G-ATRV made the first flight on 23 March 1966 as the lightest Carvair built.[1] Two days later it was purchased by British United Air Ferries and ferried to Southend. The C of A was issued on 25 March 1966. Additional checks were conducted at Southend on 27 March before going into schedule service on 01 April. The first revenue flight was made from Southend to Rotterdam, returning to Manchester, back to Rotterdam and returning to Southend the same day without incident. G-ATRV was only in British United service for a few days more than a year before being transferred in 1967 to Cie Air Transport (CAT).

Compagnie [Cie] Air Transport (CAT)

The French carrier Cie Air Transport (CAT) operated a conjunction marketing arrangement with BUAF from Le Touquet and Calais to Lydd and additional service between Nimes and Corsica and Palma. CAT initially operated leased Bristol Superfreighters prior to the Silver City merger with Channel Air Bridge. The CAT interchange agreements with Silver City were carried over and the carrier purchased the two Interocean Carvairs in 1965 to improve and expanded service.

During the off season CAT used the Carvairs for general cargo and charter flights. CAT lost Carvair four, F-BHMU in a tragic crash on 08 March 1967 at Karachi Pakistan, leaving the carrier with a single Carvair. In April a replacement option was exercised in order to maintain the conjunction agreements with BUAF. G-ATRV was overhauled and repainted at Southend in the Cie Air Transport livery before transfer. The paint scheme was a modification of the British United colors. The red British United titles were deleted and Cie Air Transport titles added along the dark blue stripe on the forward fuselage. The vertical stabilizer was repainted white with Carvair titles above two narrow bands.

As the lightest Carvair built it was designated by BUAF to operate the Deep Penetration routes and not fitted with the wing "Bob weights" used for short cross channel service to permit heavier loading. Before delivery to CAT the weights were installed changing it to an ATL-98A allowing heavier cargo loads. It made the first flight as F-BOSU on 06 May. The next day the British registration was cancelled and it ferried to Le Touquet and on to CAT operations at Nimes. Three days later the purchase was finalized becoming the fourth and final CAT Carvair.

On 10 May 1967 Carvair 15 was named "President Gamel" in honor of Pierre Gamel, a pharmacist, who practiced in Nimes. During World War II Gamel was a member of the French resistance. He was captured by the Germans and sent to Buchenwald. After the war he returned to France and his business becoming involved in politics and serving two terms as the President of the Chamber of Commerce. After his retirement he was made a Commander in the Legion of Honour. He died 29 March 1966 and was honored with his name on the aircraft. The name was not applied until it came off lease from Shell Nigeria and returned to France. Many old records show the name of the aircraft as "President Malon," which is in error. The name "President Malon" was the name given to Carvair 16, HI-168, when it was purchased by Dominicana. It appears that previous accountings confused the two aircraft and reported the name in error. The name "President Gamel" remained visible even when used for spares by TAR at Nimes-Garon France and was still below the cockpit window when it was broken up in 1972. Many people publicly expressed the feeling that the name should have been removed during the declining days of the aircraft. Leaving it visible was not a fitting end for a tribute to a noted hero.

Shell Nigeria

In 1967 F-BOSU was leased to Shell Nigeria for support work on oil projects in central Africa and the Middle East. This same airframe did oil field work in the Arabian Gulf as a DC-4 while owned by Trans Mediterranean prior to conversion. Shell also leased Carvair 11 from British Untied Air Ferries in April 1967 for oil field work. On 21 November 1967 while en route back to France over the Sahara Desert in North Africa, F-BOSU suffered a catastrophic failure of Number Four engine. The engine seized causing the propeller and reduction gear to separate from the engine. The debris ripped the cowling from the engine causing damage to the starboard wing and flap. The crew was able to maintain control of the aircraft and made an emergency landing at a small airstrip at Amenas in western Algeria. The area is very remote near the border with Libya south of Tunisia. Because of the lack of spares, temporary repairs were made in an attempt to get the aircraft back in the air. Without any facilities the crew was determined to make the aircraft airworthy in order to operate a three-engine maintenance ferry. The plan was initially to fly to the nearest airport with maintenance personnel. Once airborne,

The lightest Carvair built was painted in CAT colors after the carrier exercised a replacement option in April 1967. It was leased out to Shell Nigeria for oil field work. The new name President Gamel was not immediately applied. It was purchased three years later by TAR and used for spare parts, eventually being scrapped at Nimes still wearing the same livery and name (courtesy Terry Coxall).

minus cowling and prop, it performed without noticeable problems and the decision was made to make for France and the home base at Nimes for repair.

The Shell contract continued until 1970 when the aging Carvair was returned to CAT in France. After three years of work in remote areas under harsh conditions F-BOSU required considerable attention. CAT was experiencing financial problems and withdrew all the Carvairs from service. It had seen only brief ferry service while owned by CAT spending considerable time on lease. Shell returned it to France in the summer of 1970 and it was parked at Le Touquet.

Transport Aeriens Reunis (TAR)

Financial difficulties forced CAT to cease operations in 1970. Another French company, Transports Aeriens Reunis (TAR) purchased F-BOSU along with the two other CAT Carvairs. The Nice based charter operator inspected F-BOSU while stored at Le Touquet. Because of poor mechanical condition TAR officials decided in August 1970 to use the aircraft for spares support for Carvairs, F-BMHV and F-BRPT. It was ferried to Nimes-Garons France where it remained until 17 April 1972 when it was transferred to Air Fret. In November 1972 BAF purchased it for salvage. Maintenance engineers arrived and broke up the remaining airframe, which already had the control surfaces, engines and nose gear removed. It was still standing on the mains, but the nose gear and mountings had been cut out for use on another aircraft. The under-nose section was salvaged for possible future use. Carvair eight had suffered a nose wheel collapse at Gander the previous May and the only remaining spare at Southend had been used to repair it. F-BOSU was still wearing the Cie Air Transport livery, which was never changed after being purchased by TAR. The aircraft was not airworthy after 1970 but it remained insured until June 1974. The French registration of F-BOSU was not cancelled until 1975.

Carvair 15 saw very little scheduled commercial service during its operational life. It was operated slightly more than a year with British United before the sale to Compagnie Cie Air Transport (CAT). Shortly after the sale it was leased to Shell Nigeria. Upon return it went into storage where it remained until broken up. Seventy-five percent of the operational service life was the three-year charter work on long term lease to Shell Nigeria transporting oil-drilling equipment. Carvair 15 had a very short operational life and saw the second shortest car-ferry service of all Carvairs built.

22

Carvair Sixteen

c/n 10485-16, C-54B-1-DC, Iberia/AVIACO

Conversion 16 had an operational life of five years and 19 days serving with only two carriers before being lost in a tragic incident resulting in six fatalities on the ground in addition to the crew.

The C-54B-1-DC c/n 10485 was built at Chicago and delivered to the USAAF on 23 December 1944 with military serial 42-72380. It is the last of four C-54B-1-DC aircraft converted to Carvair standards. This C-54 was equipped with eight fuel tanks including cells in the outer wings giving it a fuel capacity of 3,740 U.S. gallons. This series C-54 had a maximum weight of 73,000 pounds and was fitted with P & W R-2000-D3 engines. It remained in military service until purchased by American Airlines in 1946.

American received civilian registration NC90417 and christened it "Flagship Monterey." In 1949 after only three years of service it was purchased by Iberia. The DC-4 carried the Spanish registration EC-AEP for the next 15 years. In April 1963 Iberia leased it to subsidiary AVIACO.

Carvair Conversion

AVIACO officials became interested in Carvair conversions and made inquiries early in the development program. Studies were done for vehicle ferry services between mainland Spain and Palma but none were ordered. Company officials eventually concluded that the market prospects were promising for mixed-traffic service and the decision was made late in 1963 to convert two Iberia DC-4s to Carvairs and take an option for five more. As part of the transaction Iberia traded-in five Bristols.[1] AVIACO officials wanted the Carvair as soon as possible to be in place for the 1964 season. ATEL recommended several options; Accept a ship that was currently in production or lease a Carvair until an Iberia/AVIACO DC-4 could be pulled from service and converted.

Carvair 15 was being converted on speculations and was available. Officials at ATEL strongly recommended that AVIACO accept LX-I0F, which was arriving as a trade-in from Interocean. ATEL viewed it as an opportunity to dispose of the surplus DC-4, which had been speculatively acquired for Carvair conversion. It was at Stansted and conversion could begin immediately. Officials at Iberia rejected both suggestions and reviewed the logistics of pulling its DC-4s from service for conversion. Iberia believed it was essential that AVIACO service begin immediately. A lease-purchased agreement was negotiated for Carvair 12 beginning in April.[2]

After review Iberia officials decided it would be in its best financial interest to convert a DC-4 from its own fleet. EC-AEP was selected to be the first AVIACO conversion. It was ferried from Barcelona to Stansted on 15 February 1964. The Interocean DC-4 LX-IOF arrived on 06 February and was slated to be ship 16. With no pending orders it was moved to position 17

When the Iberia DC-4 EC-AEP arrived at ATEL, engineers were quite surprised to find it to be in excellent condition requiring no retro work before conversion (via Jennifer M. Gradidge).

to allow for the Iberia/AVIACO conversion. Disassembly began and much to the surprise of ATEL engineers unlike previous conversions this C-54 was in excellent condition and did not require any major retro work.

The original C-54 airframe was built with the double rear cargo doors. When it was converted from military C-54 to commercial DC-4 standards after the war, the rear half of the double door and hinges were removed and skinned over. There was no trace of the rear half of the door. AVIACO did not require the double doors further expediting conversion time making it one of the few ships without the double side cargo door. There were fourteen older airframes used for conversion. This was the fastest conversion of any airframe because of the excellent condition of the later model DC-4.

The engines on the Iberia DC-4 had the short ram-air carburetor intakes, which were left intact and not converted to the long non-ram type. This DC-4 arrived with eight fuel tanks verses six tanks of other Carvairs. The carrier opted to keep the eight tanks, however sometime after delivery to AVIACO two of the tanks were removed by Iberia maintenance reducing it to six.[3]

The nose arrived by road from Stansted on 02 February 1964. The nose loading door was the earlier type equipped with two landing lights which was not upgraded to the di-electric patch. Iberia accepted the raised ceiling and relocated control runs in the cargo bay in order to exploit the Spanish horse transport market. The raised ceiling was offered in regular production beginning with ship thirteen. Work progressed so well that ATEL engineers issued a completion date of May 1964. It was rolled out and the new empty weight recorded as 41,331 pounds. It flew for the first time 04 June 1964 and received the C of A on 17 June after 6.53 hours of test flights.

EC-AXI was accepted by Iberia on 20 June 1964 and in turn transferred to AVIACO the same day but did not ferry to Spain until 24 June. The aircraft was registered EC-AEP before

EC-AXI was in service with AVIACO for only four years when declining market share forced the carrier to abandon car-ferry service and place the aircraft up for sale (courtesy Transportraits/Richard Goring).

conversion and was re-registered in Spain as EC-AXI on 19 May 1964 before the first flight as a Carvair. It was temporarily re-registered EC-WXI for the ferry flight to Barcelona. The registration of EC-WXI was only used as a transport call sign for delivery. The lettering was not applied to the aircraft and EC-AXI remained on the aircraft from first flight until leased to Dominicana. It was painted in AVIACO colors but actually registered to parent company Iberia after conversion.

Officials at AVIACO believed it was essential that service begin immediately with Carvair 12, which was leased in April to begin service. The carrier hurried into the high season market with very promising early results. With the arrival of the second ship, EC-AXI, plans for future car-ferry operations were expanded. In the first five months of operation ending in November AVIACO transported 3,300 vehicles and considerable freight volume leaving AVIACO with very optimistic expectations about the future. Company officials believed that the market would require a third Carvair. Preliminary negotiations began with ATEL to purchase the leased G-AOFW as one of the five Iberia options. During this same period AVIACO transferred a second DC-4 to Stansted for conversion. The optimism was not to last because several anticipated contracts did not materialize and traffic declined rapidly. The leased Carvair (G-AOFW) now carrying the Spanish registration EC-AVD was returned to Aviation Traders at Stansted and placed in storage. AVIACO considered canceling the order to convert the second Iberia DC-4 to Carvair eighteen. Reluctantly the contract was honored and the conversion was completed.

AVIACO struggled with the car-ferry and general cargo flights for nearly four years. After continually losing market share to the roll-on roll-off sea ferry operations the service was withdrawn in 1968 and Iberia began looking for a buyer. EC-AXI had been flown only 4123 hours since converted to Carvair standards.

Compañía Dominicana de Aviación (Dominicana)

The Cargo division of Dominicana was shopping for at least four additional aircraft to service the newly approved Haiti and Curaçao authority as well as the existing Miami–Santo Domingo routes. The Carvair appeared to fit the need because they were relatively inexpensive to operate and had a high bulk cargo capacity. AVIACO was anxious to dispose of them and agreed late in 1968 for Dominicana to lease the Carvairs with option to purchase.[4] Leasing allowed Dominicana to participate in a proposed plan of economic aid to Latin countries with minimum financial exposure. The rather questionable aid scheme was devised to transport fruit and commodities for struggling Latin countries, which included the Dominican Republic.

Dominicana leased EC-AXI for the aid flights on 01 February 1969. It was transferred with only 4127.52 hours on the airframe since conversion and training began. After being assigned Dominican registration HI-168 it ferried from Barcelona to Santo Domingo on 20 February and was christened "President Malon." Twenty-four days after the lease was signed, HI-168 received the Dominican C of A. The pair of Carvairs went into limited service based in Santo Domingo but the aid program was very short lived.

Dominicana is not an IATA carrier, therefore the quality and thoroughness of its aviation program is not known, however it is generally considered to have been lacking.[5] The Latin nation economic aid plan failed leaving the possibility of the Carvairs returning to Spain. Undisclosed concessions were negotiated with Iberia and Dominicana exercised the option to purchase in May. The Carvair operated limited cargo service to Miami for less than four months.

Dominicana acquired the AVIACO Carvairs under an economic aid program designed to help struggling Latin countries. The program failed and only four months after delivery to Santo Domingo (MDSD) the aircraft above — HI-168 — was lost in a tragic crash at Miami (courtesy Eddy Gual).

Fatal Flight of HI-168 (NTSB-AAR-70-17)

HI-168 ferried empty from Santo Domingo to Miami on 16 June 1969 arriving at 18:00. Scheduled for a revenue cargo flight on 22 June 1969 the aircraft was being positioned early for weight and balance check. The passenger compartment and galley equipment had been removed at Santo Domingo the previous week to increase the capacity for pure freighter operations. The inbound flight crew did not report any discrepancies upon arrival at Miami. The co-pilot stated the flight was routine and the aircraft performed well. The inbound flight crew deadheaded back to Santo Domingo later that evening on another Dominicana aircraft.

HI-168 was de-fueled on the morning of 17 June for the aircraft weight check. After the weight check and re-computation was completed a total of 926 gallons was pumped back on the aircraft and it remained parked until 21 June. Fuel records indicate that an additional 934 gallons were pumped on-board that morning, which totaled 1860 gallons in the tanks weighing 11,160 pounds. It should be noted that the aircraft was modified from the original configuration of eight fuel tanks, four main tanks and four auxiliary tanks. The two inboard auxiliary tanks numbered two and three were of fabric construction. They were removed by Iberia sometime before delivery to Dominicana reducing it to six tanks. The cockpit fuel selector valves and refueling caps had been blocked off. The fuel gauges had also been removed from the instrument panel.[6]

Early on the morning of 22 June a total of 15,368 pounds of cargo consisting of auto parts, plumbing pipe fittings, empty plastic cosmetic containers, and cartons of aerosol deodorant spray cans were loaded by Dispatch Services Incorporated of Miami. Dominicana flight 401 was scheduled to depart Miami at 1700 that evening. The flight crew arrived from Santo Domingo and consisted of the same captain and engineer that had ferried the aircraft to Miami on 16 June. The co-pilot was another airman assigned to Flight 401, who had arrived earlier.[7]

The flight crew consisted of Captain Jorge E. Bujosa age 42, F/O Carlos A. Brader age 30, and Flight Engineer Carlos M. Gonzales age 31. The ATL-98 Carvair was designed for a two man cockpit crew. Although an engineer was not required an additional airman was often assigned to reduce the workload on long flights. His duties include aircraft pre-flight inspection, checking fuel load and cargo load security, and to assist with engine run-up. Dominicana flight standards also state the flight engineer will perform feathering of props during in-flight emergency on command of the captain.[8] Dominicana had employed Captain Jorge Bujosa since 29 June 1947. He had been on vacation in Santo Domingo when he was called back early because of a labor dispute. He held U.S. Air Transport Pilot Certificate 1392657 with a type rating in the C-46. He also held Dominican Republic Air Transport Certificate Number One with type ratings in the DC-3, DC-4, and DC-6B. He was the designated pilot examiner for the Dominican Republic Department of Civil Aviation. His last U.S. medical certificate was issued 19 December 1968. It stated he must wear corrective lenses while flying. Captain Bujosa had 13,736 pilot hours, 500 were in the DC-4. It is not known how many of these hours were in the Carvair. However, it was relatively low since Dominicana had only flown the aircraft 94.48 hours in less than five months. Captain Bujosa had flown a total of 175 hours in the previous 90 days in several aircraft types.[9]

It should be mentioned that Captain Bujosa had deficiencies noted in his flight training records regarding emergency procedures. Records from Flight Safety Incorporated of Miami indicate that in October 1964, after completing 64 hours of ground training, he was given initial instrument and equipment check flight in the DC-4. At that time he was graded unsatisfactory in engine out procedure, engine fire, attitude, application, minimum speed maneuvers and emergency procedures. He was re-checked and received a satisfactory grade. However the following remarks were made on the check form.

> Captain Bujosa did not show up for two scheduled training flights and was late for this one, which accounts for the low grade in attitude and application. Needs work in emergency and

normal procedures. Recommended four hours Dehmal (ground training device) before further flight training.[10]

Other records show that in April 1964 he had four DC-6 simulator training flights in preparation for his initial DC-6 checkout. After the third flight it was noted on the record that all emergency maneuvers were weak because he did not memorize the Phase I procedures. Captain Bujosa was given a progress evaluation in May 1966. The instructor noted his general aircraft procedure work limited. However, he did receive an overall performance of satisfactory. Despite an apparent record of procedure deficiencies in training he was scheduled to begin jet training in Madrid the next month. Dominicana had employed F/O Carlos Brader since 08 August 1964. He possessed U.S. Commercial Certificate 1581389 and Dominican Republic Commercial Certificate 103 with co-pilot type ratings in L-749A, C-46, DC-3, DC-4, and DC-6B. He also held U.S. multi-engine, land and instrument ratings. His U.S. medical certificate was issued 23 July 1968.[11]

Co-pilot Brader had a total of 2,333 hours, 476 in the DC-4. It is not known how many of these DC-4 hours were in the Carvair. He had flown 200 hours in the previous 90 days. Records indicate that Brader completed his DC-4 ground and flight training at Flight Safety Incorporated at Miami. The records indicate that he was very proficient receiving 14 above average grades in categories including, engine out procedure, engine out maneuvers, feathering procedures, and engine fire.

Carlos M. Gonzles was the Flight Engineer. Dominicana hired him 20 November 1967. He held a Dominican Republic Flight Engineer's certificate type rated in the L-749A and DC-6B. Gonzles received 6.6 hours of ground school on the DC-6B and 14 hours of training in the DC-6B simulator in November 1967. He also received 10 additional hours of DC-6B training in January 1968. He had been a part-time Dominicana employee until January 1969 when he was upgraded to full-time.[12]

Also on board in the cockpit was Major Cesar Molina, a Dominican Republic military pilot reported to be one of their top jet fighter pilots. He was anxious to return home a day early and was granted non-revenue cockpit jump-seat authority. Major Molina had been in Miami on personal business for a week and had a return revenue ticket on a Dominican passenger flight for Tuesday 24 June.

Dominicana 401 was scheduled to depart at 17:00 22 June 1969. Captain Bujosa cancelled the departure because of possibility of arrival in Santo Domingo after 20:00 hours, which is prohibited by Dominicana company policy. The flight engineer verified the fuel in the tanks with a dipstick and advised Captain Bujosa of the fuel on-board. The captain requested a fuel truck at the aircraft the next morning to top off the tanks. Flight 401 was re-scheduled for a 09:00 departure 23 June.

Early the next morning additional cargo consisting of three air conditioners weighting 390 pounds was loaded making the total cargo load 15,758 pounds. An additional 202 gallons of fuel was added per the captain's request, 151 gallons in Number Four Main and 51 gallons in Number One Main. This brought the total fuel on board up to 2062 gallons (12372 pounds).

The original pre-planned load sheet of 22 June indicated the aircraft to be only 330 pounds from (ATOG) maximum takeoff gross of 68,600 pounds with a (ZFW) max zero fuel weight of 59,000 pounds. This is calculated with a dispatch fuel of 1600 gallons or 9600 pounds. The aircraft now had 2062 gallons or 12372 pounds of fuel.

	June 22 *Pre-planned*	*June 22* *Actual*	*June 23* *Actual*
Dry wt. 41,840	41,840	41,840	
Crew 480	640	640	
Cargo 16,350	15,368	15,758	

June 22 Pre-planned	June 22 Actual	June 23 Actual
ZFW 58,670	57,848	58,238
Fuel 9,600	11,160	12,372
TOG 68,270	69,008	70,610 (block wt.)
		-300 (taxi fuel)
		70,310 (actual TOG)

If the fuel records are correct the aircraft already had 1860 gallons (11,160 pounds) on board. That is 260 gallons more than the dispatch fuel of 1600 gallons, putting the aircraft 1,230 pounds over weight. The additional 151-gallon top off of tanks requested by the captain over fueled the aircraft by 462 gallons (2772 pounds). The planned fuel burn from Miami to Santo Domingo is 6000 pounds (1000 gallons). At Maximum Takeoff Weight the Carvair will burn 275 gallons the first hour, and 250 gallons the next two hours and 220 gallons thereafter. With this rate of burn the aircraft had more than twice the required fuel on board.[13]

It has never been determined why the captain requested more fuel. There were no weather conditions that would require extra fuel. It does appear that Number One Main contained more fuel than Number Four Main and the additional fuel was an effort to asymmetrically balance the aircraft. However, if this were the reason it could have been accomplished by adding 100 gallons to Number Four tank. It has also been speculated that they were ferrying fuel because of supply shortages in the Dominican Republic, however this was never determined. It would appear the captain had the tanks topped off to bring the aircraft up to XTOG but failed to consider the 90-pound per degree Fahrenheit high temperature penalty.

The takeoff was attempted at maximum gross weight in high humidity at a temperature of 89 degrees Fahrenheit (31 C). The weight and balance manual states that 90 pounds is to be subtracted from the indicated takeoff weight for every degree Fahrenheit above I.S.A. Using this formula relative to the humidity and temperature the aircraft could have been as much as 1260 pounds overweight.

The load sheet prior to the flight indicated takeoff gross of 68,270 pounds. Assuming the fuel company record is correct, and the actual cargo load and jump seat passenger is added, the recalculated takeoff weight would have been 70,610 pounds (less taxi fuel). The crew should have realized the aircraft was 1410 pounds over the MTOGW (max take-off gross weight). The NTSB recalculated the gross takeoff weight after the accident as 69,710 pounds, which is still 510 pounds over gross.

At 13:16 EST, 23 June, Dominicana 401 departed the northwest cargo ramp at Miami International for departure to Santo Domingo. Shortly thereafter at 13:25 it returned to the ramp reporting high oil pressure (135 psi) on Number Four engine. The normal operating pressure is 80 psi. The engines were shut down and flight engineer Carlos Gonzales went to the Aerocondor office to request a mechanic.

Maintenance dispatched two Aerocondor mechanics, Rodrigo Zapata and Narciso Camargo, who were natives of Columbia to check the Number Four engine. Zapata stated that when the throttle was increased the engine oil pressure went up to 135 psi. He stated that the oil pressure regulator was loose, so he and Camargo tightened it and adjusted the pressure back to normal. Captain Bujosa and the two mechanics ran up the engine and it checked out "ok" stated Zapata. The crew and mechanics went to lunch and were back on board the aircraft by 15:00 for a second departure. Zapata had been performing Dominicana turn around maintenance for the last year and a half. He knew all the crew well, in fact engineer Gonzales and the two mechanics had split $4,000 they won at the dog track the night before.[14]

It is assumed that the flight crew conducted the preflight run-up and again the Number Four engine checked out. The engine should have been run up to 2200 rpm and magnetos checked at 30 in Hg. If the oil pressure had not been 80 psi, the mechanics were standing by.

The weather conditions as observed by the weather bureau at Miami International were as follows:

> 3,000 feet scattered clouds, higher light broken clouds, Visibility 10 miles, pressure 1013.4 millibars, temp 89 Degrees F., wind 170 degrees at 7 knots, altimeter 29.92 inches. Thunderstorms southeast with cumulus build-ups to the west.

At 15:30 Flight 401 was cleared to taxi from the cargo ramp to Runway 9 left and to Runway 12 for departure. Runway 12 is 9,604 feet long and 150 feet wide. Flight 401 taxied into position at 15:39 and was instructed to hold. The aircraft was cleared for takeoff 15:41.05.

Dominicana 401 began takeoff roll and the aircraft lifted off and crossed the intersection of Runways 35 and 12 at an altitude of approximately 15 feet. The ground controller observed several puffs of black smoke coming from Number Four engine. At 15:42.30 the tower observed the smoke coming from the Number Two engine of the Carvair and advised:

"Dominican 401 heavy white smoke, Number Two engine, you're cleared to land any runway." The flight replied: "Four zero one."[15]

The tower controller alerted the Airport Fire/Crash rescue. The fire equipment proceeded to positions along runway 27. When the aircraft impacted the ground they left their positions and took approximately four minutes to reach the scene.

The captain of a commercial flight that had just landed on 9 left and was taxiing toward the tower also saw the puffs of black smoke from Number Four. He stated that as the aircraft flew over the intersection of Runways 12 and 9 right heavy white-yellow smoke was streaming from Number Two engine. Another pilot one-quarter mile southeast of the departure end of runway 12 observed 401 as it crossed the outer airport boundary at approximately 100 feet. It was flying directly toward him and all four engines were turning and heavy white smoke was trailing Number Two engine. As 401 passed over his position he observed Number Four propeller go into feather position and stop.

The smoke from Number Two subsided momentarily and the Number Two propeller appeared to windmill. The aircraft continued to climb to an altitude of 300 to 350 feet on the runway heading. The Carvair began to turn northward and started losing altitude as it turned.

The aircraft passed directly over another witness located approximately two miles east south east of the airport. He stated the aircraft was at 125 feet and the landing gear and flaps were up. The Number Four prop was feathered and Number Two was windmilling and trailing heavy black smoke. Engines One and Three sounded as if they were at maximum power. The heavy black smoke changed to heavy white as it passed his position and began a shallow turn. Then it disappeared from his sight.

The crew was probably using the load sheet that was calculated at 68,270, however the aircraft at this point probably weighed 70,310. The Carvair operating manual states: "Assuming infinite (sufficient) runway length, the limiting factor which restricts take of gross weight will always be the second segment takeoff climb performance of the airplane. (in this case 69,200 pounds) This is the rate of climb in feet per minute with one engine inoperative and it's propeller windmilling, takeoff power on the other three engines, landing gear retracted, wing flaps at 15 degrees, and cowl flaps in trail position. This should produce a rate of climb of 250 feet per minute."[16]

The climb performance data shows that with two engines inoperative at this weight (69,200 pounds) level flight cannot be maintained. We know the aircraft was approximately 70,000 pounds. In fact with maximum continuous power on the other two engines you can expect a rate of descent of 170 feet per minute. The maximum weight which level flight can be maintained is 58,400 pounds. The maximum fuel dump rate of 2,280 pounds per minute would require five minutes. At only 300–350 feet of altitude there would not be enough time and the possibly of the fuel being ignited would have eliminated dumping as an option.

The ATL-98 manual states that the Carvair should not attempt a bank turn into a dead inboard engine. The aircraft because of the large nose has a tendency to lose altitude more rapidly when banking into a dead engine on the left side. This is a documented characteristic covered in Carvair engine out procedures. In original flight test by Aviation Traders it was noted that the most critical engine loss would be on the left. Test proved that to maintain directional control considerable rudder movement, counteracted by trim is required. To continue to climb, a speed of 140 mph must be maintained. The aircraft could not have made the field in straight level flight. The left turn virtually insured an increased rate of descent. As the aircraft was turning back toward the airport and heading west at 15:44.15, Dominicana 401 advised the tower:

> "Four zero one is gonna land on niner right." The tower replied, "You're cleared to land four zero one straight in."[17]

The aircraft was east of the airport heading toward Runway 27 right. This was the last radio transmission.

The tower controller stated that after the aircraft made the first turn it began a slow descent. As the aircraft neared the ground it was in an extremely nose high attitude. Nose high attitude in steep turns is a characteristic noted by the manufacturer. Considerable trim is needed to keep the nose level.

Miami police officer Robert Archer saw the plane coming down and radioed, "We're going to have a plane crash here!"[18] Approximately one mile east of Miami International Airport the aircraft, now on a heading of 260 degrees, struck power lines and severed three power poles along Northwest 36th street and 33rd avenue. It ripped the roof off of the Ryon Medical center, then the tops of buildings on the south side of 36th street. Forty-two vehicles were hit along 36th street before it came to a stop in a single story building housing an auto body shop. The fuel ignited and exploded a large amount of auto paint and thinner stored in the building. The building and the aircraft were destroyed. The right wing separated from the aircraft and came to rest in the middle of 36th street. The Number Four engine was still attached to the wing by the lines and cables. It took firemen with 14 fire trucks 45 minutes to bring the fire under control. All four on board the aircraft perished along with six persons on the ground and twelve persons were seriously injured.

The Miami tower visually followed Dominicana 401 from the time the controller observed smoke from the Number Two engine until impact. The controller stated that the aircraft altitude was never more than 300 feet and it continued to lose altitude as it made a turn westward back toward the airport. It disappeared from view east-northeast of the airport and then flames and two large columns of smoke were observed.

Post Crash Investigation

The crash site covered an area approximately 740 feet in length and 180 feet wide. Because of impact and intense ground fire no reliable data could be obtained from the cockpit instruments, pedestal or upper center panel. The flight control system showed no evidence of malfunction, however because of fire damage no trim settings could be obtained from the control surfaces. The wing flap actuators were recovered and indicated the flaps to be in the full up position. There was no evidence of electrical or hydraulic failure and the gear hydraulic cylinders were in the retracted position. The Number One, Two, and Three engines were separated from the aircraft by the impact. Engines One and Three were partially disassembled and showed no evidence of failure.

The Number Two engine was disassembled and examined for failure. The engine cowling was found and showed no evidence of penetration. The Numbers Six and Eight cylinders had separated from the crankcase and were found in the wreckage. The front row master rod and all the front row cylinders link rods were broken. The rear row of cylinders was intact, although the Number Five and Seven cylinders were damaged on impact. There was evidence of valve pounding marks in the heads, however all cylinder rings moved freely in their grooves and had a film of lubrication.

All of the front row pistons were extensively damaged and sections of number Eight and 14 cylinders were ground and pounded away. The front row master rod "I" section was broken and ground off, as were the link rods. Only a small portion of each link rod remained attached to the master rod assembly. There was pronounced rotational damage in this area indicating pre-impact failure of Number Two. The master rod attaching bolts were severely damaged with one bolt broken away.

The crankshaft bearings showed no evidence of damage and were adequately lubricated. The spark plugs, cams and rollers, left magneto (right magneto not recovered), and propeller governor showed no evidence of pre-impact failure.

The Number Four engine showed no evidence of internal distress or rotational damage. All bearings were adequately lubricated, as were the cylinder walls. Ground fire damaged both magnetos precluding testing, however all spark plugs were serviceable and capable of firing. The carburetor assembly was badly burned and magnesium throttle body burned away. The Cable operated throttle valve was in the fully closed position. The cable operated manual mixture control valve was set in the "auto rich" position. The fuel strainer and carburetor inlet showed no evidence of blockage.

The Number Four engine oil pressure relief valve, which had been a pre-departure problem, was found securely in place on the engine. The valve assembly was removed and tested at the Dominicana maintenance base in Santo Domingo. The valve was installed and operated on two different engines of HI-172, the other Carvair operated by Dominicana. The following results were obtained as stated in the NTSB report:

1. Oil pressure relief valve installed in Number Two engine. (TSO 1,173 hours) of HI-172

    (a) At 800 rpm., oil pressure 25 psi.
    (b) At 1,000 rpm., oil pressure 35 psi.
    (c) At 2,000 rpm., oil pressure 55 psi.
    (d) At 1,500 rpm., oil pressure 45 psi., propeller control levers then moved to the high-pitch position and, when governing action took place, oil pressure declined 35 psi. and remained steady.

2. Oil pressure relief valve installed in the Number Four engine HI-172 (TSO 20 hours)

    (a) At 1,000 rpm., oil pressure 30 psi.
    (b) At 2,000 rpm., oil pressure 60 psi.
    (c) The same test were performed as in (d) above with the same results.

Conclusion

Examination of the captain's training records does not suggest a high degree of proficiency in emergency procedures. His training records indicate that he was very weak in the knowledge and execution of engine out procedures.[19]

Whether or not the captain was highly proficient in emergency procedures and had sufficient confidence in them so that he might have had time for further analysis of the situation is difficult to determine. The available facts suggest he was not.[20]

The failure of the Number Two engine occurred shortly after lift off. It is difficult to explain why it was not secured and propeller feathered in a reasonable amount of time. All indications are that it remained unsecured until impact. Conversely the Number Four engine examination disclosed only the oil pressure relief valve problem, but no other evidence of failure. The propeller was feathered shortly after takeoff and remained feathered until impact.

Since post crash testing of the oil pressure relief valve indicated it was consistently below operating parameters, it is conceivable that during takeoff it dropped below 50 psi. If the oil pressure falls below 48 psi, the low oil pressure limit warning light located on the panel is activated. The illumination of the light should not precipitate action until the oil quantity, oil temperature, and cylinder head temperature are noted to assess the significance of the light.

There is no doubt that the cockpit crew was focused on the Number Four engine because of pre-flight oil pressure problems. The failure sequence of the Number Two engine probably began at the same time of the illumination of the Number Four low oil pressure light on the panel.

The safety board concluded that the probable cause of the accident was confused action of the crew while attempting to cope with a catastrophic engine failure (Number Two engine) during takeoff. In their confusion and preoccupation the wrong engine was shut down. If only the Number Two engine, which sustained the catastrophic failure, had been shut down, the aircraft although overweight was capable of maintaining flight long enough to return to the field. Hard left bank into the dead engine compounded the mistakes and increased the rate of descent.

Post-mortem examination of the flight crew revealed no evidence to indicate any pre-existing disease that would affect the performance of their duties.

The accident was of a survivable nature for the cockpit occupants. Three of the four aircraft occupants did not sustain significant impact injuries. Only one occupant died of impact trauma. The other three died of smoke inhalation and/or burns.

Aircraft Statistics

HI-168 Carvair 16 c/n 10485 had 4,222 total flight hours since converted to Carvair standards. Dominicana logged only 94.48 of those hours. The aircraft was powered by four Pratt & Whitney R-2000-D3 radial engines. (Carvair standards specify the R-2000-7M2. Provisions allow for the substitution of the -D3) The -D3 has a takeoff rating of 1,450 hp and normal rating of 1,200 hp at 5,000 feet. The engines were last overhauled and zero timed by Talleres Aeronáuticos de Barajas of Madrid Spain and installed by Iberia maintenance. The average engine TSO for U.S. certified carriers is 1,600 hours. Both Iberia and Dominicana is 1,700 hours.

The total engine hours were as follows:

#1 eng. 701.39 hrs. s/n 109052 Installed 19 August 1967

#2 eng. 635.04 hrs. s/n 109161 Installed 12 September 1967

#3 eng. 867.51 hrs. s/n 107770 Installed 21 June 1967

#4 eng. 1282.29 hrs. s/n 109050 Installed 14 June 1967

The aircraft was equipped with four Hamilton Standard three blade model 23E50 propellers.

The last 200-hour inspection was performed by Iberia Airlines on 30 December 1968. Dominicana had performed a 50-hour inspection at their maintenance base at Santo Domingo on 19 April 1969, only two months before the crash. Examination of the maintenance records revealed no discrepancies that would indicate the aircraft not airworthy.

The aircraft remained on the Dominican aircraft registry for six more months until it was cancelled 01 Jan 1970.

23

Carvair Seventeen

c/n 18342-17, C-54B-5-DO

Carvair 17 is the third of four aircraft converted from C-54B-5-DO airframes. It was built at Santa Monica in block one of 100 C-54s of this series manufactured in California. This group had two of the cabin fuel tanks moved to integral tanks in the outer-wings giving a total fuel capacity of 3,740 U.S. gallons. It was fitted with Pratt & Whitney R-2000-3 engines rated at 1,100 hp at 7,000 feet and had a maximum gross weight is 73,000 pounds.

It was delivered to the USAAF on 19 July 1944 with serial 43-17142. After 15 months with the Air Force it was declared redundant and transferred on 18 October 1945 to the Reconstruction Finance Corporation for disposal. The Skymaster was re-purchased by Douglas, assigned civilian registration NC30042, and returned to the factory with only 617.56 airframe hours to become the ninth conversion to DC-4 civilian standards.

The cargo hoist and remaining fuselage fuel tanks were removed and rear half of the cargo door was bolted shut. On 5 January 1946 the engines were sent to Pacific Airmotive where they were overhauled, magnafluxed and re-configured to R-2000-11 standards rated at 1,100 hp at 7,500 feet with takeoff power of 1,350 hp. On 23 February 1946 the carburetor air induction was re-worked. The military model had hot and cold plus filtered air. The filters were removed from the non-ram air scoops on top of the nacelles and blocking plates installed.

United Airlines purchased the DC-4 from Douglas on 12 February 1946 for $90,000 USD. It was configured for 44 passengers and delivered on 20 March. After many years of passenger service with United it was converted to cargo configuration and named "Cargoliner Rainier." It was inspected and the C of A renewed. The engines were recorded as Pratt & Whitney R-2000-D5 series. The -D5 is rated 1,200 hp at 6,400 feet and a takeoff rating of 1,450 hp indicating the engines had been upgraded by United. After more than ten years of service it was sold on 28 December 1956 to Transocean Air Lines, which maintained a close association with United.

Transocean was the creation of Orvis M. Nelson, who learned to fly in the Army Air Corp and resigned his commission to work for United Airlines. He flew for United from 1931 to 1943 becoming a senior pilot. While flying contract missions for the Air Transport Command during World War II, he and several other officers developed a plan to form an airline after the war. When Nelson learned United Airlines was seeking a sub-contractor to operate military charters to Hawaii he formed Transocean with two surplus C-54s incorporating on 01 June 1946. Transocean established an excellent safety record as a major supplemental carrier and grew to a fleet of 114 aircraft. The carrier relied on Matson Air Transport for maintenance and conversion of its C-54s to DC-4s. Matson attempted to acquire the U.S. Hawaii routes that United eventually dominated and considered purchasing Transocean in its declining years. Five of the DC-4s that later became Carvairs (2, 3, 7, 13, 17) were either owned or leased by Transocean. Two more Skymasters used for conversion nine and 17 were owned by The Babb Company of

New York, a division of International Aircraft Sales and subsidiary of Transocean Corporation of California. The Transocean and Babb Company DC-4s not only operated military charters but also did movie work and transported refugees. Nelson also used some of the DC-4s to transport $6.2 million in gold bars from Tokyo to New York. Transocean lasted for 14 years as a charter and sub-charter for other carriers. It ceased operations in April 1960 and filed for bankruptcy on 11 July.

Transocean operated N30042 for two years before leasing it to Lufthansa in 1958. On 14 January just prior to the Lufthansa lease the airframe had accumulated 34,466.42 hours. The Skymaster was returned to Transocean in 1959 and recorded on 20 October as having R-2000-7M2 engines with 44,295.56 hours on the airframe.[1] It was re-purchased by United Airlines on 21 October and re-sold the same day to The Babb Company for $193,000 USD. No reason is given as to why it transited United Airlines to The Babb Company, possibly financial and bookkeeping since Orvis Nelson was closely associated with both of them.

The management at Intercontinental Airways planned expansion in Europe and applied for Luxembourg registration LX-IOF in March 1960. Associate company U.S. Transport Corporation purchased N30042 from the Babb Company on 21 April 1960 for $215,000 USD. Lawrence Kesselman of Intercontinental Airways (Interocean) headed The U.S. Transport Corporation. The day before the sale troop benches were installed forward of the wing by Flight Enterprise of Hartford Connecticut and the airframe hours were recorded at 45,403.20.[2] In August 1960 it was leased to New York based Intercontinental Airways, which shared the same management (Kesselman-Goldberg-Peck) of Luxembourg based carrier Interocean. It was then transferred to associate company Interocean for work in Europe. At this time it had both U.S. and Luxembourg registrations.

On 02 February 1961 during annual airworthiness maintenance inspections, the periodic re-sealing of the fuel tanks was being performed and cracks were found in the wing spar near

The DC-4 in Intercontinental colors at Newark in 1961 before transfer to associate company Interocean. It is the last of the three aircraft ATEL acquired from the carrier for conversion (via Jennifer M. Gradidge).

the Number Three fuel tank. Servicios Aerotecnicos Latino carried out the repair in San José Costa Rica. Although it was being operated by Intercontinental with a Luxembourg registration the owner was still listed as U.S. Transport Corp.[3]

After repairs U.S. Transport sold N30042 (LX-IOF) to Interocean on 18 April 1962. That same year Intercontinental sold Aviation Traders two DC-4s as part of a purchase agreement for Carvairs four and five. The FAA notified the Luxembourg Minister of Transportation that the U.S. registration of N30042 was cancelled 15 June 1962. Aviation Traders purchased it on 04 February 1964 when Interocean reduced its fleet. ATEL attempted to sell LX-IOF to Iberia/AVIACO for Carvair conversion but the carrier opted to use its own DC-4s and lease Carvair 12 while waiting. Intercontinental — Interocean is a major part of Carvair history because three of its DC-4s (7, 10, 17) were converted for other carriers and it purchased Carvair four and five.

Carvair Conversion/Storage

The Interocean DC-4 LX-IOF arrived at Southend on 04 February 1964. The carrier was still operating Carvairs four and five at this time and appeared to be consolidating its fleet by selling off older DC-4s. Within a year the airline was in decline selling the Carvairs and reverting back to a DC-4 fleet. It is the only conversion candidate known to arrive equipped with a radar nose. There were nine older airframes used for conversion. It arrived on 06 February and there were no current orders for Carvairs pending. Consequently the first AVIACO DC-4 that arrived in February was placed ahead of it as ship sixteen.

LX-IOF entered the production line on 10 February 1964 and the nose assembly arrived by road from Southend on 22 February. The decision was made to proceed with the conversion on speculation with no orders pending. Manpower for conversion was allocated between other projects. Carvair eleven was still in early stages of conversion at Southend as a long-range ship and 12 through 16 would soon fly as Carvairs (except 15, which did not fly until 1966).

Carvair seventeen was built as an ATL-98A with the 250 pound wing bob-weights in each wing for cross channel work. It is unique that G-AXAI was built with this option with no potential buyer and no plan for it to enter cross-channel service. Since ship 17 was built on speculation, the last of the earlier produced noses without the raised control runs was used. It was built without the "Rolamat" floor system but years later in 1983 a DC-8 roller system was adapted and installed.[4] Since it was being built on speculation, cost cutting was a factor of it being built like the older ships. When Iberia/AVIACO purchased Carvairs it took an option for five additional conversions and ATEL considered the possibility that it would eventually be sold.

After completion in 1964 it was placed in open storage beside hangar four at Stansted less engines and instruments. It was not painted but left in bare metal with some areas in zinc chromate or "sozzled" as described by ATEL personnel. Over the next two years parts were removed to keep other Carvairs flying.[5] Aviation trade magazines listed LX-IOF as the next BUAF Carvair in June 1966. Aviation Traders speculated that it would be sold, however the next four conversions were for carriers that supplied DC-4s, one more for Iberia and three for Ansett. The purchase did not materialize and conversion 17 remained in storage almost three more years.

The Carvair assembly line at Stansted had long been shut down when the decision was made in December 1968 to complete Carvair 17, which created somewhat of a logistical nightmare. It was not possible to move the completed airframe to Southend by road. The last Carvair, ship 21, had been completed at Southend for Ansett and flown 10 months earlier. In a strange twist of fate the last Carvair to fly was not the last one built.

After years of storage Carvair 17 received British registration G-AXAI on 19 February 1969. It was removed from storage on 17 March and towed into hangar four at Stansted to be completed. Since this ship never flew after completion in 1964 many parts had been borrowed, creating a considerable task to bring it up to flying status. It made the first flight as a Carvair from

Stansted on 02 April 1969, five years after it was built. Later that afternoon BAF training Captain Bob Langley made a low pass over Southend for everyone to see. The flyby was an anti-climatic event since Carvair production ended ten months earlier at Southend. Stansted production had ended prior to that with ship 20. The C of A was issued on April 04 after requiring only 6 hours and 53 minutes or trials.

British Air Ferries (BAF)

Carvair 17 was delivered to BAF on 05 April 1969 painted in the third BUA/BAF livery of a two-tone blue stripe with black BAF titles. It is the second BAF Carvair that did not receive a "Bridge" name. While in storage for nearly five years it missed the dark blue stripe red titles British United and the BUA and BAF sandstone and blue schemes. It received the character name "Fat Albert" prior to being painted in the last BAF dark blue color scheme in 1972.

Beginning in 1972 when the remaining BAF Carvairs were being repainted in the final dark blue and white livery G-AXAI became the focal point of advertising campaigns. Since it was the last Carvair completed and the newest in the fleet, photos graced magazine ads with copy stating the BAF fleet now had upgraded interiors with music systems increasing passenger comfort. Vehicle ferry bookings had declined to a point that G-AXAI was configured with a 65-seat passenger cabin. By the beginning of 1975 the BAF fleet was reduced to six Carvairs, -AREK on long term lease, -ASDC in cargo configuration, -ASHZ; -AOFW; -ASKN in 55-seat and -AXAI with 65 seats. Still in inventory was G-ASKG but it was sold to SOACO in February.

Carvair 17 was completed in 1964 and placed in storage. Production had long been shut down when it first flew in April 1969. It is the last conversion to fly since the last Carvair (21) was completed for Ansett ten months earlier. G-AXAI never received a Bridge name and eventually became Fat Albert (courtesy AAHS/Francisco Gual).

It was a forgone conclusion the Carvair was long past its prime and would soon be phased out. One by one individual Carvairs were withdrawn from service or put up for sale. G-AXAI along with the other two car-ferry ships received a reprieve in the summer of 1975. An unexpected high seasonal increase in traffic prompted BAF to re-introduce the old Southend — Basle service that had been dropped in 1967. Demand for the Basle car-ferry service ended by the fall of 1975. BAF now had three newly acquired Handley Page Heralds for passenger service. Carvair 17 was withdrawn and put up for sale. The British registration was cancelled 29 December 1975 in anticipation of disposal.

Secmafer/SF Air

The Nice based French cargo carrier Secmafer S.A. a division of Aerial Transportation operating as SFAir reviewed the purchase of an all cargo Carvair. Secmafer was formed to fulfill a contract to transport fabricated steel products to Iran. The carrier agreed to purchase Carvair G-AXAI from BAF in December 1975 and a DC-4, EC-BCJ, from Spanish operator Freezes Europe. The DC-4 was acquired and re-registered F-BVQK. British Air Ferries stripped G-AXAI of the 65-seat configuration and all paint to maximize payload. It made a debut in bare metal still carrying the British registration G-AXAI with a French flag on the tail.

The French registration was applied to the aircraft then removed. The British registration was reinstated on 05 January after the purchase stalled because of certification and licensing problems with SFAir. On 07 January 1976 the Carvair was ferried to Nice as G-AXAI with the French registration assumed to be forthcoming. After the licensing and certification was resolved the French registration F-BVEF was reinstated on 27 January. The British registration was cancelled the same day but title to the aircraft was not transferred to SFAir until March. The associate company SFAir was formed in March 1976 to operate six cargo aircraft, the Carvair, a DC-4, and four DC-6s. Two other BAF Carvairs, G-ASDC and G-ASHZ were already in all cargo configuration, which could have been made available. Secmafer opted to purchase the newer aircraft for transfer to SFAir.

French cargo carrier Secmafer acquired F-BVEF (G-AXAI) to operate under SFAir. Arriving at Nice in July 1976, it was purchased to transport fabricated steel products to Tehran, Iran (courtesy Zoggavia/Paul Zogg collection).

Shortly after arriving at Nice, F-BVEF was painted with two parallel red stripes running the length of the fuselage along the windows. It was otherwise left in bare metal finish and operated at least through July 1976 without the SFAir titles in red on the tail. Twin antennas were also installed along the top of the hump behind the cockpit. It is the only Carvair painted in this livery and to have twin antennas on top of the cockpit. After going into service with SFAir it was based in Tehran for an unknown time on the cargo contract. Secmafer tendered an offer for a second Carvair, G-AOFW, in July 1976, which was not completed.

After returning to Nice from Tehran, SFAir made the decision to repaint F-BVEF in the same livery as the carriers DC-4 and DC-6 aircraft. The new livery consisted of white upper fuselage and bare metal bottom with a parallel red and blue stripe the full length of the fuselage with "CARGO" titles in white. A large blue circle was applied to the forward fuselage with "SFAir" titles in white. Late in 1978 F-BVEF was removed from service and stored at Nice France. SFAir moved its base of operations from Nice to Nantes and Bordeaux in 1980 and became the first European company to operate a Lockheed Hercules in civilian cargo service.

Airtime/Aero Union/Kodiak Western Alaska Airlines

Airtime Corporation of Portland Oregon, an associate company of Aero Union, purchased Carvair 17 on 26 November 1979. Aero Union submitted the application for U.S. registration N55243 on 24 December 1979, which was granted on 18 January 1980. The French registration was cancelled on 07 January 1980 with the recorded serial number as 18342/17F indicating the French had designated it as a pure freighter. It remained in the SFAir livery for nearly a year after it was sold in the United States. It was seen on 07 February 1980 at Chico with no registration on the fuselage. Even after receiving the U.S. registration, the SFAir color scheme was not changed. The airframe hours were recorded on 05 September 1980 as 62,646.3.

Dick Foy and Dale Newton established Aero Union, originally known as Western Air industries, in Redding California. They began with a single B-25 in 1961 contracting to the U.S. Forestry Service for fire fighting missions. The fleet was expanded on 28 July 1961 with two B-17s. The company name was changed to Aero Union in 1962. From this modest start the

Aero Union purchased N55243 via Airtime Corporation in 1979 for support of its aerial tanker fleet. The titles were changed to Aero Union but the livery was otherwise not altered, and it remained in SFAir colors (via Jennifer M. Gradidge).

company has become the largest air tanker company in the United States. Chico Aviation was purchased in 1971 and the company base moved to Chico California. The company specialized in fire retardant contracts by maintaining a fleet of tankers. It was expanded in 1973 when the State of California contracted with Aero Union to construct a prototype tanker. In 1980 the carrier began contract cargo services to augment the seasonal fire control contracts since the tankers return to Chico each winter for maintenance.

The Aero Union maintenance facility provided contract maintenance and aerial fire fighting systems service and development. It expanded into Aerospace manufacturing specializing in the development of equipment and actual aerial fire fighting. Primarily an operator of large piston aircraft, C-119s and DC-4s, it provides crews, aircraft, and ground support to both government agencies and private operations. The company in 2004 had more than 240 employees.

Carvair 17 only transited Airtime Corporation and was transferred to Aero Union on 27 February. It was not immediately put into service and remained in the SFAir livery for nearly a year. Maintenance records indicate a pair of Benson 525-gallon auxiliary fuel tanks were installed in the fuselage on 05 January 1981 for ferry flight positioning. Another maintenance log entry indicates that a pair of Benson tanks was removed on 02 April 1981.[6] This is just prior to the lease to Kodiak Western in April 1981. Aero Union removed the SFAir lettering and logo about the same time. The new livery consisted of the name "Aero Union" above the letters "CARGO" on the forward fuselage on the same color scheme SFAir applied in Nice. The SFAir fleet designator "F" on the nose gear doors was not removed and was still visible when the aircraft was scrapped in 1993.

Charles Willis Jr. (Pacific Aerolift) leased N55243 and became part of a dispute during the same time he was attempting to acquire Carvair N83FA from Ruth May. Ms May wrote in her journal, "Chas Willis is bad news. He had Dick's plane (Dick Foy) on lease, Just got it back. Dick used his own crews."[7] Willis attempted to purchase Carvair five from Ruth May for a contract that he stated he had in Panama, the Philippines, and Hawaii.[8] This could account for the log entries of fuel tanks being installed on N55243 before the Kodiak lease.

The Aero Union Carvair was used on cargo flights and support work for its fire fighting operation. In April it was leased to Kodiak Western Alaska Airlines, formerly known as Air Fortynine, for work in the remote airfields of Alaska and the northwest. Kodiak Western, a subsidiary of Gifford Aviation of Anchorage also leased Carvair five from Ruth May in February 1981 with an option to purchase. Gifford later exercised that option when N55243 was returned to Aero Union. Kodiak Western Alaska Airlines and other Gifford subsidiaries were not financially sound and fell into bankruptcy in 1983. Rocky Mountain Helicopter of Utah attempted to purchase Gifford Aviation on 06 May 1983.

After N55243 was returned from Kodiak Western in the summer of 1981, Aero Union put the Carvair to use in Alaska transporting cargo to remote areas. It flew various ad-hoc charters and was in Montreal on 11 June 1982 on a cargo charter.

Pacific Air Express/Kemavia/Philippine Air Lines

Louis Khem attempted to purchase N55243 under the company name of Pacific Air Express (PAE). He applied for transfer of registration in 1982. The purchase was not completed and registration was not taken up because of marginal financing of the start-up company. Pacific Air Express, a Honolulu based company, was incorporated in Nevada on 01 June 1982.[9] Records indicate Kemavia Incorporated as the parent company of Pacific Air Express with Torrance California as the corporate address. Louis Khem is listed as President of PAE and Kemavia Inc. Corporate papers list A.P. Fairchild as Vice President of Pacific Air Express.[10] Fairchild first became involved with the Carvair in 1963. He flew this aircraft when it was a DC-4 with Interocean before conversion and also flew Carvairs four and five with Interocean in the Congo operation. After leaving PAE he became Director of Operations with Carvair operator Hawaii Pacific Air (HPA). Fairchild was also a corporate officer with Pacific Aerolift (PAL) and the first attempt to import and operate the two ex–Ansett ships in Hawaii. (See ships 20 and 21 for details on Fairchild and other Hawaiian Carvairs.)

Maintenance records dated 13 May 1983 indicate that Winters Aircraft Engineering installed a cargo roller system in N55243 for Aero Union on 20 May. The oak floor panels were removed and a DC-8 roller system was installed for the use of 88-inch pallets. Louis Khem believed this would be an advantage in the Hawaii inter-island market and attempted to purchase N55243 a second time.

The FAA records indicate that it had two 525-gallon Benson tanks installed in the cargo compartment on 21 May 1983 and removed on 26 May. Restrictions state, "For ferry flight positioning only. Must be removed for regular cargo operation." In all probability they were installed for the ferry flight to Hawaii. Pacific Air Express completed the purchase of N55243 from Aero Union on 27 July 1983 for $325,000. It arrived in Honolulu painted in the SFAir colors and operated in them for a short period.

No reason is given as to why FAA records indicate Kemavia Incorporated purchased it back from PAE in August 1983. Considering the fact that Louis Khem was president of both Kemavia and Pacific Air Express, leaves the possibility of a paper transfer to realize tax benefits by leasing it back through another corporation or some other technical reason. Kemavia had also purchased Carvair seven in December 1982 and leased it to PAE. Records show that N55243 was owned by Kemavia and leased to Philippine Air Lines in September 1983. There is no evidence that the carrier ever acted on the lease or it flew any service for Philippine Air Lines. The Hawaii business registration indicates the corporation name as a "Master Name for a Foreign Profit Corporation" possibly indicating the intention to operate for Philippine Air Lines.

Pacific Air Express purchased the Carvair a second time from Kemavia in 1984. It was partially repainted in Hawaii with white upper half and green stripe. The stylized flying "P" logo was added to the vertical stabilizer but never applied to the sides of the forward fuselage. The SFAir fleet identification "F" remained on the nose doors. The PAE titles were not immediately applied possibly because of Kemavia ownership or the lease to Philippine Air Lines. FAA records of 01 May 1985 indicate Kemavia held a PAE mortgage of $190,400 on the Carvair.

The carrier was in financial trouble by early 1986 and parked its DC-6 fleet. The Carvair could not generate enough revenue to cover operational cost and was operated for a short time on an ad hoc basis before also being withdrawn in 1986. As late as 28 October 1987 N55243 was in storage in Honolulu. It was in Pacific Air Express titles but the full livery was never completed.

Hondu Carib/Custom Air Service

After the demise of Pacific Air Express, N55243 remained stored at Honolulu. During negotiations with Frank Moss for Carvair seven, Louis Khem offered both PAE Carvairs. Moss informed Khem that he was not in the financial position to purchase both aircraft. Khem wanted to dispose of N55243 and proposed that Moss could take possession of it and not make any payments for two years or earlier if he made a profit. Moss and his partner Lee Mason agreed to the arrangement but no papers were signed. Moss absorbed the cost of ferrying it to Naples Florida and Mason applied for transfer of the registration to Great Southern Airways.[11] Two 525-gallon Benson auxiliary fuel tanks were installed in the cargo bay on 21 March 1988 and N55243 departed Honolulu 24 March to San Jose California with an additional stop at Griffin Georgia en route to Naples Florida. The stop at Griffin adds a strange twist since after it was broken up the salvage parts returned to Griffin. FAA documents reflect that it was sold to Fred Lee Mason under Great Southern Airways between 1986–88. However, there are no records of the terms or purchase. The plan was to operate the two Carvairs N55243 and N103, purchased in September 1987, under the Hondu Carib (Honduras Caribbean) certificate. Moss had a working arrangement with Fredrick Lee Mason who owned Great Southern Airways.

After arriving in Florida the aircraft was parked awaiting the papers to formerly complete the transaction. When they arrived there were major discrepancies such as serial numbers on engines did not match. Corrected papers never arrived. The aircraft had been parked in Honolulu for two years and needed work. It continued to deteriorate at Naples and upon inspection major corrosion was discovered that had been hastily covered up. A letter to the FAA dated 03 April 1988 and signed by Louis Khem requested a change of location to Pacific Air Express at Williamson Georgia indicating he maintained ownership and a sale had not taken place.[12] The aircraft remained at Naples and as a condition of the agreement Frank Moss removed the engines as need for use on the other Carvair. He had paid a portion of the purchase to Khem in cash and incurred the expense of ferrying the aircraft to Naples. Mason and Moss became embroiled in a dispute because Mason had not made any payments or contributed to the ferry cost. Naples Airport placed a lien on the aircraft on 24 July 1991 naming the owner as Lee Mason, Great Southern Airways Orlando Florida. Charges owed continued to accumulate until the Naples airport advertised in April 1992 that it would be sold at auction on 15 May 1992 with no avionics and two engines.[13] It was also stipulated it must be removed from the airport. It was transferred to the City of Naples for disposal on 05 June 1992.

Bob McSwiggan purchased N55243 in June for transfer to Custom Air Service that operated two other Carvairs under the same Hondu Carib certificate. He intended to put engines on it to ferry it back to Griffin Georgia.[14] It appeared to be an opportunity to add another Carvair to the Custom Air Service cargo fleet. The aircraft had been parked for years at Naples with two unserviceable engines remaining. The plan to put four new engines on it and fly it back to

N55243 sat for five years at Naples, Florida (KAPF), deteriorating during the ownership dispute between Khem, Moss and Mason. It was eventually confiscated, sold and broken up for spares (courtesy Keith Burton).

the Carvair base at Griffin Georgia never materialized. At this particular time McSwiggan did not have four spare engines. Two of the engines from Carvair N89FA were used to support N83FA. The R-2000 engines at that time were going for $30,000 each, which is a considerable expense for a small operation like Custom Air Service with limited capital. The Naples airport authority was pressing McSwiggan to move the derelict Carvair or it would be confiscated and scraped. In August 1992 the City of Naples Florida Airport Authority made good on their promise and started proceedings to take possession of N55243 for the second time. McSwiggan reluctantly made the decision to salvage it for parts.

The break up began on 11 February 1993, five years after it had arrived at Naples. After careful inspection it was revealed that the corrosion was even more severe than it appeared confirming he had made the right decision. Frank Moss assisted in breaking up the airframe along with aircraft mechanic George Dyess who had performed regular work on the other Carvairs for McSwiggan. Dyess believed at first that they were chopping up a good airplane. Upon inspection he found the corrosion was so severe that he punched a pencil through the skin in several areas of the outer wing.[15] The area where the wing bob-weights were installed was extremely deteriorated and he feared the weights would fall out.

Carvair 17 was still equipped with a DC-8 roller floor system installed by Aero Union. The system was salvaged and transported back to Griffin with the plan to eventually install it in either N89FA or N83FA. There was major corrosion of the wings but the fuselage and other salvage parts could support the Griffin operation. The rudder, gear, center wing, nose door, cargo door frame and roller floor system were stored at Griffin for many years. The nose door is the

The cockpit of Carvair 17 in 1996 with Aero Union titles visible through the faded paint. After the aircraft was broken up at Naples, Florida, the parts were salvaged to Griffin, Georgia, for spares support of Custom Air Service (author's photograph).

original and was never replaced during the operational life of the aircraft. The cockpit was traded to an aviation restoration company located at Griffin and is planned for restoration and display. Currently it still sits beside their hangar at Griffin. When examined closely the Aero Union lettering can still be seen through the faded white paint of the PAE livery.

The registration N55243 was cancelled on 11 January 1996. The remains of the aircraft were still stored at Griffin in July 2000. With only three Carvairs left in existence most of the salvage parts will probably never be used and eventually sold as scrap. HawkAir purchased the nose gear doors in 1997 to replace the doors on Carvair 20 after it suffered a nose gear collapse at Wrangell Alaska on 30 June 1997.

24

Carvair Eighteen

c/n 18340-18, C-54B-5-DO, Iberia/AVIACO

Carvair 18 is the last of four ships converted from C-54B-5-DO airframes and like aircraft 3, 11, and 17 it was originally built at Santa Monica in block one of the 100 produced. This series has two cabin fuel tanks and two extra integral fuel tanks in the outer wings giving the aircraft a fuel capacity of 3,740 U.S. gallons. It has a maximum aircraft weight of 73,000 pounds and is fitted with P & W R-2000-D3 engines. The -D3 engines are rated at 1,200 hp at 5,000 feet (1,524 m) and a takeoff rating of 1,450 hp.

It was built one line number behind C-54 c/n 18339 USAAF 43-17139 that became Carvair three. The fact that these two C-54s with consecutive Douglas construction numbers and USAAF serial numbers became Carvairs three and eighteen is noteworthy and quite a coincidence, since conversion candidates were chosen completely at random. It was delivered to the USAAF 12 July 1944 with serial 43-17140. The two sister ships were separated after being declared redundant by the Air Force. The earlier of the two went to Braniff and this aircraft was purchased by American Airlines. Almost 20 years later they were brought back together when converted to Carvairs. In addition this airframe is only two construction numbers behind the C-54 that was converted to Carvair 17.

When c/n 18340 was surplused in 1944 it was converted to DC-4 civilian standards and

The second Iberia DC-4 was also found by ATEL to be in excellent condition, not requiring retro work before conversion. The short ram-air carburetor scoops, only used on the AVIACO and Aer Lingus aircraft, are visible (via Jennifer M. Gradidge).

sold to American Airlines late in 1945. American received U.S. registration NC90403 and named it "Flagship Phoenix" operating it until 1949.

Both Iberia Carvairs were DC-4s purchased from American Airlines. This DC-4 was registered in Spain as EC-AEO and had the letters applied to the rear fuselage in very large block letters. Iberia named the aircraft "115," which was printed in a circle on the vertical stabilizer. This is more of a fleet identification number and not a name. It continued in Iberia service until 05 November 1964 when it was transferred to ATEL for the second Carvair conversion for subsidiary AVIACO.

The Iberia domestic and European network was flown with the non-pressurized DC-4s and Convair 440s and long haul routes with pressurized Constellations. Early in the Carvair development the carrier reviewed the DC-4 Carvair for use in the European network while at the same time looking to upgrade to pressurized equipment by ordering Caravelles. On 21 November 1960 Iberia began passing the older equipment to associate company AVIACO. In 1964 two Carvairs were ordered by Iberia for AVIACO to begin a combined car-ferry cargo service. The next year Iberia Constellations were transferred to AVIACO. Iberia introduced Caravelles in 1962 assuring the phase out of all non-pressurized aircraft.

Conversion/AVIACO

The DC-4 arrived at Stansted on 05 November 1964. After clearing customs and examination by engineers it was repositioned on 10 November in an area for fuel tank draining where disassembly began. In anticipation of the conversion the nose was transported to Stansted on 08 November two days ahead of the aircraft. Like the previous conversion Iberia/AVIACO ordered it with the relocated control runs and elevated ceiling in the cargo bay.

The military C-54 airframe was fitted with double rear cargo doors. The rear half of the cargo door and hinges were removed and skinned over when converted from C-54 to DC-4 for American. The rear half is not needed in passenger configuration and the gaskets tend to leak or rattle with age causing drafts. As with the previous Iberia Carvair the double cargo doors were not reinstalled, reducing conversion time. It was fitted with the earlier design nose door with only the two landing lights and was never upgraded to the di-electric patch for radar.

Similar to the Aer Lingus ships the DC-4s arrived with engines fitted with the short ram-air carburetor intake scoops. Iberia opted not to convert to the revised non-ram air system with long fairings along the nacelles. Only AVIACO and Aer Lingus Carvairs were fitted with these small carburetor intakes scoops.

The conversion was structurally complete by mid February and flew for the first time on 12 March 1965 with a new Spanish registration EC-AZA. The new empty weight was recorded as 41,224 pounds. There are only seven older airframes used for conversion. Despite it being an older airframe, no retro work was required because of the excellent condition maintained by Iberia. The C of A was issued 15 March 1965 after only 5.25 hours of flight trials and delivered to Iberia/AVIACO on 26 March. It was registered to Iberia although it was painted in AVIACO colors.

Carvair 18 is the second Iberia DC-4 converted for AVIACO and the third Carvair operated by Iberia/AVIACO. G-AOFW was leased to begin car-ferry operations in April 1964 while waiting for delivery of ship 16. AVIACO had grand plans for service in France, Spain and the Balearic Islands, which is evident by the original order for two Carvairs with an option for five more. AVIACO reviewed canceling the order for a second conversion because of declining business, unrealized contracts, and competition with the sea ferries. The contract was reluctantly honored allowing it to be completed and delivered.

Aviation Traders Carvair 12, which was leased to AVIACO during the summer season of 1964, operated car-ferry services between Palma, Barcelona, and Valencia until Carvair 18 was delivered. Carvair-AOFW, which was originally intended for BUAF was placed in storage in November 1964. It had been returned to ATEL pending the expiration of the AVIACO lease.

Both of the Iberia DC-4s converted for AVIACO had three Spanish registrations in a very short time. This DC-4 arrived at ATEL registered to Iberia as EC-AEO with the registration of EC-AXY reserved for it after conversion but was not taken up. AVIACO purchased it from parent company Iberia prior to the return to Spain. It was registered EC-WZA for the ferry flight to Barcelona in March 1965 however, the registration painted on the aircraft was EC-AZA. The registration EC-WZA was used as a call sign only for the ferry flight and never applied to the ship. It was delivered to AVIACO on 26 March 1965 and took up the Spanish registration of EC-AZA in April. The shuffling of registrations and transfer of ownership between the two carriers caused EC-AXY not to be used.

AVIACO added Nimes to its car-ferry services in an exchange agreement with Compagine Air Transport (CAT) but struggled to show a profit with declining bookings until 1968. Eventually Carvair 18 was withdrawn from service although Compagine Air Transport (CAT) operated two Carvairs in some of the same markets during this period with some success.

Compañía Dominicana de Aviación (Dominicana)

Dominicana was formed in 1944 as the national airline of the Dominican Republic. As a small island country it was dependent on carriers of other countries for all air transport needs. Dominican citizens were immigrating in to Miami and considerable numbers to San Juan and Madrid beginning a close aviation relationship with Spain. With large numbers of surplus military transports available, the carrier began service with DC-3s and soon acquired DC-6s. Dominicana suffered its first crash in January 1948 when 28 passengers perished in the crash of a DC-3 at Santo Domingo. This marked the beginning of a very problematic safety record and financial situation, which plagued the carrier until it ceased operations in 1995. During the 1950s company officials believed there was great domestic potential expanding to La Romana, Puerto Plata, and Santiago. The carrier acquired C-46s for passenger and cargo service and embarked on a major expansion plan in the 1960s believing it could be the central point for commercial air operations in the Caribbean. The airline entered the jet age and acquired DC-8s for long haul with DC-9s and 727s for Miami and the Caribbean.

Dominicana was party to an economic aid scheme late in 1968 to provide cargo service to struggling Latin nations. The plan found some favor in Spain with the belief that new export markets could be opened while boosting the growth of poor Latin economies in the Caribbean. Dominicana officials were anxious to exploit the aid scheme and obtain additional aircraft while looking at the cargo potential from Miami, which was much closer than Spain.

Iberia/AVIACO officials were anxious to sell the two Carvairs that had been losing revenue for several years. As part of the Latin American aid scheme, Dominicana entered into a lease with option to purchase agreement believing the two aircraft were better suited for an all cargo roll. The first Carvair leased by Dominicana registered HI-168 operated some service around the Caribbean and between Santo Domingo and Miami.

Dominicana originally leased both AVIACO Carvairs, 16 and 18 and initial cargo profit projections were good. With the potential for success Dominicana exercised the option to purchase in May 1969. It appears that HI-172 was originally obtained in the economic aid relief program on 22 March 1969. Iberia/AVIACO was financially supportive but details are very vague as to who and how the aid program was operated. The purpose of the program was to distribute commodities and freight to the struggling Latin governments in and around the West Indies. As with most poorly planned and funded schemes it soon failed with the details of the operation still unclear.

Dominicana encountered serious economic and morale problems after the crash of Carvair,

HI-172 in the second livery of Dominica NA (National Airlines) at Santo Domingo (MDSD). After the crash of HI-168 the crews became disenchanted and some refused to fly it (courtesy VIP/Nicky Scherrer).

HI-168, at Miami in June 1969 and the crash of a DC-9 at Santo Domingo where 102 perished. The DC-9 crash killed boxing champion Carlos Cruz and his family and the Puerto Rican women's national volleyball team. Even though the publicity brought attention to the Dominicana safety, maintenance, and financial situation, the carrier acquired and briefly operated a 747 to Madrid.

Dominicana continued to operate Carvair 18 until 12 January 1974 despite reports that it was withdrawn from service and parked at Santo Domingo. Crews were disenchanted with the Carvair after the Miami crash and the situation continued to deteriorate with some pilots refusing to fly it. It still appeared complete and serviceable in July 1976 while parked at Santo Domingo. Sometime during this period the livery was changed, possibly in an attempt to return it to service after a re-organization attempt. The title "Dominicana" in gold was changed to "Dominica NA" in blue. The color was changed on the circled logo on the tail and the red and blue stripes were removed from the vertical stabilizer. The registration moved to the top of the stabilizer and a white strip was added below the red stripe the length of the fuselage.

After the Carvair served rather sporadically from 1968–74 combined with the negative pilot reaction, HI-172 was withdrawn from service and remained at Santo Domingo until 1978. During the time it was in storage, on or about 30 September 1976, HI-172 was repainted again for unknown reasons. The Dominica NA titles were removed from the fuselage along with the "D" logo on the tail. The twin red and blue stripe along the windows was extended up the tail. The stripes break and alternate colors about half way up the vertical stabilizer. No reason has been established for the change of livery but there were unidentified purchase inquiries. It was not airworthy and had been stored in a deteriorated condition for approximately two years.

It was surplused for scrap late in 1977 and never flew again. It received a last reprieve when

Above: After being in deteriorating condition for years, Carvair 18 was disassembled in November 1978 and moved to downtown Santo Domingo to become a bar and restaurant (courtesy Stephen Piercey/Peter Marson collection). *Below*: The Carvair was converted to a restaurant-bar at the El Embajador Hotel in Santo Domingo. The display appeared as more of a transportation junkyard than a theme restaurant (courtesy Stephen Piercey/Peter Marson collection).

the wings were removed in November 1978 and the fuselage transported to the Embajador Hotel on Avenue Sarasota in downtown Santo Domingo. It was place beside the hotel with plans of converting it to a theme restaurant. The restaurant never opened and it became a nightclub named "DC-4 Piano Bar." It was painted with a red and blue stripe similar to the last livery with an entrance marquee awning and landscaping. At a later date the display was changed again for a proposed restaurant. It was painted white and combined with grass huts and a number of buses that formed entryways into the aircraft. The engine cowlings were mounted upright on top of the fuselage in a row. Transparent domes were placed in the front opening of the cowlings to form skylights. The outer wings were mounted vertically and outlined in neon lights. The restaurant never opened and it was rumored to have been a disco in February 1984. Actually the last change transformed it from a novel bar to a transportation junkyard. The busses and other clutter took away any chance of identifying it as ever being a flyable aircraft. It was still on display in Santo Domingo in 1995.

Carvair 18 was owned and operated by only two carriers and flown approximately 12 years. The profitability of the operational service of this Carvair was marginal at best. It is unfortunate that it was idle during most of the service life and was allowed to deteriorate. The remains were displayed in the city of Santo Domingo for many years. It is one of the least documented Carvairs with few if any outstanding missions or cargo loads. Despite the mundane operational record, the airframe remained in tact longer than the majority of the fleet with only three ships surviving it.

Dominicana continued on a deteriorating path through the 1980s adding service to Caracas, Bogota, and Quito. By 1987 the carrier was reduced to only a few aircraft. The San Juan route, which sustained the carrier, was dropped and U.S. officials began to restrict the carrier in Miami. By the 1990s Dominicana only operated to Miami with three 727s and ceased operations in 1995. Very little is left of its existence except the Carvair restaurant in Santo Domingo.

25

Carvair Nineteen

c/n 42927-19, DC-4-1009, Ansett-ANA

Carvair 19 is the first of two Carvairs converted from a postwar DC-4-1009 airframe. Ships 19 and 20 are true DC-4s while all other conversions were originally C-54s that were upgraded to DC-4 standards. Only one later model airframe was used for Carvair conversion. This aircraft was originally built at Santa Monica as the 23rd of only 79 postwar civilian DC-4s. All of the DC-4-1009s were built as civilian transports using the large inventory of partially completed C-54 airframes and parts left after the cancellation of military aircraft orders at the end of World War II. The -1009s were designed as passenger transports with a single passenger entry door. There were other design features that increased the aircraft weight to 2600 pounds more than the war time version C-54. Cabin pressurization was available yet none were built with the option.

Scandinavian carrier SILA Svensk Interkontinental Lufttrafik AB purchased the DC-4 from Douglas and took delivery on 17 May 1946. It was registered SE-BBD, and named "Monsun." In 1947 the aircraft was transferred to SAS. A request for transfer of registration was filed with the new name "Stybjorn Viking." The registration was not taken up or name used and the transaction was cancelled. In August 1948 SILA merged with SAS and the registration was transferred to SAS and it was re-named "Sigmund Viking."

It remained in service with SAS until 23 February 1954 when it was sold to Japan Airlines, registered JA6008, and named "Zao." Japan Airlines re-configured it for 98 passengers and operated it for nine and a half years. Ansett-ANA of New South Wales purchased it on 06 August 1963. The Japanese registration was cancelled 07 August 1963 and Australian registration of VH-INJ issued the same day. It departed Tokyo on 09 August commanded by Captain John Adams and assisted by Captains D. Baker, and Ian Strother with navigator W.C. "Bill" Kennedy. The ferry flight staged via Manila and Darwin arriving Melbourne (Essendon) on 12 August.[1] Ansett converted it to a 60-seat configuration before it was ferried Essendon-Sydney on 12 December 1963 beginning service the next day. It was wet leased to Airline of Southern Australia (ASA) from 24 February to 23 March 1964 while the airline was having aircraft overhauled, returning back to Ansett at Adelaide on 24 March.

When the DC-4 was purchased from JAL and received Australian registration VH-INJ it was repainted in Ansett colors of red, white and black. The title of NSW was applied to the tail fin with Airlines of NSW above the windows. It remained in this livery nearly two years and even after Australian National Airways (ANA) was acquired in 1957. Eventually it was repainted in the first Ansett-ANA livery. It was not determined at this time that VH-INJ would be one of the DC-4s converted to Carvair standards. If it had been known it is quite possible it would not have been repainted since it was transferred a short time later to ATEL for Carvair conversion.

Carvair Conversion

The DC-4 VH-INJ departed Melbourne 11 May 1965 for England. The crew consisted of Captains Rod Lapthorne, Kevin Hants, John Coakley, Flight Engineer N. Johnstone, and Navigator E.W. "Pat" Adams. The six-day flight made stops at Perth, Cocos-Keeling Islands, Colombo, Karachi, Cairo, Athens, and Marseilles arriving at Southend on 16 May 1965.[2] After clearing customs and inspection it ferried to Stansted on 17 May becoming the first of the original Ansett order for two all cargo Carvairs. The nose section had been delivered to Stansted on 26 November 1964, nearly seven months before VH-INJ arrived for conversion. The last group of nose units was delivered to Stansted without firm orders. When Carvair production ceased, two nose units were still on hand at Stansted and never used.

When VH-INJ arrived, Carvair 17, which was completed in October 1964, remained unsold and in storage less engines at Stansted. Officials at ATEL suggested that Ansett purchase or lease this aircraft, which could be immediately available instead of converting VH-INJ. This would allow Ansett to continue operating without reducing its cargo fleet. Unfortunately for ATEL, Ansett was restricted from purchasing or leasing Carvair 17 because of a ruling by the Australian Directorate of Civil Aviation. The Ministry was undecided as to whether the Carvair could be flown under the DC-4 certificate or if it would be considered a new aircraft. A directive was issued that any Carvair operated by Australian carriers must be converted from postwar DC-4s, which is 2600 pounds heavier automatically reducing cargo capacity. Some of the weight could be made up with the deletion of the passenger cabin and windows. The ministry also reduced the standard takeoff gross (ATOG) by an additional 1000 pounds to a maximum of 71,000 pounds.

Ansett ordered two cargo conversions under this ruling with an option for a third. If the Carvair was successful it was assumed there would be an eventual requirement of eight aircraft.

VH-INJ was painted in the first Australian livery of Airlines of NSW after purchase from Japan. It was repainted in the Ansett-ANA colors before delivery to Stansted for the first of three conversions for Ansett (courtesy R.N. Smith).

Ansett placed an order for two Carvairs providing the only two postwar DC-4-1009 airframes used in the program. The Australian Directorate of Civil Aviation dispatched an aircraft surveyor to monitor and evaluate the conversion (courtesy Lindsay Wise).

Ansett ordered the "Rolamat" floor system designed to handle seven of the international 88" × 108" freight pallets. Although ordered as a pure freighter without the rear passenger cabin and rear windows, the roller system ended at station X794 leaving the floor area the cabin occupied as space for bulk cargo. The DC-4-1009 series were originally built as civilian passenger transports with a single entry door on the left rear fuselage. Ansett required the double opening rear cargo door for side loading. The car-ferry ships had the double doors installed with the rear half permanently closed and secured with steel plates to utilize the passenger cabin.

The "Rolamat" floor system for Ansett is slightly different from systems installed on previous Carvairs. It is designed primarily for pallets but the tracks are spaced to allow for vehicles to be loaded. To accommodate the larger pallets the forward entry doorframe was squared and widened at the bottom. It is similar to the Interocean ships, which were also notched at the top for loading large vehicles and widened at the bottom. Ansett also required the relocated control runs and high ceilings in the cargo bay behind the cockpit to increase volume. The wing "Bob Weight" feature ATL 98/5150 was also fitted to increase cargo payload by 1000 pounds.

The Australian registration VH-INJ was changed from DC-4 to ATL-98 on 22 June 1965. The Conversion was completed in four months making the first flight as a Carvair on 14 September 1965. In a ceremony the following day at Southend ATEL technical director Bob Batt handed over the logbook to Ansett Captain John Withecombe. The Australian C of A was issued 17 September 1965.

The first Ansett ship was put through nine days of final checks and testing to satisfy Australian authorities. It departed the U.K. on 24 September 1965 for the more than 50-hour flight home to Australia. It was at Stansted on 23 September having shuffled over to Southend a number of times prior to leaving for Australia. The return crew consisted of Captains John Adams,

John Withecombe who accepted the logbook from ATEL director Bob Batt, John Coakley, Navigator W.C. "Bill" Kennedy, Radioman Rob Gale, and Engineer N. Johnstone.[3] The crew was comfortable with the aircraft prompting a decision not to ferry empty but to take on a freight load plus spares for the return flight to Melbourne. The cargo included a replica of a Santos-Dumont Demoiselle aircraft built for the film "Those Magnificent Men in Their Flying Machines." The aircraft was to be used to promote the film in Australia.

The new Carvair departed for Athens continuing across the Middle East to Damascus, Bahrain, and Masirah Island Oman. Because of a break out of hostilities the normal ferry route to Calcutta and Karachi was altered. It continued on to Colombo, Singapore, Denpasar and Darwin arriving home at Melbourne/Essendon on 30 September 1965 after six days en route.

Ansett-ANA

More testing and flight trials were conducted in Australia before VH-INJ operated the first scheduled revenue flight on 25 October on the Sydney — Melbourne — Adelaide route. The aircraft performed well and went into regular cargo service. In June 1966 it made its first visit to Cairns where it was met by the press and a large crowd. By mid 1969 VH-INJ was scheduled for periodic resealing of the fuel tanks and overhaul at Hong Kong Aircraft Engineering Company (HAECO). Early in the program Aviation Traders licensed HAECO to actually produce Carvairs in the event ATEL could not keep up production at Stansted but they were never ordered in great numbers and the problem never occurred. The Carvair made a rare visit to Hong Kong for heavy maintenance in 1969. The ferry under the command of Captain Neville Currey with Captain V. Leinstead, F/O Robinson, and navigators W. C. "Bill" Kennedy and L.

Alan Fraser martials VH-INJ on the first visit to Cairns in regular cargo service. A large crowd and the press assembled for the event. Fraser loaned his camera to Max Harrison to photograph the event (courtesy Max Harrison/Maurice Austin).

P. Kettner, VH-INJ departed Melbourne/Essendon 01 July. The routing took them via Darwin, Jesselton (now Kota Kinabalu) on the northern tip of Borneo and across the South China Sea to Hong Kong.[4]

After overhaul the Carvair returned by the same routing to Melbourne and returned to regular cargo service. On 02 October 1970 while flying near Adelaide it was struck by lightning. Damage was confined to a small area. VH-INJ returned to service after repairs and testing.

Ansett-ANA became Ansett Airlines of Australia (Ansett) on 01 December 1968 but VH-INJ remained painted in the old Ansett-ANA livery until April 1971. The new Alpha livery with cargo titles was quite simple compared to the old Ansett-ANA look. It consisted of a white upper fuselage with bare metal bottom separated by a red and black stripe and Ansett titles. The red tail fin had a stylized "A" triangle logo in a white circle. Two months after repainting, it was struck by lighting a second time on 01 June 1971 near Devonport. The damage was repaired and it operated 14 months in the new livery before the last flight to Melbourne/Tullamarine on 09 June 1972. It was withdrawn from service and placed in storage. A tentative sale to South East Asia Air Transport (SEAAT) was negotiated in April 1973 pending finance. Eventually all three Ansett Carvairs were parked near gate 27 at Tullamarine airport. VH-INJ had its titles removed on 22 June 1973 and AAS markings applied.

Australian Aircraft Sales/Jack M. Garfinkle/South East Asia Transport Service (SEAAT)/Air Cambodge

Australian Aircraft Sales (AAS) of Sydney, a transition owner, purchased VH-INJ from Ansett on 23 June 1973. A Bill of Sale was filed with the FAA on 25 June 1973 indicating an American, Jack M. Garfinkle, purchased Carvair VH-INJ, and an Ansett DC-4 c/n 43071, VH-INL, for $150,000. He applied for U.S. registration N33AC the same day, which was granted on 04 July 1973.[5] Garfinkle's address at that time was recorded as Van Nuys California. The aircraft was purchased for the purpose of leasing out through a subsidiary co-owned by Garfinkle known as South East Asia Air Transport (SEAAT), which was involved in relief and supply flights in war torn Southeast Asia.

South East Asia Air Transport (SEAAT)

Bob Ferguson, Jack Garfinkle and Cecil (Cy) Wroten loosely ran SEAAT. The trio supplied Air Cambodge with crew and aircraft in full livery through SEAAT in an effort to supply Phnom Penh. All the SEAAT aircraft maintained U.S. registrations and crews that were mostly FAA licensed Americans. Later in the operation just prior to the fall of Phnom Penh some of the SEAAT Carvair and DC-4 co-pilots were said to be Taiwanese. At least they held Taiwan certificates but there were many U.S. pilots that had their tickets pulled. SEAAT also became involved in other relief work and some clandestine supply missions in Southeast Asia, primarily Cambodia and to a lesser extent Laos.

A second Bill of Sale dated 25 June 1973 for $100,000 was filed with the FAA for Carvair 19 between Australian Aircraft Sales and Jack M. Garfinkle. It listed his address as Encino California with other documents listing his address as Tarzana California.[6] Ownership is quite hazy during this period with no explanation as to why there are two sets of documents for the same date with different address for Garfinkle. The puzzle can be partially explained by the nature of the business and the political situation in Southeast Asia at the time.

The application for U.S. airworthiness certificate for Carvair 19 was submitted on 03 July 1973. VH-INJ was removed from the Australian Aircraft Register on the same date with the total airframe hours recorded as 51,159. The aircraft was re-weighed on 04 July and empty weight certified at 41,603 pounds, up 299 pounds from last weighing. Jack Garfinkle signed the application for Airworthiness as owner of the aircraft and Cecil Wroten signed as Vice President of Operations for SEAAT.[7] The Airworthiness Certificate was issued the next day, which is quite puzzling. All three of the Ansett ships were built with certain features different than those required by the FAA for U.S. registration. In 1990 when the other two Ansett aircraft were purchased by Hawaii Pacific Air (HPA) the FAA required major modifications to bring them up to U.S. standards. They could not be operated in the U.S. until these standards were met although they previously had U.S. registrations. It has been suggested that the FAA was pressured to turn a blind eye to operations of U.S. civil registered aircraft operated in Southeast Asia.

When the FAA inspected the aircraft operating in Cambodia in 1974; they were shocked at the deplorable condition of some of them and threatened to shut down the entire operation. The U.S. Military attaché in Phnom Penh was instrumental in soothing the FAA and allowing the operations to continue.[8] Because of the volume of cargo required to keep Cambodia supplied and some clandestine activity that was taking place the military was willing to look the other way to maintain the status quo for these cargo operators.

During the 1970s a number of air transport operators were active in the area. Some were government sponsored and others were strictly opportunists who flew for quick profits. Because of the political instability and military presence in the area a clear history of air operations is vague at best. Although not officially sanctioned, some carriers operated under the umbrella of international organizations with missions that were undoubtedly of a military nature. Several U.S. military pilots during the war in Southeast Asia stated that the Carvair was operated for the military. This is highly suspect since no records were located to support the claims. SEAAT did transport anything but these reports are probably a result of seeing N33AC in Saigon on the Air America ramp parked among military aircraft.[9] Also not confirmed and unlikely is the report that Bird Air crews flew the Carvair on some of these missions.

The familiar operators in Southeast Asia, primarily Vietnam were Air America (AAM), Continental Air Services (CASI), Tri 9, and Bird Air. Lesser-known operators supplemented them. Southeast Asia Air Transport (SEAAT) operated a ragged fleet of DC-3s and DC-4s primarily in Cambodia and Laos with some flights to Saigon and Bangkok. Individuals like Jack Garfinkle, Bob Ferguson, William Bird, John Morley, Jim Zeigler, and Stan Booker were buying up old airliners for use in theater.[10] (Booker later surfaced with Nevada Airlines, flying tours over the Grand Canyon. He perished in a Lodestar crash in 1984 while allegedly running drugs in New York.) These individuals set up a number of companies that both operated and leased aircraft. Some of the companies were shells to lease aircraft back to themselves for either financial or liability reasons.

Carvair 19 with U.S. registration N33AC was scheduled for delivery to Singapore on 07 July 1973 staging through Alice Springs and Darwin. The day before departure Australian Customs grounded Garfinkle's Carvair N33AC and DC-4 N32AC believing the aircraft were to be used for the war effort in Cambodia. Authorities refused to release the aircraft until they received assurances that it would not be used in the war. They were parked on a taxiway near gate 27 where they remained nearly a month until the authorities were satisfied. Permission for export was granted on 01 August. The next morning at 10:00 N33AC commanded by ex–Ansett Captain Presgrave departed Melbourne headed northwest to begin operations in Southeast Asia. The Carvair was formally leased to SEAAT on 02 August and sub-leased to Air Cambodge on 07 August 1973.

The "Rolamat" floor system was removed by SINGAS and the standard floor installed with cargo tie-downs before going into service with Air Cambodge. The livery was also altered at Singapore with the vertical stabilizer painted white to match the upper half of the fuselage and the red and black stripe changed to a solid red band. The port side titles were "AIR CAMBODGE" and the starboard was in Cambodian characters. The name "Barb" in script was added below the cockpit windows after it went in service in Cambodia. Barb has never been positively identified but is believed to be the wife or lady friend of "Cy" Wroten. Other SEAAT aircraft were also named such as "Debbie" and it is speculated all the names were wives or girl friends.[11]

Royal Air Cambodge was established in 1956 with the help of the French. Air France provided financial support along with management and technical advice in exchange for a 34 percent stake in the airline. Royal Air Cambodge began service with DC-3s and later operated DC-4 and DC-6 aircraft. One of the DC-4s Air Cambodge operated was N32AC (ex–Ansett) and like Carvair N33AC it was owned by Jack Garfinkle and leased through SEAAT. The Carvair was operated between Phnom Penh's Pochentong airport/Siemreab, Battambang, Poipet, Kampong Saom, Kampong Chhnang, and Bangkok. On occasion it was operated to Vientiane, Laos and Saigon along with DC-4 N32AC. The DC-4 was in Manila in November 1973 loading engines and spares from Stan Bookers spares supply. Air Cambodge also operated routes to Kuala Lumpur and Singapore. Most maintenance was conducted at Phnom Penh by Filipino mechanics.[12] On occasion the Carvair received maintenance at SINGAS in Singapore, which was SEAAT's base of operations. SEAAT also used HAECO contract maintenance at Kai Tak for some of its fleet.

In 1970 Royal Air Cambodge became Air Cambodge after a Coup de' tat changed the Cambodian government. Lon Nol came to power when he deposed Prince Norodhem Sihanouk who was visiting Moscow. Air Cambodge did not lease the Carvair until August 1973. The operation and particulars of Carvair 19 from 1973–75 is quite vague as it flew in and out of Cambodia and around Southeast Asia. Little is recorded about the cargo activity and it can only be speculated as to what was being transported in the Cambodian war zone. On at least one occasion wounded were transported from Kampong Chhnang to Phnom Penh.[13]

The arrival of the Carvair did contribute to some bizarre attempts at cargo handling. When locals at Battambang became aware of the original car-ferry operations in England they

Jack Garfinkle's SEAAT Carvair N33AC shortly after arriving in Cambodia. At this time it was still in good condition and had not been named Barb. It was often flown with the overwing exits open to vent fumes from fuel drums or the smell of pigs (courtesy Bill Ernst/Paul Howard collection).

Locals at Battambang, Cambodia, made a rather ingenious and dangerous attempt at car-ferry operations using trucks and planks as ramps. Some small trucks were transported to Phnom Penh to support air operations, but the car-ferry effort was short-lived (courtesy Paul Rakisits).

attempted a short-lived car-ferry. With no cargo loaders available an ingenious attempt was made to use plank ramps to back a car into a truck. It was then placed near another truck with the front facing the nose of the aircraft. The second truck had a set of ramps placed over the cab to the cargo floor. A rope was tied to the front of the car and the other end pulled by several locals inside the aircraft. They were successful in loading a few cars with this dangerous maneuver but further attempts were considered too risky.[14]

On another occasion Chief SEAAT Pilot Paul Rakisits was scheduled to fly N33AC empty from Battambang back to Phnom Penh. Stan Booker had a Lincoln automobile that he wanted transported. They attempted to lift the car, which was thought to weigh 5000 pounds. Several forklifts of adequate capacity were placed around the car but it continued to tip them. They tried putting weights on the lifts and even had crewmembers stand on the back of the loaders to add weight but nothing could lift it. Finally it had to be left behind. It was later revealed that the doors and inner panels were lined with gold making it so heavy it could not be lifted.[15]

Captain Batt Masterson hired Paul Rakisits to fly for SEAAT. Rakisits had gotten Masterson a job with Interocean in 1962. Other pilots were surprised when Rakisits arrived in Cambodia with a Carvair rating. During 15 months with SEAAT he flew throughout Cambodia from the Phnom Penh base transporting anything. Flying N33AC he picked up wounded at Kampong Chhnang; made three flights to Vientiane Laos for live pigs and a trip to Saigon to pickup badly needed spares from Air America. Engine failures were common while operating in these primitive conditions. Rakisits and F/O Jimmy Jacks were returning from Vientiane when on climb-out they took a hit from ground fire on Number Three engine. The shell hit the number 12 cylinder at the exact moment it was at the top of the stroke. The hit and compression created a double explosion that blew the top of the cylinder off. Rakisits executed engine out procedures and turned back to the field as Jacks feathered it. After three days on the ground waiting for parts or to be rescued they realized they were on their own. It could not be repaired so they made a three-engine ferry back to Phnom Penh. Rakisits did not know at the time that this was his last flight in the Carvair.[16]

Paul Rakisits made three other trips to Saigon for spares in a Convair 440. On the last flight in August 1973 he was stopped at the end of the taxiway by a man in a jeep waving his arms. He had people standing on board but called back to drop the stairs. The man scrambled in and Rakisits started taxiing for the runway with the stairs still down. As soon as the stair lights went out he advanced the throttles. Just as they cleared the treetops they took a dozen hits to the tail. Rakisits held it steady and made for the Mekong Delta where he could fly at ground level.

The man they picked up stood by the radio-rack all the way back and identified himself as Prince Panya of Laos. To show his appreciation for being rescued he offered Rakisits a position training pilots in Laos. He decided that after taking a hit in both the SEAAT Carvair and Convair it was best to move on. He took the position and went to Paris to pickup two pilots and fly them to the Seychelles for training on two DC-4s purchased for Vientiane — Peking service. The Chinese protested and would not allow the route be flown if any Americans were involved in the operation. Rakisits returned to Brussels, until he found work in Saudi Arabia and discovered Bob Ferguson there involved in a rather nefarious deal buying aircraft parts. Ferguson barely escaped with his life. Rakisits flew Convairs for the Saudis and eventually returned to Carvairs in 1983 flying for PAE out of Honolulu.[17]

The situation in Cambodia deteriorated to such a point in 1975 that more flights were needed. The refugee crisis along the Cambodian border with Thailand was supplied by relief flights sponsored by major world charity organizations. Another Carvair, LN-NAA was en route to Pochentong but arrived too late to make any contribution and was grounded at Bangkok in June. The war in Vietnam and Cambodia made Air Cambodge operations difficult. Khmer Rouge forces had secured nearly half of Cambodia by the time the war in Vietnam ended. As they continued to expand, Phnom Penh was being surrounded and slowly strangled. With no place to fly, Air Cambodge was virtually confined to intra–Cambodia service. South East Asia Transport (SEAAT) opted to continue flying for Air Cambodge in the Carvair, DC-4s and Convair 440s.

The possibility of a massive re-supply of the countryside by the Phnom Penh government was thought to be imminent. More than 20 contract carriers descended on Phnom Penh to get a piece of the relief work. The effort was being coordinated out of the U.S. Embassy in Phnom Penh and Singapore. The airlift operation became a tremendous undertaking with extremely dangerous flying conditions but there was no shortage of adventure seeking pilots. The pilots were known as the "Pig Pilots" of Phnom Penh Association or PPPPA.[18] The name was derived because of all the pigs that were transported into Phnom Penh. The shippers were never concerned about weight and would crowd as many pigs as would physically fit on the aircraft. It was impossible to determine the cargo weight so the SEAAT crews came up with ingenious ways to determine the aircraft gross. They would take a pack of cigarettes and place it against the shinny area of the main strut. If it would fit then the aircraft was not considered overweight. The second test was the amount of manifold pressure required to get the aircraft rolling off the blocks. Anything under 30' Hg was considered safe but if it got to 35' hg cargo was removed. The unofficial PPPPA headquarters became the Tropicana Bar in Phnom Penh where stories were compared of the insane conditions and incredibly overweight flights to remote airfields. Some of the fields were actually roads that had to be cleared for the aircraft to land.[19] Whatever the mission the Carvair performed well but was poorly maintained and operating in questionable circumstances.

By April Pochentong airport was under daily rocket attacks. The Carvair was actually doomed at Kampong Saom a week earlier. It had been dispatched to the seaport for critical supplies where it developed an engine problem. The crew opted to fly the 200 km back to Pochentong on three engines. Because of the cargo load and strain on the other three engines N33AC lost another engine before limping into Phnom Penh. It was grounded at Pochentong with two unserviceable engines and no spares available. With the war raging it was not possible to change them. By 10 April the field was under constant mortar attacks with aircraft being destroyed on the ramp. It was now apparent the field would be over run within days if not hours. On 01 April 1975 Lon Nol resigned and his wife escaped in a Caravelle under intense mortar fire. In less than two weeks the country fell.

Khmer Rouge forces over-ran the field on 13 April 1975. Three C-46s, two DC-3s, three Convairs, and a Lodestar made it to Singapore. One of the DC-3s, N64422, took a hit on departure but managed to stay airborne and escape. SEAAT crews abandoned a Convair and got a DC-4 to Bangkok. An Air Cambodge Caravelle XU-JTB and DC-7 N774R also made Bangkok. A DC-4 owned by Cecil Wroten presumed to be N11117 c/n 3077 was also left behind (See Matson, Carvair five). Carvair N33AC didn't make it out and was reported hit with mortar fire. The Air Cambodge DC-4, N32AC, also owned by Jack Garfinkle was among the other aircraft left behind. The majority of airport personnel perished in the attack.[20]

Air Cambodge and SEAAT operations immediately came to an end with the fall of Phnom Penh. An unconfirmed report stated that Carvair N33AC had been damaged by an earlier mortar attack in addition to the engine problems and was last seen on the ground as the last aircraft departed. The extent of the damage will never be known and no reports on the condition or disposition ever surfaced. It was presumed destroyed although a date has never been established. The aircraft did sit for sometime on the military ramp at Phnom Penh and was reported to have been used as a decoy for mortar fire.

All of the SEAAT staff vanished with the fall of Phnom Penh. They were assumed to be in Bangkok or Singapore. In those final months the situation in Southeast Asia intensified and many of the operators did not trust each other. This is supported by a lien on Carvair N33AC that Robert Ferguson filed on 26 August 1974 for $75,000 against Jack Garfinkle, his partner at SEAAT. The lien filed in California stated it was for maintenance services performed between August 1973 and August 1974.[21]

Jack Garfinkle filed a letter with the FAA dated 28 April 1980 requesting the following aircraft be de-registered and to delete his name as owner. The list included Convairs N102KA/N103KA, DC-3s N82AC/N83AC, DC-4 N32AC, and Carvair N33AC. He stated the aircraft were abandoned at Phnom Penh Cambodia.[22] The assumption was the aircraft were destroyed, but later Convair N102KA mysteriously appeared at Bangkok. It remained there derelict for many years. Garfinkle filed a second request with the FAA, which he signed 31 December 1980. It stated the Carvair was destroyed and scrapped. The U.S. registration N33AC was cancelled 13 March 1981.[23]

It is of interest that the FAA sent a letter to the U.S. Embassy in Singapore at the request of James A. Cunningham of Bathesda Maryland. The letter dated 15 June 1978 stated that Carvair N33AC was registered to Jack Garfinkle on 4 July 1973. Cunningham purchased Carvairs 20 and 21 in March 1978 and this appears to be an effort to prove that the U.S. had previously certified Carvairs. Cunningham has never been linked to Carvair 19. The reason for this letter is not established, however Garfinkle and Ferguson spent a lot of time at the Embassy in Singapore and received mail and conducted business there. Cunningham became involved in establishing a country of registration for Carvairs 20 and 21. There is no evidence that he assisted in acquiring registration for Carvair 19 (N33AC).

A captain for a contract Australian cargo operator Interstate Parcel Express Corporation (IPEC) flying for World Vision Relief spotted the forward fuselage of N33AC on the scrap heap at Phnom Penh (PNH) in the mid 1980s. The captain was not familiar with the Carvair and could only identify it as a "Guppy Type." The entire aircraft was not visible and could not be determined if the fuselage was complete.[24] When he was there again in 1992 he looked for it and reported it was no longer visible. Another pilot did report spotting a "Guppy" on the scrap pile in October 1997 that could have been the remains of N33AC but it could not be verified. The report is highly suspect since it is doubtful that it was in existence in the 1990s. It is believed to have been dragged into the brush and used by the homeless for shelter.

J.M. (Jack) Garfinkle

The tension between the three SEAAT partners Garfinkle, Ferguson, and Wroten intensified after the collapse of the Cambodian operation. Ferguson was known as "Mister Southeast Asia" during the SEAAT Phnom Penh operations. He ran his SEAAT interest and sub-companies from his headquarters at the Nana Hotel in Bangkok, a small office in Phnom Penh, and Seletar at Singapore.[25] While N33AC was being researched a Jack Garfinkle was contacted in Tarzana, California. He signed and returned an inquiry denying any knowledge of the operation or owning the aircraft stating he was not "that" Jack Garfinkle. Several years later Jack Garfinkle's signature was found on FAA documents. When compared the signature was the same, verifying that the correct Jack Garfinkle had been located. It was reported that Ferguson was a partner in Rocky's Pizza parlor in Singapore. It has never been confirmed. He later turned up flying Convairs in Saudi Arabia.[26]

26

Carvair Twenty

c/n 42994-20, DC-4-1009, Ansett-ANA

The second Ansett Carvair was converted from one of the 79 postwar civilian DC-4-1009s built at Santa Monica. The cabin doors on the first group were a single type but not like those fitted to the DC-6 with four rounded corners. They were the forward half of the double doors with the rear frame deleted.

The DC-4 was purchased by Norwegian operator Det Norske Luftfartsee Iskap Air Service (DNL) on 24 June 1946, registered as LN-IAE, and named "Olav Viking." When DNL merged with SAS on 01 August 1948 the DC-4 was transferred to the new carrier and remained in operation for eight years and two months. Japan Airlines purchased it on 27 October 1956 assigning registration JA6012 and re-naming it "Mikasa," after a Japanese mountain. JAL operated it for six years and two days then leased it short term to Korean Airlines on 29 October 1962 receiving registration HL4003. Korean Air returned to JAL in 1963 and the previous Japanese registration was restored. JAL declared it redundant in 1964 canceling the registration on 08 February. It was parked pending sale on 21 February 1964 to Ansett for freighter service.

The DC-4 received Australian registration VH-INK before being ferried to Hong Kong where it was idle two months for repainting and conversion to cargo configuration. The ferry to Australia on 22 April was commanded by Captain J. Adams and assisted by Captain S. Telford, First Officers Johnston and C. Musche, Navigator W.C. "Bill" Kennedy and Engineer R. Searle. The flight from Hong Kong staged through Manila, Jesselton (Kota Kinabalu), and Darwin before arriving at Melbourne/Essendon where it began service with Ansett for the last year of operation as a DC-4.[1]

Ansett-ANA/Carvair Conversion

Aviation Traders officials were convinced the Carvair was well suited for the carriers cargo needs and held detail talks with Ansett to demonstrate the benefits of a Carvair cargoliner. Officials at Ansett considered the idea of purchasing conversion kits from ATEL to convert two DC-4s. After careful review officials decided to have two DC-4s converted by ATEL rather than do the conversions in house. Carvair sales were very sluggish at this time with Carvair 17 still unfinished and stored at Stansted. The conversion of two more DC-4s was considered a good sign for the Carvair program and could possibly generate more orders to keep engineers and craftsmen employed.

The DC-4 had been in limited cargo service for a year before being selected for Carvair conversion. VH-INK departed Melbourne/Essendon on 20 June commanded by Captain S. Hayward assisted by Captain Brett. The First Officer was J. Blair and Engineer B. Butterfield. The Navigator was W.C. Kennedy, who was a member of the crew that delivered the DC-4 from Hong Kong when purchased form JAL. It followed the standard route from Perth, Cocos-Keeling

The second DC-4-1009 VH-INK arrived at Stansted in June 1965. It is seen being stripped for conversion with the right outer wing and rudder removed (courtesy Transportraits/Richard Goring).

Islands, and Colombo. The balance of the route was not verified but believed to be Karachi, Cairo, Athens, and Rome (Ciampino) direct to STN. It arrived at Stansted on 25 Jun 1965, just six weeks after the first Ansett DC-4, VH-INJ, was delivered for conversion.[2]

It is the second of only two postwar DC-4-1009s converted to Carvairs and the newest model airframe used in the program. Nineteen conversions were from C-54 type airframes that were upgrade to DC-4 standards. The two Ansett DC-4s VH-INJ and VH-INK were converted simultaneously to ATL-98A standards. Essentially the only difference in the "A" and standard Carvair (ATEL modification 98/5150) is the installation of the 250-pound ballast "Bob-Weight" in each outer wing.

The Australian registration of VH-INK was changed from DC-4 to Carvair on 19 July 1965. The conversion took five months and made the first flight as a Carvair on 27 October 1965. This ship, like Carvair 19, was built as a pure cargoliner without rear passenger cabin, large rear windows, or rear emergency exit. The "Rolamat" floor like the previous ship was slightly different than other Carvairs with the rollers ending at station X794 leaving the cabin floor space for bulk loading. The aircraft also has the re-routed control runs and raised ceiling in the cargo bay. The front cargo door bulkhead is squared with modified opening at the bottom for loading of 108 × 88 inch palletized cargo. The lower alignment pin is moved up some 18 inches from the floor unlike the standard door pin, which is about four inches up the doorsill.

Only five Carvairs were originally built as pure freighters. The Interocean ships four and five were later re-configured to car-ferry standards but the three Ansett Carvairs were built as cargoliners and never re-configured. Since DC-4-1009 series were built with a single entry door on the left rear fuselage the double opening rear cargo doors had to be installed during conversion. VH-INK is the latest airframe used for conversion and is one of only two 1946 model airframes used. It is the last Carvair built at Stansted before the line was shut down.

The new Carvair departed Southend 04 November 1965 for the return trip across the Middle East and Southeast Asia to Australia. Captain S. Hayward was in command of the new Carvair with stops at Athens, Damascus, Bahrain, Karachi, Colombo, Cocos-Keeling Islands, and Perth. After seven days en route it arrived at Melbourne/Essendon on 10 November.[3] Unlike the first Ansett-ANA ship VH-INK was given the name "Kasby I." The livery is the same as it wore as a DC-4 prior to conversion. The only difference being the red stripe behind the DC-4 cockpit window now ran around the nose because of the high cockpit windows on the Carvair.

Ansett-ANA became Ansett Airlines of Australia on 01 December 1968 prompting a new color scheme but VH-INK remained in the old livery for eight more months. On 12 September 1969 it was rolled out in the new "Alpha" color scheme with air cargo titles. It is the second Ansett-ANA Carvair built but the first to be repainted in the new livery. It was another year and a half before the other was repainted. The Carvair was transferred to Ansett Transport Industries on 25 November 1969 and continued in cargo service until 09 March 1973 when it was withdrawn and stored at Melbourne/Tullamarine airport.

Air Australia (Singapore)—Australian Aircraft Sales (AAS)

The redundant Ansett Carvair was purchased 01 January 1974 by Air Australia (Singapore) also known as Australian Aircraft Sales. The Australian registration of VH-INK was retained and the livery altered. The Ansett titles were removed, the vertical stabilizer painted solid red, and a small winged "AAS" Air Australia (Singapore) logo was applied behind the cockpit window. On 11 February 1974 it was tentatively sold to Jakarta based Seulawah-Mandala Air service. The sale collapsed when financing did not materialize and it remained idle.

Seulawah-Mandala Air Service

Indonesian carrier Seulawah-Mandala Air Services of Jakarta entered into a lease-purchase for Carvair 20 on 01 January 1975 from AAS. The carrier applied for Indonesian registration and some records show a registration with a PK-prefix, however it appears that no registration was ever assigned. Because of financing problems VH-INK remained at Melbourne. It was seen at Sydney on 22 March and in May both VH-INK and VH-INJ were parked at Melbourne. VH-INK operated a test flight on 03 July from Melbourne to Broken Hill and back to Melbourne in anticipation of a renewed deal with Seulawah.

The Seulawah titles with Indonesian flag were applied in April to the sides of the forward fuselage. The small winged AAS Air Australia Singapore lettering was retained behind the cockpit windows. It departed Melbourne on 17 July and in August 1975 application was made to the Australian authorities to de-register VH-INK because of export to Indonesia. Financing collapsed in December and it appears the Jakarta based carrier never operated the Carvair during the five months of possession. The lease was cancelled and it remained registered to AAS. It was at Singapore in December of 1975 still wearing Australian registration VH-INK. Conflicting reports indicated it officially returned to Singapore–Seletar on 16 January 1976 and was withdrawn from service. The terms of the lease-purchase agreement were not revealed.

The aircraft had been at Jakarta and technically without a country of registration. It remained there for some time before AAS officials became concerned they may lose both Carvairs. In a daring move both aircraft were unlawfully flown unregistered to Seletar at Singapore. It was selected since it was the closest airport where AAS had a presence, Australia was too far and the aircraft did not have a country of registration. Also there was concern about getting enough fuel to leave Jakarta without arousing suspicion.[4]

Bayu Air/Air Express Australia

Some records have brief notations of a tentative sale to Bayu Air in May 1975. It appears that Bayu Air, also an Indonesian carrier was associated with Seulawah and became involved after the failed 1974 attempt to purchase by Seulawah Air. Sometime in 1976 after the return to AAS at Seletar, it was again reported sold, this time to Air Express Australia. The carrier applied for authority to operate scheduled cargo services. Authorization was never approved or sale completed and the aircraft remained idle at Singapore. (See Carvair 21 for more detail on Air Express Australia and Bayu Air.)

Dwen Airmotive

Dwen Airmotive, an Auckland New Zealand company, advertised the two ex–Ansett Carvairs with considerable spares in October 1976. The advertisement gave the appearance that Dwen was the owner but it was actually AAS. Twenty years later Dwen Airmotive crossed paths with the operator of this same Carvair. Prior to the purchase of Carvair 20 in 1996, the HawkAir Bristol Freighter C-FTPA was severely damaged while landing on the gravel airstrip at the Bronson Creek Mine. The aircraft was broken up and parts transported back to Terrace British Columbia to be used to re-build Bristol C-GYQS. HawkAir purchased a pair of overhauled Hercules engines from Dwen Airmotive and props from Straits Air Freight Express (SAFE) an associate company under the Air Holdings corporate umbrella. SAFE operated Carvair 11, G-APNH, for a few months in 1965 on a trial basis.

With the Bristol out of service and a backlog of freight the HawkAir group began searching for additional suitable aircraft. They settled on purchasing N5459X (ZK-NWA), which is the same Carvair Dwen Airmotive advertised for sale 20-years earlier. As of this writing Dwen remains an aircraft broker, and maintenance provider. Although Dwen offered the two Carvairs for sale in October 1976, AAS owned the aircraft and had possession. They were stored at Singapore–Seletar under the custody and maintenance of Singapore General Aviation Services Pte. Ltd. (SINGAS). Again in August 1977 AAS advertised the two Carvairs with six spare QEC engines and a large amount of spares.

Car Haulaways Limited became interested in the two Carvairs in early 1978 through New Zealand broker Dennis Thompson Limited. The carrier had plans of operating car-ferry service between the north and south islands with the creation of Nationwide Air. Car Haulaways negotiated the purchase directly with AAS but because of registration problems the sale could not be completed. The Carvair transited a series of other owners before certification and registration was established.

James A. Cunningham

The overhaul and registration of the two Carvairs was left to SINGAS, which is the only company at Seletar with facilities and the capability to do the work. Both aircraft had been stored without registration for sometime and were under some suspicion since they had been ferried to Singapore from Jakarta rather covertly without registration. The biggest problem was the aircraft lacked a country of registration. In 1977 SINGAS Engineering Manager Keith Gordon was given the task of acquiring the NZ registration. This involved a considerable process because this was the first time for the type in New Zealand.

Gordon determined that acquiring a New Zealand C of A for a new type that did not have a country of registration would be impossible. After some research, he determined that the FAA had standards for the modification of DC-4/C-54 to Carvair standards. It would have also been impossible to put them back on the Australian register. The answer was to get them registered in the United States, then transfer the registration to the NZ register. A considerable amount of work was required to get them airworthy since neither had flown for quite some time. SIN-GAS had a good relationship with the FAA office in Honolulu so Gordon worked through them to acquire the U.S. registration. British Air Ferries got involved by supplying technical data and BAF owner Mike Keegan even used the certification process as an excuse to take a holiday in Singapore.[5]

There were many problems with the mods that had been approved when manufactured for Australian service. The FAA certification was held up by such absurd items as ashtray installations (SINGAS removed them). Considerable corrosion repair was required along with a number of other items. The FAA protested regularly and did not want to certify them claiming that Carvairs had never been registered in the United States. Gordon reminded them that he could provide the name of the FAA inspector and the Phnom Penh bar (Tropicana) where the U.S. certification was signed for Carvair 19 (N33AC). He also pointed out that two U.S. registered Intercontinental — Interocean Carvairs N9757F and N9758F had operated in the Congo in 1962. The FAA reluctantly relented.[6]

Temporary U.S. ownership was set up to acquire registration for transfer to New Zealand. On 17 March 1978 Carvair 20 was purchased from Air Australia (Singapore) by James A. Cunningham through American Aircraft Sales also AAS. It is quite confusing since Australian Aircraft Sales, Air Australia Singapore, and American Aircraft Sales all use the same winged AAS letterhead and logos. All official documents were routed through American Aircraft Sales while Cunningham applied for and received U.S. "temporary" registration of N54598 on 30 March. This temporary registration indicated a "for profit" purchase and Cunningham had no interest in operating the aircraft. He never physically took delivery since it was strictly a paper work shuffle to receive U.S. registration for which he was compensated.[7]

The purchase was for the Nationwide operation brokered by Dennis Thompson between AAS and Car Haulaways. Cunningham also stated in a letter to the FAA that the aircraft were not flown between 25 August 1975 when de-registered in Australia, and 17 March 1978. The exact date of the unregistered flight from Jakarta to Singapore is not known but was in all probability after 25 August 1975 and possibly early 1976. Cunningham most likely was unaware since he was not a party to the unauthorized flight. As of 10 April 1978 it was at Singapore Seletar for work that progressed until September. The engineers carrying out the overhaul work at SIN-GAS included ex–Air America trained Thai mechanics.

James Cunningham filed an Application for Export Certificate of Airworthiness with the FAA and Department of Transportation dated 25 May 1978. The application stated "Test flight awaiting FAA inspection at Singapore." The purchaser is recorded as Nationwide Air Limited.[8] It is falsely reported that after the inspection and test flight it was flown to Brisbane on 17 September. The date of 30 September was used by SINGAS for STC modification. It was held at Seletar awaiting payment through BCCI bank and did not leave for Brisbane until at least October or later. It did not arrive in Auckland until November, which coincides with the second U.S. ownership of Leonard Lundy.

The registration of ZK-EKY was reserved in New Zealand for the tentative sale to Nationwide but was not taken up. The FAA issued only an export C of A and ferry permit to New Zealand. Since the goal was to get the aircraft to New Zealand to begin the process of NZ registration, the FAA conditions were accepted. The FAA required a placard in the cockpit stating the aircraft had been modified to U.S. standards and the STC number recorded. In the end the inspector was only interested in the placard before issuing the certification. There was no date on the placard so SINGAS Engineering Manager Keith Gordon quickly wrote in his birthday.[9] A New Zealand crew commanded by Nationwide Air Chief Bob Gilbert was dispatched to Singapore to do the test and the ferry flight.

American-trained ex–Air America Thai mechanics overhauled N54598 at Singapore Seletar in 1978 for Nationwide. The FAA required a placard in the cockpit stating the aircraft had been modified to U.S. standards and the STC number recorded. There was no date on the placard, so SINGAS Engineering Manager Keith Gordon quickly wrote in his birthday (courtesy Keith Gordon).

Leonard Lundy

The problems were still not resolved and officials were aware the U.S. registration was acquired only as a transition country. The answer to the problem was quite simple a second U.S. owner was set up. On 13 September 1978, prior to the collapse on paper of the first sale to Nationwide, Philadelphia attorney Leonard Lundy applied for transfer of U.S. registration of N54598. Lundy completed the purchase of the two Carvairs on 28 September and the registration transferred to him on the 29th. Lundy owned both aircraft less than two months. An application was filed with the FAA on 01 November 1978 requesting U.S. de-register N54598 for export to New Zealand.[10]

Credit Finance Air (Auckland NZ)

The Carvair was purchased by Credit Finance Air Limited of Auckland New Zealand in care of Michael L. Curtis as the mortgage holder for Nationwide Air International. The registration of ZK-EKY, which was not taken up by Nationwide, does not appear to have ever been applied to the aircraft even for a ferry flight. On 01 November 1978 Credit Finance was assigned registration ZK-NWA for the Carvair. This registration is believed to have been reserved prior to sorting out the country of registration issue.

Nationwide Air International

Nationwide Air of Wellington was successful in its second attempt to acquire Carvair 20 completing the purchase on 28 November 1978. It was painted in the Nationwide livery by SIN-GAS in the spring of 1978 proving that Cunningham and Lundy never intended to operate the aircraft. It only transited them on paper to acquire the needed U.S. registration of N54598 that it carried for a short time after arriving in New Zealand. The NZ registration of ZK-NWA was acquired by Credit Finance for Nationwide, which was formed in 1978.

The two primary owners were Trevor Farmer a prominent New Zealand business leader and trucking tycoon Matt Thompson. Thompson transported new automobiles throughout NZ from the assembly plants and started an airline to circumvent the government owned Cook Strait ferry monopoly. He reduced the fare from NZ$85 to NZ$70 per car and supplemented loads with palletized bulk freight. Almost all cars were south bound from Wellington to Blenheim with a few cars from the British Leyland plant at Nelson transported north.

The parent company, Car Haulaways, merged the two subsidiary companies of Akarana Air & Air North to form Nationwide. The Carvairs were delivered in November 1978 after the failed attempt in September to purchase them. In 1978 the carrier operated service between Christchurch — Nelson, and Paraparaumu Airport Wellington — Blenheim. Operations were restricted in May 1979 because of a fuel shortage. The fuel supply was embargoed on the pretext that Nationwide did not have an allocation for the following year. The actual reason was political. The carrier was competing against the railroads and Cook Strait Ferry, which were owned by the New Zealand government. In order to eliminate competition the government cut off the fuel supply forcing Nationwide to pay cash for fuel before each flight. Limited flights were operated into June with the airline shutting down on 16 July. The carrier did not have the financial strength to purchase fuel on a cash basis for an indefinite period. The Carvairs were stored at Nelson NZ with an uncertain future.

After the collapse of Nationwide Matt Thompson moved to Africa and became the BAE 146 factory representative. He had been a prominent and influential business leader in New Zealand but soon became despondent over the financial ruin of his company and eventually took his own life.

James Air (James Aviation Ltd.)

The Nationwide Carvairs were acquired from receivership in October 1979 by James Air Limited at a considerable discount considering monies owed for storage and maintenance. It was strictly a speculative transaction with no intention of operating them. James Air is the operating division of James Aviation based at Hamilton New Zealand. The parent company is owned by Ozzie James a pioneer in aviation engineering. James was also a major shareholder in aircraft manufacturer Pacific Aerospace, manufacturer of the Fletcher FU24 and Victa Airtrainer. James Aviation became a pioneer in aerial topdressing (crop dusting) of farmland. He also operated B-99 charters, maintained a fleet of aircraft and helicopters and became the Beechcraft representative.

Air Cargo Panama

In 1980 negotiations were entered into with Air Cargo Panama. The talks do not appear to have progressed very far. No serious proposals were made or contracts signed. It appears that it was no more than a group of investors attempting to operate under Panamanian registration. Without serious commitment it never proceeded past the planning stage. (See Carvair 21 for more details.)

Turner Aviation Limited

Turner Aviation became involved with Carvairs in 1982. The Carvair had been evaluated for Hawaii inter-island cargo service in 1962 when Hawaii State officials and representatives from Hawaiian and Aloha Airlines visited Aviation Traders to review the aircraft. James W. Turner president of both Turner Aviation and Love's Bakery Honolulu recognized the need to create a dependable inter-island air transport network. All bakery products for Hawaii are baked in Honolulu and distributed to all outer islands daily. The two major bakeries were dependent on less than reliable cargo service for distribution at that time. Turner purchased Carvair 20 in December 1982 in a planned attempt to set up a cargo network to transport bakery products within Hawaii. (See ship 21 for details)

Pacific Aerolift

In 1982 Charles Willis Jr. with either the financial backing of James Turner or a lease agreement for two Carvairs formed Pacific Aerolift (PAL). A report on the condition of the Carvairs dated October 1983 states; with a cargo mean density of seven pounds per cubic foot they would

Hawaii operator Pacific Aerolift requested U.S. registration N406JT for ZK-NWA. Retraction checks were conducted before mounting the outer wings and outboard engines during complete overhaul by Pacific Aerospace at Hamilton, New Zealand (courtesy Pacific Aerospace/Murray Dreyer collection).

allow for economic transportation of bread in Hawaii. In addition former Nationwide ship ZK-NWA had the U.S. registration of N406JT reserved with the "JT" indicating Turner as the owner and it was named Ruth I. At the same time Pacific Air Express, another Hawaii cargo carrier founded in June 1982, owned a DC-4 registered N301JT and was negotiating for two different Carvairs. The "JT" suffix indicated Turner had interest in both operations. Whatever the arrangement, none of the principles located would comment. (See ship 21 for details.)

Both Carvairs were inspected at Nelson and discovered to have severe corrosion because of being parked for more than two years with exposure to sea breezes. After a ferry flight from Nelson to Hamilton with the gear down in 1983, Carvair 20 was transferred to James Air, an associate company of Pacific Aerospace, for complete overhaul for PAL. The engines, outer wings, and empennage were removed. The airframe was stripped of all paint and all areas were inspected for damage and corrosion. The vertical fin was removed because of corrosion of the mounting saddles. In addition the nose gear was replaced because of heavy rust. The bolts in the QEC engine packages were so corroded they vibrated loose and wore half way through on the ferry flight.[11] Since the aircraft was being overhauled for work in Hawaii all the de-icing boots, cockpit heater, and anti-ice supply tank for props and windshield were removed. Years later this became a problem for HawkAir when ship 20 operated in Alaska and Canada.

The cargo floor roller system was extended from station X798-X891 to utilize the floor area the passenger cabin occupied on Car-Ferry ships. Up until this time it was used for bulk cargo. The extension increased the capacity to seven positions for pallets. The Carvair is nose heavy and requires compensation CG loading to take advantage of forward cubic capacity. The area designated "Bay 8" could be used for palletized ballast or handle an additional LD3 container or 88 inch pallet up to 3300 pounds. While operating in Hawaii an 88-inch pallet with a block of concrete attached was used for ballast in this aft position.

All systems that were inspected and repaired are too extensive to be listed. An unreported and unapproved doubler was found on the spar section. It had corroded to the point that the piece behind it was completely missing. New spar cap extrusions were made in the U.S. and shipped to New Zealand for all four outer wings on both aircraft. In producing it a mistake was found in the Douglas production drawings that had gone undetected for forty-years. The fuel tank panels were found with more than 70 damaged rivets. They were drilled and re-sized and the tanks cleaned and re-sealed. The overhaul was completed and aircraft test flown in September but never delivered to PAL because of financial problems.

An October 1983 Pacific Aerospace condition report notes the aircraft were converted by ATEL under S.T.C. SA2 which approves combi operation with a 23-seat passenger cabin. It further states, "We have seats for only one aircraft. The moveable bulkheads for above are not available but we do have the manufacturing drawings." These aircraft were built as pure freighters without the passenger cabin windows. Until this time it was not known that Pacific Aerolift or a new owner was considering converting them to Car-Ferry configuration. The report recorded N406JT total airframe time in October 1983 as 53,613 hrs.[12]

Interest began accruing from 04 November 1983 for monies owed, which is noted in a December 1983 report compiled by First Hawaiian Bank titled, "Evaluation of Hawaiian Inter-Island Air Cargo Service Using Two Carvair Aircraft." The report does not indicate whom it was compiled for however, James Turner president of Love's Bakery is noted as owner of the aircraft. After the collapse of Pacific Aerolift the aircraft remained parked at Hamilton through 1987. A variety of brokers attempted to generate interest but because of the limited market and high asking price both aircraft remained idle until Hawaii Pacific Air (HPA) expressed interest. A copy of the Evaluation of Hawaiian Inter-Island Air Cargo Service report previously mentioned was obtained from HPA files after that company ceased operations. The affiliation between Pacific Aerolift and Hawaii Pacific Air is not defined yet some of the same individuals appeared in both operations.

After the collapse of Pacific Aerolift, both Carvairs were repainted in two-tone orange stripes in antic-ipation of resale. They remained stored at Hamilton for six years before being purchased by a second Hawaii operator (courtesy Peter Berry).

ZK-NWA was still painted in the Pacific Aerolift livery in March 1984. It was repainted in the two-tone orange stripes possibly in 1985. In 1986 Captain Jack Ellis of Australia attempted to purchase the aircraft for a proposed cargo operation out of Bacchus Marsh under the name of Gold Crown Aviation. The Victorian country airfield was mostly used for gliding and never suitable for the operation. He was unable to secure authority to operate and plans were dropped. In November 1987 ZK-NWA remained stored in a two-tone orange scheme with severe dirt and mold on fuselage. The New Zealand registration was cancelled 19 May 1990.

Hawaii Pacific Air/Air Cargo Hawaii

Hawaii Pacific Air (HPA) was registered as a Hawaii corporation for Domestic Profit on 24 July 1989 naming George Crabbe Jr. as president. The corporation purchased the two New Zealand Carvairs from Turner Aviation on 08 February 1990. The value of the two Carvairs was set at $1,407,800, which included charges incurred while stored in New Zealand. The cost of the aircraft was set at $1,232,800 not including the interest after 04 November 1983 and NZ$1500 per month engine run cost. The report indicates that Turner Aviation had already expended $500,000 on the two aircraft of which $199,100 was paid to James Aviation. It is also noted that ship 21 was not completed when the report was compiled.[13]

A.P. Fairchild was named Director of Operations for Hawaii Pacific Air and applied to the FAA for U.S. registration on 12 March 1990.[14] Fairchild had been a captain with Interocean from 1961–63 during the United Nations Carvair operation in the Congo. While with Interocean, he flew the DC-4 that became Carvair seven, which eventually came to Pacific Air Express (PAX) where he was Vice President. He was also involved with the forming of Pacific Aerolift (PAL) but his position and title is not known.

The crew at Pacific Aerospace turned out as ZK-NWA makes a final pass before departing Hamilton to new owner Hawaii Pacific Air in Honolulu. It staged through Pago Pago, where an engine was replaced (courtesy Pacific Aerospace/Murray Dreyer collection).

The aircraft ownership changed on 25 June 1990 when N5459X was re-registered to Air Cargo Hawaii Limited. At this point the status of ownership by the group of companies becomes very confusing. Air Cargo Hawaii was registered as a corporation in Hawaii on 02 May 1990 with Harry Snow as President. The registration expired on 30 August 1990. The corporation was registered again on 21 September 1990 with an expiration of 22 January 1991 then cancelled. Records indicate it was registered a third time as Air Cargo Hawaii Limited Partnership on 07 November 1990. Harry Snow was previously president of Air Cargo Enterprises that held interest in Pacific Air Express.[15]

The Air Cargo Hawaii registration noted for the purpose of Inter-island Air Cargo Transportation d/b/a as Hawaii Pacific Air (HPA). The partners included Robert Iwamoto (Owner of Roberts Hawaii Inc tours), Larry Mehau (Hawaiian Businessman), Al Harrington (Entertainer and AH Investments), and Alvin Baron. During this period HPA had considerable problems with the aircraft meeting FAA standards. They were able to obtain a C of A for crew training from 18 November to 15 December 1991. The Carvair remained registered to Air Cargo Hawaii until at least 21 January 1992 when it was recorded as having 53,732.8 total airframe hours.[16]

In early 1992 the Carvair was sold back to the associate company Hawaii Pacific Air. Air Cargo Hawaii Limited Partnership purchased N5459X from HPA again on 15 April 1992. The general partnership had changed since formed in November 1990 now showing George Crabbe Jr. as the Sole General Partner. The inter-island cargo market was changing with the introduction of 737 and 727 cargo liners. The cargo business declined for the Carvair and Air Cargo Hawaii sold it back to HPA in 1993.[17]

Roberts Hawaii/Joseph Mallel

Roberts Hawaii (Robert Iwamoto, once a partner in Hawaii Air Cargo) purchased N5459X from Hawaii Pacific Air on 02 June 1993. Registration was transferred to Roberts Hawaii on 13 September 1993.[18] HPA ceased operations and the Carvair remained parked at the far end of the Diamond Head ramp at Honolulu International for several years. Ownership was changed between members of the partnership and associated companies, which were all under the same corporate umbrella.

Roberts Hawaii is a major tour operator in Hawaii with little need for cargo aircraft. The Honolulu Airport Authority implemented plans for major improvements to the area of the airport where Carvairs 20 and 21 were parked, which necessitated moving them. They were a financial burden to Roberts Hawaii and the partnership and had been allowed to deteriorate. An advertisement in December 1995 with a Honolulu contact of John Wessner got the attention of Aircraft Traders Belgium owned by Joseph Mallel and André Pierre Vigano. Airline Marketing Consultants of Miami purchased the two Carvairs on behalf of Mallel. During the summer of 1996 Mahalo Airlines and South African mechanics were observed working on the two HPA Carvairs still registered N5459X and N5459M. Inquires to their disposition were met with short answers that they were being returned to service.

Mahalo Air, a small inter-island carrier incorporated 29 April 1993, is owned by a group of investors including Robert Iwamoto Jr. The carrier challenged Aloha and Hawaiian Airlines for a piece of the lucrative $30 million a month Hawaii inter-island business. Mr. Iwamoto who holds majority control of the major tour operator in Hawaii was the primary financial force behind Mahalo Air. Officials at Mahalo Air denied any work was being done by its mechanics and security was increasingly tight around the Carvairs with individuals working on the aircraft refusing to comment. Unofficially the story at the Honolulu airport was that neither Roberts Hawaii nor Mahalo Air would be operating them. Instead an unknown "Dutchman" (actually Belgian) had purchased the two aircraft for $125,000 each. Both aircraft had been purchased on 01 August 1996 by Giuseppe (Joseph) Mallel through Airline Marketing Consultants of Miami. A temporary operating certificate was issued on 01 August with an expiration of 27 October 1996.[19] The temporary certificate set the urgency to bring the aircraft back to flying status and explains why the lead mechanic had been offered a $5,000 bonus to get them airworthy by a specified date. Further inquires revealed that the two Carvairs were being purchased for proposed African relief work.

After several failed attempts N5459X commanded by Bruce McSwiggan departed Honolulu in November via Georgia and Lakeland Florida en route to Africa. A South African mechanic was on-board who had fitted it with a 350-gallon auxiliary oil tank in the cargo bay. It was equipped with an electric pump that connected to the existing oil line manifold to supply the nacelle mounted tanks. Two 150-gallon fuel tanks were mounted in the cargo bay and connected to the inboard aux-tank selector.[20] It arrived at Griffin, south of Atlanta, and joined N5459M and the two Custom Air Service Carvairs, N83FA and N89FA. The four Carvairs together marked the final event in the Carvair era and a historical moment in aviation history. Only five Carvairs were still in existence and there had not been four at the same field since the 1970s. These four ships had never been together before. The two HPA Carvairs were the last two built for Ansett. The other two Carvairs, five and nine, began with Interocean and BAF respectively. The remains of Carvair 17, N55243, was also present on the field with the cockpit sitting beside a hangar and the wings, empennage, nose door, and cargo door assembly in a storage area. McSwiggan commented on the historical perspective that five Carvairs, one quarter of the fleet, were at the same place for probably the last time.

After maintenance at Griffin Georgia, Carvair N5459X was scheduled to continue on to Florida. It was en route to relief work in the Democratic Republic of the Congo, formerly Zaire.

An important moment in aviation history occurred at Griffin, Georgia, in November 1997. Four of the last five Carvairs in existence were at the same field for the last time. Never again would there be even two Carvairs together at the same field (courtesy R.D. McSwiggan).

A tentative lease-purchase agreement had been negotiated with Avia Air Charter of South Africa. The aircraft suffered an engine failure after departing Melbourne Florida for San Juan and diverted to Lakeland Florida. A spare engine was on-board but not used awaiting a decision from the new owner. Eventually the sale was switched to the other aircraft N5459M. Bob and Bruce McSwiggan had returned to Griffin awaiting the decision. They flew N5459M, ship 21, from Griffin to Florida where the spares and ferry equipment were transferred to it for the flight to Africa. Carvair N5459X remained at Lakeland.

HawkAir

HawkAir, based at Terrace British Columbia Canada, was awarded a three-year cargo contract to supply the gold mining project at Bronson Creek British Columbia. Officials of HawkAir were in desperate need of a larger aircraft and inspected several DC-4s in Arizona. After learning of the available Carvair, Joseph Mallel was contacted and coordinated with HawkAir to have Canadian transport authorities inspect the aircraft. It was returned to Griffin Georgia where additional maintenance and painting was performed.

Paul Hawkins, Rod Hayward, Dave Menzies, and Don Vienneau founded HawkAir at Terrace British Columbia late in 1993. The group had been with Trans Provincial Airlines (TPA) when the carrier ceased operations in March 1993. They recognized the need for another carrier to take up the void left by Trans Provincial. The original HawkAir plan was to operate Bristol Freighters in western Canada to service some of the same contracts and remote projects.

HawkAir had many problems to overcome and did not have any long-term contracts for security. The carrier struggled with on call work transporting gold ore from remote mine operations and hauling fuel and supplies to the sites. Don Vienneau sold out his share in the first year while HawkAir survived by servicing light aircraft. Finally a contract was secured from Royal Oak Mines to remove the support equipment from its Wind-Craggy project. HawkAir service was visibly better than the other operators. The carrier proved its ability on this contract with the Bristol Freighter generating enough cash flow to keep the company going into 1996.

Ad-hoc work was picked up at Bronson Creek with the Bristol Freighter flying segments that Air North was unable to operate. Air North, based at Whitehorse Yukon, was contracted to Prime Resources for the mine work. It's DC-4, C-FGNI, was having trouble keeping up because of high maintenance. On 14 August 1996 it departed Bronson Creek en route to Wrangell Alaska with a full load of gold ore. After takeoff Number Two engine had a fire warning. It was reported but never verified that the starter failed to disengage, burning up and starting an engine fire. The engine separated from the wing taking out the Number One prop. In an attempt to return to the field the Captain made a hard right causing the port wing to strike a rock outcropping in the rough terrain. Unable to make the field he put it down in the Iskut River near the end of the runway. All three of the crew escaped into waste deep water. The co-pilot and engineer made the 40 foot swim to shore. The Captain was swept into the swift current and disappeared.[21] His body was never recovered. The owner of Air North was so devastated by the loss of his close friend that he gave HawkAir the contract. During the following winter Buffalo Airways operated a DC-4 into the mine. Ninety-day permits were issued to Brooks Fuel and Northern Air Cargo to transport fuel in but gold ore was not transported out.

Prime Resources agreed to a three-year contract with HawkAir to service Homestake Canada's Snip Mine at Bronson Creek specifying DC-4 equipment. HawkAir tried to convince Prime Resources that it would be better with two Bristol Freighters but agreed to acquire a DC-4 to secure the contract. The contract required materials and fuel transported in and the 3000-pound bags of gold ore flown out for shipment to Japan. HawkAir only had one operational Bristol with a second not in flying condition. To operate twelve round trip flights a day from Wrangell Alaska to Bronson Creek Canada would be impossible with one aircraft.

While looking for a DC-4 the L-100 version of the C-130 was reviewed and found to be far too costly. On 24 April 1997 the HawkAir Bristol with a cargo of 1700 gallons of diesel fuel on board was destroyed when the right main gear horizontal strut failed on landing at Bronson Creek. The gear separated at the fitting and the aircraft slid off the runway into the grass wrinkling the fuselage, making repair impossible. Although diesel fuel was leaking the crew escaped safely and there was no fire. A few weeks later the aircraft was broken up and parts salvaged to get the other Bristol airworthy.

At this point it was imperative to locate a replacement aircraft since the second Bristol was not ready. After inspecting DC-4s in Arizona and the HPA aircraft in Florida the decision was made to purchase the Carvair. The operating cost is the same as the DC-4 but it can handle cargo up to 80 feet long and considerable more volume. The experience with the nose loading Bristol proved that straight in loading was essential. Transport of pipe and odd size cargo at these remote mine sites is impossible with the side loading DC-3 and difficult for DC-4 aircraft leaving only the Carvair, which can handle a cargo of 2000 gallons of diesel inbound and five 3000-pound bags of ore outbound.

The aircraft, N5459X, had been tentatively sold to an African operator. Because of an engine failure on the first leg of the ferry flight the other HPA Carvair N5459M was substituted for the African sale.[22] HawkAir purchased Carvair 20 in April 1997 and the U.S. registration N5459X was cancelled on the 14th. The engine failure en route to Africa caused HawkAir to get a DC-4-1009 Conversion rather than C-54, which later proved to be a problem.

Work began immediately to get it certified in Canada leaving HawkAir without an airplane for May and most of June 1997. Paul Hawkins, the name behind HawkAir, mortgaged his home to cover the expense of getting the Carvair on line. Extensive work was done at Griffin Georgia to bring the aircraft up to Canadian standards, which included mandatory prop overhauls and complete inspection and repair of engines and airframe. Many control cables were replaced and the aircraft was repainted. The HawkAir crew headed by Dave Menzies and Rod Hayward accomplished the work along with Sam Knaub.

Frank Moss, owner of Carvair seven, which was based in Alaska and filling in on the mine contract, was called in. The dedicated group worked 16-hour days for six weeks to complete maintenance. During this same period the pilots had to be trained and certified. The problems intensified because of increased scrutiny by the FAA and Canadian transport inspectors after

Carvair five, N83FA, crashed only yards away on the night of 03/04 April 1997 at Griffin. The airport authority and the Griffin community also became concerned about the Carvairs operating from the 3700-foot field.

Most of the work was completed by 25 May 1997 and the FAA inspector was at Griffin to review the aircraft and to check-ride Dave Menzies of HawkAir. The FAA inspector was reviewing the logbook and manuals with Menzies on a picnic table outside the FBO. Attached to the cover of the logbook was a photo of Carvair 20 in Hawaii Pacific Air livery taken by the author in September 1995 at Honolulu. It is still a mystery as how this un-circulated photo became part of the logbook.

By mid-morning the crew was ready for engine run up and high-speed taxi runs. If everything checked out the Carvair now C-GAAH was set to fly in the afternoon. Frank Moss took the left seat and began starting each engine, however Number Three engine would not fire off. After about 45 minutes of frustrating attempts with Menzies and Frank Moss switching places in the right seat, the other engines were shut down. Tension was high as the engine stands were pulled up and the frustrated crew made adjustments to Number Three. Several more attempts were made pausing to allow the starter to cool. With the other engines running after several more attempts Number Three fired off with a big ball of smoke. Most of the spectators cheered and whispered "I told you they could do it." After warming up, the engines smoothed out as Moss began to taxi to the north end of the field.

The FBO radio operator Kerry Jones advised the remaining spectators that if the aircraft checked out they would take off. However, if there were problems the runway was too short to abort. They would continue takeoff and divert to La Grange where the runway was longer. Frank Moss was at the controls and quite familiar with La Grange since he had lived there rebuilding his Carvair N103. As the aircraft taxied past the FBO shack it was obvious this would be a high-speed taxi, the gear pins were still in. It sat on the end of the runway for about 20 minutes before the engines came up to power. The brakes were released and C-GAAH did a high speed run getting light on the gear. Just as it passed mid field the power was cut and it began to slow down approaching the FBO radio shack when the Number One engine popped and got a stack fire. Frank Moss throttled up the engine as he hung his head out the open cockpit window. He had a smile on his face as the fire blew out. The Carvair taxied back to the parking ramp and turned across the skid marks burned into the pavement where Carvair five had crashed seven weeks earlier. As the engines were shut down it was obvious C-GAAH would not fly today. It was already mid afternoon and the decision was made to begin again the next morning.

C-GAAH did fly a few days later. It received a temporary certificate and departed Griffin for the 15-hour flight to Vancouver on 02 June 1997 then ferried to Abbotsford where Chief Pilot Dave Menzies completed his check ride. It was moved to home base at Terrace British Columbia to begin service to the gold mine operation at Bronson Creek. HawkAir received the official Canadian C of A for C-GAAH on 05 June 1997.

The aircraft suffered a lot of problems at first, which could have ruined HawkAir. An incident occurred on 30 June 1997 when the nose wheel collapsed on landing at Wrangell Alaska. After selecting gear down the transit light is illuminated for a set period of time. When the gear is down and locked the three green lights are illuminated. The transit light was only on for a few seconds before the green light came on indicating the gear down and locked. The pilot, who did not have a lot of Carvair experience, was not aware that when the transit light glows for only a few seconds there is a problem. The same condition had occurred on a previous landing when the gear locked at the last second on touch down. The collapse could have been avoided since the strut had been over serviced and would have been discovered on inspection if the previous incident had been reported to maintenance. The aircraft carried on the main gear a considerable distance before it settled on the nose gear, which collapsed, crushing the nose doors and sliding to a stop. The Number Two prop contacted the runway, taking about one inch off

A costly mistake caused the nose gear to collapse on landing at Wrangell, Alaska, on 30 June 1997. The gear malfunctioned on the previous landing, locking on touchdown due to an over inflated strut. The collapse could have been avoided if maintenance had been advised of the earlier incident (courtesy Paul Hawkins collection).

each blade. The main oleo struts were also pumped too high in error, which prevented props one, three and four from contacting the runway. There were also a few dents in the skin but the aircraft sustained no major structural damage.[23]

Bob McSwiggan in Griffin Georgia was contacted for another set of nose doors, since he had possibly the only stock of Carvair parts left in the world. The nose doors supplied to HawkAir were salvaged from N55243 (G-AXAI), Carvair 17, which had been broken up at Naples Florida. The Number Two prop was removed and replaced with the belief there was no other damage. Shortly afterward Number Two engine began to experience vibration and major prop problems. The replacement prop was removed and sent out for overhaul. It was determined that the spider was several thousands out of true proving the nose gear collapse a much more costly incident than originally thought. After the prop was overhauled and replaced the engine problems ceased and the Carvair returned to work at the mine.[24]

During the winter C-GAAH lost a piece of fabric from one of the ailerons. A repair was made but an out of balance condition was soon evident. Several other repairs were attempted but the conditioned persisted. A pair of new ailerons with fresh fabric was located in Vancouver from old SABENA DC-4 parts stock. After they were installed the problem went away.

The next problem encountered was the jamming of the "Rolamat" floor system originally thought to be an asset to the operation. After the first flight of gold ore from Bronson Creek, it was obvious the roller floor could not handle the weight of the 3600-pound pallets. The only alternative was to remove the "Rolamat" system and replace it with a reinforced plate floor adding 2000 pounds to the aircraft dry weight. When the floor was removed it was discovered that being converted from a DC-4-1009 rather than a C-54 airframe and having the "Rolamat" system there were no drilled attachment points or mounts for the major cargo tie-down fittings in the main floor members. HawkAir purchased the salvaged C-54 fuselage of N44909 from

Hawkair's C-GAAH at the Bronson Creek Mine during the winter of 1998. The wings and engines are covered with tarps to ease the snow removal (courtesy Paul Hawkins collection).

Brooks Fuel to obtain the needed fittings, plates, and hardware.[25] If not for the engine failure on N5459X en route to Africa that caused the other HPA Carvair to be substituted for the sale, HawkAir would have acquired N5459M, which was converted from a C-54 that had the attachment points in place.

HawkAir eventually overcame most of the break-in problems, getting more than a month of service from the aircraft between major mechanicals. This is quite an accomplishment considering the aircraft was built in 1965 from a DC-4 airframe manufactured in 1946. Even more impressive is the fact that minimum problems can only be expected in long haul service yet it was performing on very short segments. In the early Carvair days operators experienced engine failure on short segments fostering the opinion the Carvair was unreliable. The Carvair is a high maintenance aircraft but the dedication of the crew at HawkAir proved it is reliable with proper care.

Even with the start up problems, the Carvair still logged 800 hours in the first year of service to the mine flying the 20-minute segments between Wrangell and Bronson Creek. Despite the increased lift of the Carvair the short segments began to take a toll on engines, airframe, and crews. Even the dedicated group at HawkAir had to face the fact that a long-range aircraft was not suited for short segments. C-GAAH blew a cylinder on the inbound segment 05 June 1998 forcing the crew to overnight at the Bronson Creek mine. They spent the night replacing the cylinder returning back to Terrace British Columbia the next afternoon. HawkAir resolve is very impressive transporting more than 21 million pounds of cargo in and out of Bronson Creek with the Carvair and the Bristol. This type of performance on an aged aircraft is a credit to Douglas Aircraft, Aviation Traders and operators like HawkAir that continue to produce revenue from an old workhorse.

On 31 May 1999, after ten years, the Homestake Canada mining operation at the Bronson

Creek Snip Mine ceased operations ending the contract with HawkAir. The Carvair was parked at Terrace in June 1999 leaving the clean up operation to continue at Bronson Creek through October 1999 with the Bristol Freighter. HawkAir officials concluded the Carvair was never suited for the short 20-minute segments serving the mine. A DC-4 was a condition of the contract and completed the job in spite of the high maintenance required to fly the short segments. The smaller Bristol was the ideal aircraft for the job with front-end loading making quick turns at the remote mine easy. The high nose door on the Carvair made it impractical for the remote mine work.

C-GAAH sat idle at Terrace British Columbia after mid June 1999. The aircraft log indicated 52,204 hours on the airframe and 2,129 cycles since Carvair conversion. It had flown up to six 25 minute flight segments a day to the mine and experienced a number of engine problems. Without any new contracts its future was uncertain. Paul Hawkins liked the "Old Girl" but said it became a maintenance hog that really did not pay its way until the last.[26]

The last HawkAir charter and one of the most bizarre missions of any Carvair occurred in 1999. HawkAir was contacted by a group of ranchers in Red Deer, Alberta, Canada in early 1999 who were planning to move 300 Reindeer from Umnak in the Aleutians to Red Deer. HawkAir was quite busy at the time with the clean up at the mine as the contract was about to expire. The ranchers were persistent and the number of animals to be transported changed several times. Finally a contract price was reached and the date was set for September. As part of the deal the ranchers also wanted 13,000 pounds of horse feed transported on the outbound flight to Fort Glen on Umnak Island.[27]

Seven pens were constructed of plywood with snow fencing tops to prevent the reindeer from jumping out. The pens were constructed in such a way that they could be broken down. On 21 September 1999 the pens, 13,000 pounds of horse feed, and 32 five-gallon containers of Shell W100 motor oil were loaded on C-GAAH. After the long day of loading the Carvair departed from Terrace BC at 18:30 for the first leg to Wrangell Alaska.[28]

After departure the gear would not retract. A quick check determined that in the haste to get airborne before dark the gear pins were not removed. The Carvair returned to Terrace and with engines running Steve Hawkins slid down the ladder to remove the pins. Only two of the pins were still installed. One had vibrated out and was never found. Once again the reindeer charter was airborne with Dave Menzies in command. The remainder of the flight went without incident and Captain Menzies made an IFR approach to Wrangell. The crew cleared customs and fueled up as a big low-pressure system was moving over the Gulf of Alaska. The decision was made to remain at Wrangell for the weather to clear.[29]

On 23 September at 09:00 the Carvair departed Wrangell out over the Pacific into clear blue sky. The group arrived at Kodiak at 13:45 and took on 500 gallons of fuel. The Number One engine although operating well was a real oil burner. With a useable capacity of only 20 gallons, it took 18 gallons to top it off with the other engines taking 5 gallons each. After servicing, the Carvair was once again in the air en route to Cold Bay. The stop at the desolate field at Cold Bay only lasted long enough to take on 988 gallons of fuel at the exorbitant $3.00 per gallon. The crew noted that at least fuel was available.[30]

With full tanks the group departed Cold Bay flying out over the Aleutian chain on the last leg crossing Dutch Harbor at 7000 feet as the volcano smoke was visible against the setting sun. Dave Menzies made a perfect landing on the old volcanic cinder runway on Umnak Island. Fort Glen is an old World War II U.S. Air Base that has long been deserted where the wind blows across from the Bearing Sea to the Pacific at 100 mph. There is an old hangar at the base; the remains of a P-61 and runway lights still in place that do not work. Paul Hawkins stated; "This is a place you do not want to stay any longer than is absolutely necessary."[31]

The next morning the horse feed was unloaded and the pens assembled. The pens were placed along the starboard side starting aft of the cockpit ladder leaving a walkway to the rear cargo door. The reindeer were small at about 100–120 pounds each. With no facilities for loading the reindeer were loaded 12 at a time into a high-sided trailer and towed to the aircraft then walked up a narrow ramp and into the pens. They were not enthusiastic about being loaded and resisted with force bucking and jumping with some escaping.

While the loading was taking place Paul and Steve Hawkins removed the cowling from all four R-2000 engines. After 10 hours flying inbound a post flight inspection was required. The aircraft had to be ready to depart as soon as loading was complete and the weather was holding. The Number One engine had two loose spark plugs; Number Two had a loose starter ready to fall off and a broken exhaust stack. The Number Three had leaking oil seals and Number Four had a broken exhaust stack. The Carvair would have to be able to fly all the way to Red Deer, Alberta, with no mechanical problems. Otherwise the reindeer would be stressed and begin to die.[32]

It took five hours to load 102 reindeer. Many escaped from the coral. The excited animals got so hot in the aircraft that the forward door was opened to circulate fresh air. One of the animals got loose and jumped 10 feet out the forward door, limped off the runway into the tall grass and disappeared. Condensation was forming on the gill liner walls of the cargo hold. The smell was beginning to become intolerable and even the cockpit windows began to fog up. The situation was getting tense.

Captain Menzies and Rod Hayward were in the cockpit standing by. The sun was setting and it was getting dark with no runway lights. As the chute was pulled away Captain Menzies started Number Four. Paul Hawkins stated; "We were praying the engines would all start." At 21:30 the Carvair turned into the wind and roared down the cinder runway. The "Ark" flight was airborne to Kodiak. Almost immediately the cargo hold turned to fog. A modified air vent had been installed prior to the flight, however it was not enough as the gill-liners turned to ice. Steve Hawkins' quick thinking saved the day when he removed the port emergency exit window. After a few minutes the ice and fog cleared and the animals settled down.[33]

The old Carvair climbed to 9000 feet and cruised at 200 knots for the three hour flight back to Kodiak, arriving at 00:30. The crew wasted no time taking on 1962 gallons of fuel and topping off the oil tanks. They were back in the air without any more problems. The reindeer were calm as the Carvair landed at Wrangell Alaska at 05:30. The fuel and oil tanks were topped off again and a generator that failed inbound was quickly changed.

After a stressing two hours at Wrangell the flight was back in the air at 07:30 for the final leg to Red Deer, Alberta. The other generator failed between Wrangell and Red Deer forcing the crew to shut down the cockpit heat. It became very uncomfortable for the crew but the reindeer liked the cold. The ranchers expected a 10 percent loss and the insurance rates were based on how many animals arrived at Red Deer alive. The ranchers wanted to throw overboard any animals that died. Paul Hawkins told the ranchers straight out that there would be no throwing out of any reindeer carcasses over the Gulf of Alaska or great northwest.[34]

The "Ark" flight landed at Red Deer at 14:30 to a crowd of 200 people, news media and customs. The unloading went well with no animals lost. The Carvair was not so lucky. A tarpaulin had been spread the full length of the cargo hold and blankets put down to absorb the urine. The animal's sharp hoofs had cut the blankets and tarpaulin creating a smelly mess. The pens, tarpaulin, and blankets were removed and taken away to be burned. C-GAAH ferried back to Terrace, BC where it took two weeks to clean the aircraft. The floor panels had to be removed, the aircraft disinfected, and the control cables cleaned and sprayed with a non-corrosive lubricant. HawkAir officials stated the trip was a success although they only broke even. The Carvair and the R-2000s proved their ruggedness and ability to perform a long-range mission in adverse conditions.

The Carvair remained idle at Terrace while Paul Hawkins and Dave Menzies sought new contracts in the United States and England but nothing materialized. While they were hopeful for a contract in the Caribbean or Middle East Gulf, the decision was made to place the Carvair on the market for $350,000 USD in October 1999. With no offers the price dropped to $289,000 and eventually $189,000. It stood in partial storage at Terrace British Columbia with the engines covered and sealed to keep birds and the elements out. As of November 2000 it remained unsold with a total airframe time of 55,188 hours. The engine and prop times were as follows:

Engine Hours	*Prop Hours*
#1: 799 Hrs.	#1: 1384 Hrs.
#2: 109 Hrs.	#2: 427 Hrs.
#3: 646 Hrs.	#3: 1450 Hrs.
#4: 707 Hrs.	#4: 1051 Hrs.

It was equipped with the following Avionics: Collins VHF Nav 1 51RV1, VHF Nav 2 51RV1, ADF 1 51Y-4, ADF 2 51Y-4, COM 1 618-M2, COM 2 618-M2, DME 860 E-2, XPDR. TDR 90 (Mode C), Sunair H.F Xcvr. ASB 500, Garmin GPS 150 XL, HSI and RMI on Both Panels.

Brooks Fuel

It remained parked for nearly two years until purchased by Brooks Fuel of Fairbanks Alaska on 03 December 2002 for an undisclosed amount. The engine times were recorded as 853/176/700/768. The Number One engine needed rings and was still experiencing excessive oil

Carvair 20 returned to U.S. registration with Brooks Fuel of Fairbanks in March 2003. The new unpainted rudder clearly shows the difference from a standard DC-4 unit (courtesy Keith Burton).

burn. The aircraft was brought back to flying status and ferried to Fairbanks in December. Brooks Fuel planned to return it to cargo service for work in remote Alaska. It was de-registered in Canada on 27 December 2002 and remained idle without registration for nearly a month. Brooks Fuel requested U.S. registration N898AT on 17 January 2003. The FAA records listed the hours on 04 August 2004 as 55,239.7 indicating it had only been flown 51 hours in four years.[35]

Roger W. Brooks founded Fairbanks based Brooks Fuel in 1986. After gaining considerable experience with DC-3s at Aero Transit in Idaho and manager of the DC-3 operation at Frontier Flying Service in Alaska, Brooks began his own air service with a single DC-3. Under FAR Part 91 regulations a carrier is restricted to transporting its own cargo. In essence the company buys and sells fuel and the aircraft are a means of transporting it to the remote areas of Alaska as well as military sites. Over the years Brooks has built up a fleet of two and four engine Douglas Propliners for the hazardous work. In 1997 Brooks also added FAR 125 cargo contract carrier status to service certain contracts.

The paint on the Carvair, which was hastily applied at Griffin Georgia in 1997, had long peeled and flaked off in large areas. After considerable work in 2004 it was repainted white with a wide blue stripe before returning to service in Alaska. It remained active on an intermittent basis until November 2006 when it was parked for the winter.

Crash of N898AT (NTSB ID ANC07LAD40)

It was removed from storage in May 2007 and work began to bring it back up to active status. Roger Brooks personally took it up for a 20 minute test flight on 29 May 2007. It performed well and was scheduled for return to service the next day.

Carvair 20, N898AT, on the first revenue flight after returning to service from winter storage, was destroyed near McGrath, Alaska, on 30 May 2007. The nose section broke off forward of the wing and rolled over after a wind sheer caused the aircraft to strike the threshold and cartwheel off the runway (courtesy Brian Martin).

On the morning of 30 May Roger Brooks and F/O Jonathan Hathaway departed Fairbanks in N898AT for the 269-mile flight to McGrath and additional 28 miles to the Nixon Fork Mine. The field, elevation 1510 feet (640m) FAA code AK40, is 4200 feet long and extremely treacherous with large boulders just below the threshold. Because of a hill on the south end, all landings are approached from the north regardless of the unpredictable wind conditions. The sky was clear at the mine with some convective cloud formations. Just before noon the Captain made an approach from the north with a 5–11 mph tailwind as he tried to fly as slow as possible. Just as he approached the threshold the aircraft was caught in a thermal column. The adjacent downdraft around the column caused the right main to impact the ground with tremendous force at the runway threshold. The right wing broke off and burst into flames on the runway as aircraft cartwheeled. The rest of the aircraft came to rest on the west side of the runway in small trees, brush, and rocks facing in the direction of approach. The bottom of the fuselage was ripped open from the nose gear doors to the center-wing, breaking off and rolling on to the port side. The rear half of the fuselage remained upright, breaking off just behind the wing. The cargo of fuel for the mine was contained in bladders. Fortunately they were not punctured and there was no fire allowing the crew to escape without serious injury.[36]

The aircraft was totally destroyed, leaving only two Carvairs in existence; one active in Texas and the other stored in South Africa.

27

Carvair Twenty-One

c/n 27314-21, C-54E-5-DO, Ansett-ANA

The airframe of the last Carvair was built at Santa Monica as a C-54E-5-DO c/n 27314 and delivered to the Air Force 07 April 1945 with military serial 44-9088. It saw very little military service and was declared redundant on 30 August 1945. It was transferred to the Reconstruction Finance Corporation for disposal. The total airframe time was logged at 553.22 hours on 15 September 1945. Douglas Aircraft purchased it on 14 November 1945 for modification from C-54 to DC-4 commercial transport. The Douglas records indicate the registration of NX 88881.

It was fitted with Pratt & Whitney R-2000-11 engines, which were removed by Pacific Airmotive at Burbank. Two were overhauled and converted from -11 standards to -9 and two new P & W R-2000-9 were installed on 01 May 1946. It was completed and certified as the 13th conversion from C-54 to DC-4 on 24 September 1946.

Pan American Airways purchased it for $100,000 during DC-4 conversion requesting transfer of the registration, which was recorded as NC88881 on 26 December 1945. It was christened "Clipper Kit Carson" beginning commercial service. Pan American Airways became Pan American World Airways on 05 January 1950. By June 1954 the Skymaster had logged 24,742 hours continuing in service until early 1958. During the years with Pan Am the registration changed to N88881 and it was re-named twice, first "Clipper Golden West" and finally "Clipper Red Rover."

It was sold to Japan Airlines on 10 February 1958, and assigned registration JA6015, and re-named "Amagi" on 17 February. The U.S. registration was cancelled on 27 February. Japan Airlines operated it in the Pacific until 18 March 1965 when it was sold to Ansett-ANA. It was re-registered in Australia as VH-INM and scheduled to be re-configured for cargo service.

Ansett had dispatched a crew to Hong Kong to pickup another DC-4 VH-ANF (c/n 10302), which was being overhauled by HAECO. The crew arrived on Cathay Pacific but because of an engineers strike at HAECO the work on that DC-4 was not completed. Ansett instructed Captain A. Lovell to take a commercial flight from Hong Kong to Tokyo to conduct test flights at JAL on VH-INM. After the test flights were completed to satisfaction Captain Lovell returned to Hong Kong. The ferry crew consisted of Captains Stanley Telford, Adrain Groser, Bart Chapman, Engineer C. Burns, and Navigator W. C. Kennedy. Telford and Kennedy had also ferried the DC-4 that became Carvair 20. VH-INM departed Tokyo on 23 March 1965 staging through Manila and Darwin arriving at Melbourne on 25 March. The Japanese registration was cancelled on 29 March. After only a few months with Ansett the DC-4 was leased out to MacRobertson Miller Airlines (MAA) through the end of 1965.[1]

Carvair Conversion

VH-INM remained in Ansett-ANA cargoliner service for exactly three years before it was selected for Carvair conversion. The Australian Directorate of Civil Aviation (DCA) had previously specified that only DC-4 post-war aircraft could be modified, as were the two previous Ansett Carvairs. This was a rather questionable requirement since Ansett was operating pre-war C-54/DC-4 freighters. The authorities reconsidered and allowed this DC-4 to be modified provided the conversion included any upgrading to DC-4 specifications that the airframe did not have.

VH-INM departed Melbourne/Essendon airport 22 February 1968 for the 57 hour flight to Southend for conversion. The ferry crew consisted of Captain J. Coakley in command, Captains N. Currey, W. Oldcastle, and Engineer P. Kettner. The Navigator was W. C. Kennedy, who held the same position on the previous DC-4 ferry to Aviation Traders. He was also the navigator on the delivery flight of this aircraft when purchased from Japan Airlines. The routing took the DC-4 from Melbourne to Alice Springs, Darwin, Denpasar, Singapore, Bangkok, Calcutta, and Karachi where the exhausted crew took a rest day. The ferry continued on to Bahrain, Damascus, Athens, and Marseilles arriving at Southend 01 March 1968.[2]

VH-INM is the third and final conversion for Ansett, the third ship built entirely at Southend and the last production Carvair. Two DC-4s were converted to Carvair standards for Ansett in 1965. Ansett placed the order for a third Carvair late in 1967. Cargo bookings were strong and projections of growth required additional lift as a stopgap measure until newer aircraft could be acquired.

Demand for the Carvair waned by 1967 and the Stansted assembly line was shut down. The last Carvair almost appears to have been an after thought. Three unused nose sections were in storage at Stansted along with Carvair 17, which had been completed and never sold. All three nose sections originally built at Southend had been transported to the assembly line at Stansted. Now they were transported back to Southend. This is the only Carvair to have the nose assembly travel from Southend to Stansted and back without the rest of the aircraft. VH-INM is one of only two Carvairs converted from a C-54E, which is the last series of wartime Skymasters. The only two other aircraft converted from later model airframes were the two previous Ansett ships, which were postwar DC-4-1009s.

VH-INM arrived at Southend for conversion on 29 February 1968; two and a half years after the first Ansett Carvair flew. The first two Carvairs built, G-ANYB and G-ARSD were soon to be retired and two other Carvairs had crashed. VH-INM is the final Carvair built but not the last ship completed. Carvair 17, which began conversion in 1964, remained in storage at Stansted incomplete. Aviation Traders officials encouraged Ansett to accept Carvair 17 to reduce delivery time and expedited sales. However, Australian aviation authorities remained firm. Originally it was required that any Carvair for Australian conversion must be from post World War II airframes. After the contract was negotiated with ATEL, Australian aviation authorities relented allowing a pre-war C-54 model to be used but other conversions such as Carvair 17 could not be substituted.

The DC-4 nose was removed on 13 March. Like the previous cargo conversions the large rear cabin windows were not required and it was built with the raised ceiling in the cargo bay behind the cockpit. It was equipped with the "Rolamat" floor system and squared 88-inch doorsill to accept standard international pallets. Actual inspection of the aircraft revealed that the floor system in this ship is different from the previous two Ansett conversions. The pallet lock points are in different positions and there are guide plates a few feet back from the front entry to prevent damage to the inner walls of the cargo deck. Specifications do not denote if this is an upgrade of the previous system or a one-of-a-kind for the final Carvair.

The DC-4 was removed from the Australian register on 16 April 1968 pending completion. It flew for the first time as a Carvair on 12 July 1968 and the registration VH-INM recorded as an ATL-98A on 16 July. The Carvair conversion was now a proven aircraft that quickly completed all check flights for certification. It departed Southend on 19 July 1968 for return to Australia staging through Athens and Damascus where a cockpit window cracked and blew out.

This was unusual since the Carvair is not pressurized. A new window was installed and delivery continued to Bahrain, Karachi, Calcutta, and Bangkok where the fire warning system malfunctioned. After repair and testing the flight continued to Singapore, Denpasar, and Darwin. After six days en route the exhausted crew consisting of Captains D. Elford, G. Blackhouse, R. Stacey, Engineer M. Draper and Navigator W. C. Kennedy arrived on 26 July 1968 at Essendon airport Melbourne.[3] VH-INM was named "Kasby II," given an acceptance inspection, and trials before going into cargo service. The first revenue flight operated 01 August 1968 from Melbourne to Devonport.

Ansett

On 01 December 1968, Five months after delivery to Ansett, the company reorganized and was renamed Ansett Airlines of Australia. On 25 November 1969 the registration was transferred to Ansett Transport Industries. The new company name brought a new livery but VH-INM had been in service for only a short time and remained in the old colors. In January 1971 it was repainted in the new "Alpha" livery with cargo titles. It returned to service and was struck by lightning on 12 May 1972 while on a cargo flight 85 miles west of Melbourne. It was grounded pending inspection and repair. The damaged was confined to the High Frequency antenna and surrounding area, which was replaced. After testing it was certified airworthy and returned to service.

Ansett began phasing out the Carvair for newer equipment by 1973. Only the last two ships were in service since ship 19 had been withdrawn in June 1972 and was in storage at Tullamarine airport Melbourne. VH-INM was withdrawn on 01 February 1973 and placed in storage beside VH-INJ leaving only ship 20 in service for another month.

VH-INM was repainted in the second Ansett Alpha livery in January 1971. Two years later it was withdrawn from service (courtesy R.N. Smith collection).

Australian Aircraft Sales (AAS) Air Australia Singapore/ Bayu Indonesian ALS (Bayu Air)/Seulawah-Mandala

Australian Aircraft Sales purchased Carvair 21 from Ansett 01 January 1974 after 11 months in storage at Melbourne — Tullamarine. It was returned to airworthy status and test flown 15 January 1974 in anticipation of a tentative February lease-sale to Indonesian based Seulawah — Mandala of Jakarta. Financing did not materialize and the sale collapsed. With no other buyers the aircraft was returned to storage where it remained until June 1975.

The sale to Seulawah was re-kindled in May 1975 when Bayu Indonesian ALS commonly known as Bayu Air surfaced as backing Seulawah's attempt to purchase Carvairs. Bayu Air held preliminary talks with Seulawah to lease a Carvair in 1974. Bayu was based in Jakarta but favored Singapore for the cargo operation. Seulawah planned to lease-purchase the two Carvairs then sub-lease one of them to Bayu Air. The lease was based on Bayu obtaining authority to operate throughout Southeast Asia as well as within Indonesia.

Seulawah continued the charade for sometime while hard negotiating for a better price, pressing the point that rather than the aircraft sit idle they should be allowed to lease or purchase at an extremely discounted rate. On 22 June 1975 Australian Aircraft Sales removed it from storage for a test flight in anticipation of a lease to Seulawah-Mandala Air Services. Bayu Air pulled its support when route authority was delayed causing Seulawah financing to collapse halting the purchase in December 1975. Bayu Air was eventually granted route authority in 1976.

During this same period Seulawah also had dealings with individuals running air operations in the war ravaged areas of Southeast Asia. Jack Garfinkle co-owner of Southeast Asia Air Transport (SEAAT) purchased Ansett Carvair 19 in 1973 to operate in Cambodia and had serious problems getting out of Australia. Seulawah's first attempted to purchase Carvairs 20 and

After the purchase by Australian Aircraft Sales in January 1974, the Ansett titles were removed and small winged AAS letters were applied behind the cockpit window. Seulawah titles were added in anticipation of a lease-sale agreement (courtesy VIP/Nicky Scherrer).

21 was late in 1974. In April 1973 Seulawah sold a Convair 440 c/n 187, N18837 to Tri-9 Corporation operated by John Morley. That aircraft was later destroyed at Phnom Penh along with Carvair 19. Jack Garfinkle's (SEAAT) and John Morley's (Tri-9) operations were closely associated. Garfinkle's SEAAT also purchased several Convair 340s from Saudi Arabia and later sold them to American Aircraft Sales (AAS). That same company handled James Cunningham's purchase of Carvairs 20 and 21. (See Carvair 19 for details on SEAAT)

Of further coincidence but not directly related is the 1978 Bayu Air purchase of two CL-44s in the U.K. One of the CL-44s, c/n 19, was owned by Transmeridian Air Cargo and had been leased to associate company British Air Ferries, the original Carvair operator. Both carriers were owned by T.D. "Mike" Keegan. It is also of note that Bayu purchased DC-6s from French carrier SFAir the carrier that owned Carvair 17 until 1979.

Seulawah and Mandala Air merged in 1972 and made the first attempt to purchase Carvairs in February 1974. The Indonesian passenger and cargo operator based at Jakarta, eventually leased the two ex–Ansett Carvairs in July 1975. Carvair 21 was repainted with Seulawah titles and Indonesian flag by April 1975. The Australian Aircraft Sales (AAS) letters remained behind the cockpit window. The PK-prefix Indonesian registration was applied for but never verified. The aircraft remained under Australian registration VH-INM for the short time Seulawah attempted to operate.

The lease agreement with Seulawah Air was scheduled to be completed 07 July 1975 but not finalized until 17 July. Carvair 21 departed Melbourne the same day to Jakarta staging through Alice Springs, Darwin and Denpasar. The Australian registration was cancelled on 28 August 1975. In December less than six months after leasing VH-INM Seulawah-Mandala financing collapsed and the aircraft was withdrawn from service and remained parked at Jakarta. The Carvair reverted back to Australian Aircraft Sales (AAS) without a country of registration.

AAS officials feared both Carvairs would be confiscated by Indonesian authorities or possibly commandeered and moved to an unknown location. The Australian registration had been cancelled and would have been impossible to reinstate and the Indonesian register was never approved. AAS opted to recover the aircraft. In a daring clandestine operation the Carvairs were flown without registration from Jakarta to Singapore under questionable circumstances. They were placed in storage at Seletar about 16 January 1976.[4] This date is questionable since a letter to the FAA from James A Cunningham stated the aircraft had not flown from 1975 to August 1978. The actual date AAS covertly recovered the aircraft from Jakarta to Singapore will possibly never be verified. Cunningham was probably not aware, since the purchase was to establish U.S. registration and the location of the aircraft was not important.

Air Express Australia (AEA)

Air Express, a small Bristol 170 cargo operator, reviewed expansion plans and the purchase of the ex–Ansett Carvairs. The plan called for increased scheduled cargo service. AEA was poorly financed and the plan appears to be more wishful thinking than a legitimate ability to operate the two aircraft. Australian authorities did not approve the plan and the proposed purchase was cancelled. Air Express eventually acquired two of the last QANTAS DC-4s. Both of the DC-4s, VH-EDA, "Tasmanian Enterprise," and VH-EDN "Tasmanian Tiger" were purchased 20 August 1977. The Air Express Australia operation was continually plagued with financial problems and was out of business by 1979.

The two DC-4s were later acquired by Calm Air, the Canadian carrier that also considered the Carvairs and attempted to purchase Gifford Aviation and Carvair five. This is another example of how DC-4 and Carvair ownership seems to travel in the same circle of carriers and individuals.

Dwen Airmotive

The Auckland NZ company advertised two Carvairs with considerable spares for sale in October 1976. (See Carvair 20 for more Details on Dwen Airmotive) AAS advertised the same two Carvairs VH-INK and VH-INM with six spare QEC engines and a large amount of spares in August 1977.

James A. Cunningham

James A Cunningham purchased Carvair 21 on 17 March 1978 and applied for U.S. registration on 30 March. He received U.S. registration N54596 and applied to Australian authorities to export the aircraft. Records indicate it had a total of 53,931.20 hours with engines one and four as R-2000-2SD13G and two and three as -7M2.[5] In April 1978 SINGAS at Singapore–Seletar was contracted to overhaul Carvair 21 and repaint it in Nationwide colors. The U.S. registration was applied to the aircraft, however Cunningham was a transition owner. The aircraft were only registered in the U.S. as a way of establishing a country of origin to receive the New Zealand C of A. It was impossible to re-register it in Australia and it had flown from Jakarta without registration.

The wing "Bob-Weights" were removed during overhaul at Seletar changing it from an ATL-98A to standard ATL-98 with a reduced ZFW of 60,700 pounds. This was done because it could not be confirmed what model wings had been fitted to the aircraft.[6] It was signed off on 17 October 1978 confirming that it did not depart Singapore in September as reported. The FAA would only issue an export C of A and ferry permit to NZ. Both ships departed Singapore in November staging Darwin-Brisbane-Auckland to begin service with Nationwide Air.

The last two ex–Ansett Carvairs were painted in the Nationwide livery at Singapore in the spring of 1978. Two U.S. owners were set up to establish a country of origin to obtain a type certificate in New Zealand (courtesy Dennis Goodin/author's collection).

L. Leonard Lundy

The sale was actually between Australian Aircraft Sales (AAS) also under the name of Air Australia Singapore and Nationwide with Cunningham as an interim owner to establish U.S. registration. In order to avoid any doubt and satisfactorily establish U.S. ownership the aircraft was sold on paper to L. Leonard Lundy, a Philadelphia attorney. Lundy purchased the Carvairs, on 14 September 1978 and the U.S. registration of N54596 transferred on the 29th. United States registration was firmly established and accepted by New Zealand authorities in order to obtain a full C of A. (This is covered in more detail on Carvair 20) Lundy then sold Carvair 21 along with the other aircraft to Credit Finance Air Limited of New Zealand. He requested the U.S. de-register N54596 on 29 November 1978.[7] Later when this same Carvair was sold to James Turner for Pacific Aerolift major modifications were required to re-obtain U.S. registration.

Credit Finance Air Limited

The Carvair was purchased from Lundy by Credit Finance Air Limited of Auckland New Zealand, c/o Michael L. Curtis, as the mortgage holder for Nationwide Air International. The registration of ZK-EKZ originally requested by Nationwide was not taken up and does not appear to have ever been applied to the aircraft even for a ferry flight. On 01 November 1978 Credit Finance was assigned New Zealand registration ZK-NWB. Actually Car Haulaways had made its intent to purchase the aircraft from Australian Aircraft Sales in early 1978. All the interim owners were set up in order to establish a country of registration to obtain the New Zealand C of A.

Nationwide Air

Nationwide based in Wellington was founded in 1978 as a merger between Akarana Air & Air North. Both were divisions of parent company Car Haulaways. New Zealand registration ZK-EKZ was reserved by Nationwide in September 1978 in anticipation of the purchase. Because of problems in obtaining NZ registration a U.S. registry was acquired. Nationwide requested and obtained the second New Zealand registration once the problems were cleared and the C of A was obtained.

The Carvair arrived at Nationwide carrying the U.S. registration of N54596 originally assigned to James A. Cunningham. The new registration of ZK-NWB was not applied until 28 November 1978 when the Nationwide purchase was final. The carrier operated only a few months before the state monopoly of ferries and railroads created artificial fuel shortages in April 1979. Nationwide was forced to pay cash for fuel before each flight. In June the Number One engine failed on approach to Christchurch. The next day it returned to Nelson on three engines with a cargo of metal castings used for ballast. The carrier was unable to secure a replacement engine and operations were suspended on 16 July. Nationwide was not financially capable of maintaining the two Carvairs and purchasing fuel on a cash basis. All attempts to resurrect the carrier were suspended in September 1979 and it was stored at Nelson Airport N.Z. (See Carvair 20 for detail of Nationwide demise)

James Air (James Aviation Limited)

James Aviation purchased Carvair 21 from the receiver in October 1979 and the registration was transferred to associate company and engineering division James Air in February 1980. This well established company owned by aviation legend Ozzie James operated a fleet of charter aircraft and helicopters, imported corporate aircraft, and maintained the Beechcraft dealership. (See Carvair 20 for more detail) James Aviation and Pacific Aerospace routinely combined major engineering projects, as is the case with the overhaul of ZK-NWB. Ozzie James was also a major shareholder in Pacific Aerospace where the aircraft was rebuilt. Records indicate that Turner Aviation purchased the aircraft in 1982 and $199,100 of $500,000 advance was paid to James Air on the two Carvairs.[8]

Air Cargo Panama

During the summer of 1980 James Aviation was approached by a group of investors said to be Air Cargo Panama. Negotiations were entered into and a sale was reported forthcoming. The talks did not produce a contract and no sale was recorded. It appears to be no more than an attempt to operate under Panamanian registration. Details as to who the individuals backing Air Cargo Panama were are not known. It is an interesting coincidence that an attempt by Charles Willis Jr. to purchase other Carvairs for work in Panama occurred between 1981 and 1983. It is possible this was a first attempt by Willis, who eventually formed Pacific Aerolift, to start a Carvair operation. Willis is a bit of an enigma in the Carvair story. He first surfaced in documents relating to Carvair five (N83FA). Ruth May wrote in her personal notes on 07 June 1983 that Willis told her he had a contract in the amount of $350,000 to operate to Howard Air Force Base Panama Canal Zone.[9]

Turner Aviation Limited/Pacific Aerolift

Turner Aviation became involved with Carvairs in 1982. The State of Hawaii actually evaluated Carvairs for inter-island cargo service when they were introduced in 1962. Hawaii State officials and representatives from Hawaiian and Aloha Airlines visited Aviation Traders to review the aircraft. James W. Turner president of Love's Bakery Honolulu recognized the need to create a dependable inter-island air transport network. All bakery products for Hawaii are baked in Honolulu and distributed to all outer islands daily. The two major bakeries were dependent on less than reliable cargo service for distribution at that time. It appears that Turner formed Turner Aviation Limited to exercise some control of bread transport and distribution. Turner Aviation did not actually operate aircraft but acquired them for use by cargo operators.

Turner purchased the two ex–Ansett/Nationwide Carvairs in 1982 from James Air of New Zealand. In an unspecified arrangement they were to be operated under Pacific Aerolift (PAL), headed by Charles Willis Jr. and under the direction of A.P. Fairchild. Turner was also supplying DC-4 aircraft to PAE at the same time. Both transactions occurred in 1982 with little information on which was first. The 1983 Evaluation of Carvair service in Hawaii report stated it would take at least five Carvairs to handle the 40,000 tons of cargo carried out of Honolulu.[10] This could explain why Willis attempted to purchase Carvair N55243 from Aero Union and N83FA from Ruth May in addition to the two ATL-98s Turner was providing for Pacific Aerolift.

Problems of an unknown nature occurred almost immediately with Pacific Aerolift Cargo (PAL) and A.P. Fairchild went to PAE as Vice President. The date does not determine when Turner began providing aircraft for PAE but it appears to be after Pacific Aerolift (PAL). Pacific Air Express attempted to purchase Carvair 17 in mid 1982 and the registration was not taken up. In December 1982 Kemavia, parent company of PAE, did purchase Carvair Seven. At the same time Pacific Air Express/Kemavia acquired two DC-4s and registered them as N300JT and N301JT the "JT" indicating Turner's backing. When Turner Aviation on 21 December 1982 purchased Carvair 21 in New Zealand for Pacific Aerolift the registration N407JT was reserved and it was given the name "Ruth II."

Above: The paint is stripped away from ZK-NWB during overhaul at Hamilton, New Zealand, for Hawaiian upstart Pacific Aerolift. Four liveries (two Ansett, Seulawah, and Nationwide) are visible as the layers of paint are stripped (courtesy Pacific Aerospace/Murray Dreyer collection). *Below*: Carvair 21, N407JT, incomplete at Hamilton after work stopped because of non-payment. Financial backing for Hawaii-based Pacific Aerolift collapsed after the first aircraft was completed (courtesy Pacific Aerospace / Murray Dreyer collection).

Pacific Aerospace of Hamilton New Zealand was contracted to overhaul the aircraft and repaint them in Pacific Aerolift livery. PAL was a Hawaii based cargo carrier formed by an unspecified group of individuals headed by Charles Willis Jr. It was incorporated in the State of California on 09 May 1982 with the company address as Honolulu International Airport P.O. Box 29636. Pacific Aerolift collapsed early on leaving few records of how they were financed other than Turner providing the aircraft. The carrier is one of three attempts to establish Carvairs in Hawaii. The backers secured enough financing to form the company and convince Turner to purchase the two remaining ex–Ansett ships. Although several attempts were made to purchase ZK-NWB along with ZK-NWA between April 1979 and December 1982, none of the potential buyers were well financed or got past a planning stage.

It had been stored in Nationwide livery at Nelson in September 1979. James Air purchased the two Carvairs from the receiver in October 1979 but they remained stored. After purchase by James Turner in 1982, Pacific Aerospace of Hamilton was selected to perform a major overhaul. Both aircraft were inspected and found to have major corrosion. After the inspection and maintenance the two-hour ferry to Hamilton was operated with the gear down.[11]

Pacific Aerospace at Hamilton removed the engines, outer wings, vertical stabilizer and control surfaces. The fuselage was stripped of three layers of paint revealing two Ansett, the Seulawah and Nationwide color scheme. Carvair 20, ZK-NWA, was eventually completed but ZK-NWB was not. All work was halted when Pacific Aerolift financing collapsed before the overhaul was completed.

The engines were hung and fuselage painted solid white with blue Pacific Aerolift lettering. The empennage and outer wings were not re-fitted. A 1983 evaluation report stated Turner Aviation had already expended $500 thousand on the two ships of which $199,100 was paid to James Air. It further stated that an estimated $315,870 (NZ$ 478,570) would be required to complete ZK-NWB plus $30,000 to ferry it to Honolulu. The total airframe hours for N407JT were recorded as 54,198 in October 1983.[12]

Turner Aviation maintained a relationship with PAE after the collapse of Pacific Aerolift. Turner was stuck with two incomplete Carvairs with pending U.S. registrations. The pair remained at Hamilton with an estimate of one million dollars to purchase and complete to meet U.S. FAR part 121 requirements.

Gold Crown Aviation

Carvair 21 with reserved registration N407JT was still sitting at Hamilton partially completed. When work stopped it was pushed out to a parking ramp. At some point during storage the engines were removed from positions three and four. In 1986 Captain Jack Ellis who owned Banks Aerodome near Geelong attempted to purchase both ex–Pacific Aerolift ships and return them to Australia. He planned to operate freight services under the name of Gold Crown Aviation. The DCA was not about to allow the Carvair back in Australia refusing route authority. The purchase was not completed leaving the aircraft in a semi overhaul condition.

In 1987 they were removed from storage and completed at the expense of James Air. The partial Pacific Aerolift livery was replaced with a two-tone orange stripe. During overhaul the cockpit heaters, de-ice boots, and anti-ice tank for windshield and props were removed. Gill liners were installed in the cargo bay along with an upgraded fire detection system. The "Rolamat" floor system was extended from station X798 to X891 to utilize the floor area normal occupied by a passenger cabin on car-ferry ships.[13]

A number of aircraft brokers in North America and Canada shopped the Carvairs but little interest was generated. Dusty Rivers President of Shelmar International of Darfield British Columbia made several attempts to broker it distributing technical data on both New Zealand

Carvairs. The information detailed the condition, capacity, and lift available but there were no takers. Both Carvairs sat idle through out 1987 at Hamilton New Zealand.

Hawaii Pacific Air/Air Cargo Hawaii/Roberts Hawaii

Hawaii Pacific Air was incorporated in Hawaii on 24 July 1989 and began effort to acquire and operate Carvairs in March 1990. HPA partners based their business plan on an evaluation study commissioned by James Turner and compiled in 1983. The study evaluated the feasibility of the Carvair in the Hawaii cargo market. Hawaii Pacific Air purchased Carvair 21, which was stored in New Zealand, from Turner on 24 July 1990. The exact purchase price was not disclosed but it was estimated in the feasibility study that an additional $315,870 above Turner's $500 thousand would be required to complete the overhaul. James Air absorbed that amount when the overhaul was completed around 1985. Air Cargo Hawaii under President Harry T. Snow was doing business (d/b/a) as Hawaii Pacific Air (HPA). Snow was listed in the 1983 Hawaii feasibility study as a contributor of information. Hawaii Pacific Air financial partners were listed as Robert Iwamoto, Larry Mahau, Al Harrington and Alvin Baron.[14]

It took considerable preparation to ready the Carvair for the ferry flight and complete all necessary documents. The ferry cost was estimated at $34,900 when HPA took possession of the aircraft on 25 August. Still wearing the New Zealand registration of ZK-NWB it departed Hamilton on 10 September 1990 staging through Auckland and Pago Pago where an engine was replaced. The New Zealand registration was cancelled on 17 September 1990 after arrival in

Hawaii Pacific Air collapsed in 1994–95, leaving N5459M parked on the Diamond Head ramp at Honolulu. Work began in 1996 to bring it back to flying status after being purchased by Joseph Mallel for export to Africa (author's photograph).

Honolulu. The same day A.P Fairchild as Director of Operations at HPA applied for U.S registration N5459M. The ferry oil tank was removed in Honolulu on 24 September 1990.[15]

Hawaii Pacific Air experienced problems from the outset with a 22-month delay in obtaining FAA certification for the Carvair. Once in operation HPA was continually plagued with cash flow problems and lack of cargo bookings. On 16 January 1992 aircraft ownership was transferred from HPA to parent company Air Cargo Hawaii. At that time Air Cargo Hawaii was listed as a Limited Partnership with George Crabbe as the sole partner. In March 1992 the aircraft ownership was transferred to Air Cargo Hawaii a second time listing Robert Iwamoto as president. On June 12 documents filed with the FAA stated N5459M had logged 54,237.6 hours.[16]

Beginning in 1992 ownership on N5459M becomes very confusing. Air Cargo Hawaii, Hawaii Pacific Air, and Roberts Hawaii all share corporate officers and partners of the associated companies and the Limited Partnership group. The aircraft remained idle in Honolulu through May 1993. Ownership changed back to HPA on 02 June 1993 when a Bill of Sale showing Harry Snow as president of Air Cargo was filed with the FAA. On that same day another Bill of Sale was filled transferring it from HPA to Roberts Hawaii. That document was signed by Harry Snow as vice president of HPA. Robert Iwamoto Jr. was vice president Air Cargo Hawaii and president of Roberts Hawaii. On 03 June 1993 Robert Iwamoto filed a request with the FAA to transfer the registration to Roberts Hawaii.[17]

Carvair N5459M operated an erratic cargo service until June 1995 when it was removed from service and stored with Carvair 20 on the Diamond Head ramp at Honolulu airport. In December 1995 both HPA Carvairs were advertised in trade publications with a contact of John Wessner in Honolulu.

Joseph Mallel/Airline Marketing Consultant (Status in Africa)

After over a year in storage at Honolulu International, Airline Marketing Consultants of Miami representing Joseph Mallel of Aircraft Traders Belgium purchased N5459M on 29 July 1996. Registration was transferred to the new owner on 01 August. During the summer of 1996 maintenance activity was in progress on the two HPA Carvairs N5459X and N5459M. Mahalo Airlines denied performing any maintenance but its mechanics were observed working on the aircraft. Mahalo, a small inter-island operator competed with Aloha and Hawaiian Airlines for the lucrative 30 million a month Hawaii inter-island market. The primary financial force behind Mahalo is Robert Iwamoto Jr. who also holds the majority control of Roberts Hawaii.

Other small cargo operators at Honolulu International were concerned about competition and were assured the Carvairs would not be operating in Hawaii. The two aircraft were purchased for $125,000 each to be used for relief work by Avia Air Charter in Zaire Africa. Carvair N5459M, returned to airworthy status in August 1996. Airline Marketing Consultants filled a request signed by Joseph Mallel for transfer of the registration. It was not confirmed at that time that engineers from Wonderair of Pretoria were on hand in Honolulu to assist in making the Carvairs airworthy.

Carvair 21 was transferred on 27 September 1996 to Airline Marketing Consultants of Miami retaining the registration N5459M.[18] It was returned to service on a rather speculative lease agreement to provide aircraft for proposed African relief flights. Late in October it departed Honolulu under the command of Bruce McSwiggan. A South African mechanic was on-board who had fitted it with a 350-gallon auxiliary oil tank in the cargo bay. A line equipped with an electric pump was connected to the existing oil line manifold to replenish the nacelle-mounted tanks. Two 150-gallon fuel tanks were also mounted in the cargo compartment and connected to the inboard aux-tank selector.[19] Bob McSwiggan and his sons are the larger of the last two active Carvair operators in the world. Upon arrival at Griffin on 31 October, Carvair 21 was parked with ships five (N83FA) and nine (N89FA). A month later it was joined by Carvair 20 (N5459X).

A historical event in aviation history occurred in November 1996 marking the last time that four Carvairs would ever be together on the same field. Only five Carvairs still existed and even more noteworthy is the fact that these particular ships had never been together before. The Hawaii Pacific ships were the last two Carvairs built for Ansett 33 months apart. The other two, Carvairs five and nine began service respectively with Interocean and BAF then were reunited at Falcon Airways in 1979 and now Custom Air Service.

Carvair N5459X was grounded in Florida after an engine failure while en route to Africa, Bruce and Bob McSwiggan returned to Griffin Georgia. Months passed before the sale was switched to Carvair 21, N5459M. The McSwiggans were contracted to fly it to Lakeland where all the spares and equipment were transferred from the other Carvair. The 350-gallon oil tank had been installed in Hawaii by the South African mechanics and the two 150 gallon header fuel tanks were transferred to N5459M. The fuel line was connected to the aux-inboard fuel selector, which had the tanks removed and blocked off when converted to a Carvair. Forty-one additional 55-gallon drums of fuel were loaded with a pump system that was manned by a mechanic from WonderAir. The aircraft staged through Miami and fueled up at Barbados for the 2900 mile flight to Dakar. All wing tanks were filled to 2800 gallons and the fuselage tanks and drums contained an additional 2555 gallons. They arrived at Dakar with 1000 gallons of fuel remaining. After a layover and refueling it continued on to Abidjan, Ivory Coast. It was scheduled to stop at Kinshasa (Leopoldville) for the new owner but because of unstable political conditions it was flown to Windhoek, Namibia. Mallel met the Carvair there and rode the last 900 miles to South Africa. The Carvair arrived at Pretoria's Wonderboon airport on 13 November 1996 with a major oil leak on Number Two engine.[20]

The plans for the African relief work collapsed leaving the Carvair unsold and leased to WonderAir. Still wearing the Hawaii Pacific Air titles it was parked at Wonderboon airport. It was repainted white with no titles but remained idle. The sub-lease to Avia Air Charter for relief work at Kinshasa expired and it was abandoned at Wonderboon. WonderAir defaulted on the lease in September 1997 and ownership reverted back to Airline Marketing Consultants.

It was scheduled for overhaul before beginning the Zaire relief work. It was moved to Rand Airport and remained parked from November 1996 until January 1997. It was made airworthy at Rand Airport near Germiston South Africa and on 12 March 1998 it was flown to Lanseria airport also near Johannesburg. Since WonderAir had defaulted on the contract and the aircraft was in need of overhaul, Airline Marketing opted to scrap it. It was reasoned that cost to maintain airworthiness or cost to return it to the U.S. exceeded the scrap value. On 02 November 1999 Airline Marketing Consultants of 57th Street Miami filed a Triennial aircraft registration showing the aircraft destroyed. That was followed by a request on 15 December 1999 to deregister N5459M in the U.S. due to being totally destroyed and scrapped.[21]

On 11 May 2000 it was flown to Kinshasa to be broken up for spares to support standard DC-4s but returned to Pietersberg Gateway Airport in the capital of the Northern Province of South Africa. It was advertised for sale and ready to be ferried worldwide. On 05 November 2000 it was flown to Pretoria "Wonderboon," Johannesburg International then to Kinshasa and back to Pietersburg. It is believed the flights were to show the aircraft to a potential buyer.

John Poindexter, Director and Vice President of Airline Marketing Consultants on NW 92nd Ave Miami, submitted a request to the FAA on 09 March 2001 to reinstate the U.S. registration N5459M. The request was granted with an expiration date of 14 April 2001.[22] The Carvair was parked at Pietersburg South Africa on 01 May 2001 and offered for sale still painted white with no markings.

The last Carvair arrived at Lanseria in 1998. Plans by Avia Air for relief work in Zaire collapsed, leaving it unsold. No new buyer could be found, and it was slated to be broken up for spares. Viscount G-BFZL of British World Airways, which evolved from BAF, sits in the background (courtesy Via Paul Howard).

Phoebus Apollo

Phoebus Apollo Aviation of Rand airport Germiston purchased N5459M in March 2002. It was overhauled and repainted in Phoebus Apollo titles with a black vertical stabilizer. The nose was painted with a stylized face along with the Phoebus Apollo lettering. It retained the U.S. registration after repainting until the Zambian registration 9J-PAA was obtained in April 2002. Another application was submitted to the FAA on 09 April 2002 by Andrew Cohen vice president of Airline Marketing of Pembroke Pines Florida requesting the U.S. de-register N5459M because of export to Zambia. The FAA noted there were unreleased liens against the aircraft that must be cleared. The Carvair did not go to Zambia but apparently the lien was cleared and the U.S. registration was cancelled 30 April 2002.[23]

Phoebus Apollo is a cargo and passenger carrier created in 1994 by Hendrik (Hennie) Delport with a fleet of DC-3 and DC-4s. Mr. Delport is also chairman of the syndicate that owns Rand Airport. He received his commercial rating in 1978 and the company name Phoebus Apollo came from a certificate he received after his first solo flight in 1976. The certificate referred to the Greek mythology God of the Sun as "Monarch of the Sky" and Lord of all things that fly. Mr. Delport took inspiration from this in creating a company name that people would not easily forget. The original plan was to create a cargo carrier to transport products for his other businesses.[24]

Phoebus Apollo has a complete maintenance facility able to perform in-house and contract maintenance for other DC-3 and DC-4 operators. The company is capable of complete airframe and instrument maintenance, build up and engine overhaul for R-2000 engines (completing as many as 25 per year), and operates a flight training school for DC-4 pilots.[25]

The Carvair was to operate from Johannesburg although based in Zambia. Presently the

South Africa–based Phoebus Apollo and Exclusive cigarette titles were applied to Carvair 21 while still wearing U.S. registration N5459M. In April 2002 it was re-registered in Zambia as 9J-PAA (courtesy Michael Zoeller collection).

old propliners are the only economically viable answer to South Africa's cargo needs. The DC-9 and 727s are too expensive for low usage into remote areas. The planned usage was estimated between 65–70 hours per month. The Carvair was to operate with other Phoebus DC-4s on cargo service between Johannesburg (Jan Smuts) and Lusaka Zambia, Harare Zimbabwe, and Lilongwe Malawi. It would fly some of the same routes transporting tobacco products as the EclAir Carvair 10, 9Q-CTI, did prior to being abandoned and destroyed in 1995 at Kinshasa. Phoebus Apollo is the distributor in Southern African countries for the "Exclusive" brand of cigarettes. The Exclusive lettering has been added to the lower forward fuselage. In addition to tobacco products, other cargo work includes machinery, fresh vegetables, furniture, and hunting trophies.[26]

In early 2004 Phoebus Apollo placed Carvair 21 on the market for $250,000. It had been parked for some time because of lack of work. Once again confirming that there are not enough continuous bookings of long and oversize cargo to warrant the cost of operating the Carvair. In early 2005 it was complete and taxied on several occasions. It flew from Johannesburg to Rand airport on 25 June. As of 2007 the last Carvair built remained stored in South Africa with an unsure future.

In the spring of 2007 Hendrik Delport owner of Phoebus Apollo Aviation along with 12 others was accused of racketeering, cigarette smuggling and fraud.[27] Most of his assets including vintage aircraft were seized placing the Carvair in further jeopardy of never flying again.

Epilogue

During its run the Carvair transported many unique cargo loads. The aircraft was economical and could be profitable with a load as low as 7 lbs/cu ft density. Although originally designed for transporting automobiles and their owners, over the years the Carvair played a very diverse roll. It was used in government and military operations, some of which were clandestine. It performed in disaster relief operations and recovering damaged and downed aircraft. The Carvair transported Black Knight and Blue Streak rockets, nuclear material, oil drilling equipment, boats, races horses, military equipment/troops, yacht mast, diesel ship engines, clay models for body pressings, treasury shipments, gold bars, new automobiles, race cars, automobile parts, radio towers, rabbits, snakes, geese, monkeys, pigs, sports teams, orchestras, rock groups and refugees. Some of the more unique loads were reindeer, elephants, dolphins, and a three-ton whale from RAF Binbrook to an aquarium at Nice. Carvairs re-supplied a zoo in Poland with animals, transported rocket parts to Sweden, cars to Mozambique, coffee beans in Central Africa, live chicks to Tunisia, cigarettes in South Africa, armored cars to Berlin, toilets and bath fixtures to Gabon, radio towers to Argentina and skydivers in Illinois.

British Air Ferries dispatched a Carvair to Calais in 1968 to transport a stone signpost weighing 1000 pounds. The post bore the name of the French town of "Embry." Wing Commander Basil Embry was shot down by the Germans in 1940 and escaped through the town with the same name. The signpost was of great inspiration and now stands at the RAF museum at Hendon. In 1972 BAF transported 14 baby elephants from Bombay to Stansted.

Among the more important roles was the transporting of the Concorde Bristol-SMECMA Olympus 593 turbojet engines from Filton to the aircraft factory at Toulouse. The Carvair transported many other Concorde items prior to the "Super Guppy" coming on line. It was also used in blockade running transporting fuel during the Rhodesia UDI, United Nations operations in the Congo, and some rather treacherous missions in Cambodia including transport of wounded soldiers.

Celebrities frequently used the Carvairs to transport their vehicles. The trend began with Bristol's at Lydd (Ferryfield) in 1954 evolving to the Carvair. Car-ferry service was more charming and private during the early years eventually serving film stars, politicians, diplomats and royalty. The Carvair continued the tradition transporting HRH Prince Phillip, King Hussein of Jordan, King Fiesel, and Crown Prince Gustov of Sweden. Some of the more colorful passengers included The Who, Small Faces, and Alvin Stardust. Racecar transport included Stirling Moss and the Team Mini Cooper's of Paddy Hopkirk that won the 1964 Monte Carlo Rally. Sports teams were transported, including some very rowdy rugby players that wanted to use the cargo bay as an airborne playing field.

The Carvair is a tribute to the quality and craftsmanship of American and British aircraft engineering. No other hybrid aircraft can match the dependability and longevity of the DC-4/Carvair combination. It has been more than sixty years since the DC-4 and forty years since the Carvair portion of the airframe was manufactured. Only 21 aircraft were built, yet three examples were still active into 2007 demonstrating the durability and ruggedness of the design.

Even more interesting is the number of airlines and individuals that owned and operated these airframes. All but two including the mock-up began service as U.S. military transports. The C-54/DC-4 airframes were operated by at least 50 carriers and after Carvair conversion by at least 75. Since many of the carriers operated multiple Carvairs it resulted in the fleet being collectively painted in hundreds of liveries. The list of registrations is just as impressive representing 21 countries as DC-4s and 16 as the Carvair.

The Carvair operated in many roles through the years even though it was built specifically as a car-ferry aircraft. Ansett ordered pure freighters and BAF converted a pair to cargoliners but the concept was a single purpose design for car-ferry operations. When the Car-Ferry era ended all the other uses were after thoughts. This explains why the 21 ships produced were operated by so many owners and continually being re-sold. There are not enough specialized cargo loads of pipe, car parts, ships mast, oil drilling equipment, bulk bread, or even reindeer, elephants, and whales in one region to support a continual operation.

Manufactures of specialty products now allow for transportation cost in the design and production of an oversize item. The delivery of that product on time is imperative. Therefore the difference between the cost of leasing a Carvair, 747, or AN-124 is never considered. The delivery cannot be delayed since time is money. The carrier with the newer most dependable aircraft usually gets the charter. For this reason the final frontier for the 18,000-pound payload Carvair is probably Alaska, Northern Canada and the Arctic regions or possibly South Central Africa. It can operate into small gravel fields where cost is a factor. More than forty years after conversion the delivery of oil drilling equipment, building materials, supplies, and bulk fuel to areas hundreds of miles from roads will probably be the last stand of the remaining Carvairs. The time is growing near when the Carvair will only be a fleeting memory for all who knew them.

Appendix 1:
Carvair's DC-4 Airframes

The Carvair was converted from an assortment of different series of DC-4 airframes. The following breakdown demonstrates that not only were no two Carvairs alike, they were converted from DC-4s with different characteristics.

C-54-DO   (Carvair #13) Built at Santa Monica original troop carrying 26 passenger variant of the DC-4. No cargo doors, four long range fuel tanks in fuselage. Max weight 68,000 pounds, max fuel 3,620 U.S. gallons. Engines P&W R-2000-3

C-54A-1-DC   (Carvair #7) Built at Chicago, military version carrying 50 troops or 32,500 pounds cargo. Cargo door and reinforced floor and loading hoist. Max weight 73,000 pounds, max fuel 3,620 U.S. gallons. Engines P&W R-2000-7.

C-54A-5-DO   (Carvair #6) Built at Santa Monica, Same as A model.

C-54A-10-DC   (Carvair #2, #4) Built at Chicago, Same as A model.

C-54A-15-DC   (Carvair #5, #10, #12) Built at Chicago, Same as A model.

C-54B-1-DC   (Carvair #1, #8, #14, #16) Built at Chicago, Same as C-54A except two of the fuselage tanks moved to the outer wing. Some built without cargo hoist. Max weight 73,000 pounds, max fuel 3,740 U.S. gallons. Engines P&W R-2000-3 and -7.

C-54B-5-DO   (Carvair #3, #11, #17, #18) Built at Santa Monica, same as B model.

C-54B-20-DO   (Carvair #9) Built at Santa Monica, Same as B.

C-54E-5-DO   (Carvair #15, #21) Built at Santa Monica, new fuel system, deleted fuselage tanks and replaced with fabric bag tanks in outer wings. Quick change interior from cargo (32,500 pounds) to troop transport, 50 troops or 44 airline type seats. Max weight 73,000 pounds, max fuel 3,520 U.S. gallons. Engines P&W R-2000-11.

DC-4-1009   (Carvair #19, #20) Built at Santa Monica with R-2000-9 engines, Passenger transport with no cargo door, 44-passenger crew of five. Fuel capacity 2,868 U.S. gallons. Max weight 73,000 pounds.

Appendix 2: Individual Aircraft Dated Histories

Serial Nos. DC-4 Skymaster/ATL-98 Carvair

C-54B-1-DC Douglas c/n 10480 (Built at Chicago)

42-72375	21 Dec 44	dlvr to USAAF
	05 Aug 46	pchs by Douglas Aircraft Co. 64th
		cvrsn to DC-4 at Santa Monica
NC90862	15 Jan 47	pchs by Santa Fe Skyways name: "SKY CHIEF APACHE"
PH-TEZ	24 May 48	pchs by KLM Royal Dutch Airways name: "ZEELAND"
	10 Jan 49	reg Netherlands
PH-DBZ	29 Mar 54	re-reg Netherlands
	-- --- 58	wfu
	09 Jan 59	b/o for spares
	20 Apr 59	reg cxld Netherlands

ATL-98-0

	01 Jul 59	fuselage pchs by Aviation Traders Ltd. for mock-up of ATL-98 Carvair
	02 Nov 59	fuselage arvd Tilbury Docks dlvr Southend by road
	-- Sep 61	tail cone used to repair G-ANYB

C-54B-1-DC Douglas c/n 10528 USAAF (Built at Chicago)

42-72423	22 Jan 45	dlvr to USAAF
NC88723	-- --- 46	pchs by Douglas Aircraft Co
NC59952	-- --- 47	pchs by Braniff Airways (Rato takeoff La Paz Boliva)
N59952	31 Dec 48	"C" dropped from reg
	15 May 53	dmgd in lndg incident Dallas TX
	-- Jun 53	pchs by World Airways/E.J. Daly
	-- --- 54	pchs by Air Carrier Service Corp subsidiary of California Eastern
	24 Jan 55	pchs by Air Charter Ltd. name: "ATALANTA"
G-ANYB	31 Jan 55	reg Great Britain
	16 Jun 60	trsf British United
	01 Jul 60	mgd Air Charter to British United Airways
	-- Sep 60	ferry New York — Southend by Capt McKenzie, Air Charter Chief Pilot
	21 Sep 60	frst pre-conversion test flt Capt Robert Langley
	03 Oct 60	last flt as conventional DC-4

ATL-98-1

G-ANYB	01 Oct 60	dlvr to Aviation Traders Ltd. (disassembly began)

07 Oct 60	towed into ATEL hangar	
25 Oct 60	DC-4 nose removed	
-- Dec 60	Carvair nose attached	
13 May 61	moved to flt shed	
17 Jun 61	roll-out of frst Carvair	
20 Jun 61	high speed taxi test	
21 Jun 61	FRST FLT AS CARVAIR (08:26 runway 06/2 Hr flt)	
26 Jun 61	visit to Stansted	
27 Jun 61	Filton for testing	
28 Aug 61	dmgd by forklift	
29 Sep 61	First Flt after rear fuselage rplcd	
08 Oct 61	C of A trials at Filton	
-- Dec 61	rear windows installed	
30 Jan 62	C of A	
12 Feb 62	name: "GOLDEN GATE BRIDGE" painted on nose	
16 Feb 62	dlvr to Channel Air Bridge demo flt to Ostend	
17 Feb 62	frst revenue flt Southend — Malaga	
22 Feb 62	proving flt to Basle	
01 Mar 62	frst schedule svc to Rotterdam	
11 Mar 62	Inaugural flt to Geneva	
29 May 62	proving flt to Strasbourg	
01 Jun 62	Inaugural flt to Strasbourg	
19 Oct 62	dprtd London for Singapore	
17 Dec 62	repaint at Southend, British United	
01 Jan 63	Channel Air Bridge became British United Air Ferries (BUAF)	
-- Feb 63	acft trsf to BUAF	
05 Mar 67	last flt/wfu stored Lydd Kent	
15 Mar 67	British reg cxld	
01 Oct 67	BUA renamed British Air Ferries	
04 Oct 67	trsf to BAF	
-- --- 68	vandalized at Lydd	
26 Jul 70	b/o at Lydd Kent, U.K.	
26 Aug 70	cxld from U.K. register	

C-54A-10-DC Douglas c/n 10311 (Built at Chicago)

42-72206	27 May 44	dlvr USAAF
NC57670	22 Sep 49	pchs by Federated Airlines
N57670	-- May 52	pchs by California Eastern Airways
	-- --- --	reg Transocean Airlines
	29 Oct 54	lsd by Resort Airlines
	-- --- 56	rtn to California Eastern Airways

ATL-98-2

	-- Jun 61	pchs/dlvr to Aviation Traders Ltd.
G-ARSD	14 Jul 61	reg Great Britain to Channel Air Bridge
	20 Jul 61	dlvr Stansted conversion began
	-- Jul 61	pchs by Channel Air Bridge
	21 Oct 61	nose arvd by road frm Southend
	25 Mar 62	FRST FLT AS CARVAIR STN-SEN
	28 Mar 62	began ARB certification program
	02 Apr 62	dlvr to Channel Air Bridge name: "CHELSEA BRIDGE" rcvd Certificate of Airworthiness after 9.5 hrs 1st flt to Geneva
	05 Apr 62	schld Geneva service begins
	06 Apr 62	Inaugural flt to Basle

	29 Oct 62	repaint British United livery at SEN
	01 Jan 63	mgd British United Air Ferries (BUAF)
	18 Mar 63	trspt 9 armored cars to Berlin
	20 Jun 63	trspt 20 armored cars to Berlin in seven days
	20 Jan 67	repaint British United colors
	01 Oct 67	BUAF renamed British Air Ferries (BAF)
	04 Oct 67	ferry to Lydd Kent
	06 Oct 67	wfu stored Lydd Kent
	13 Apr 68	British reg suspended
	26 Aug 70	b/o Lydd Kent, U.K. registration cxld

C-54B-5-DO Douglas c/n 18339 (Built at Santa Monica)

43-17139	10 Jul 44	dlvr USAAF
NC88709	-- Nov 45	pchs by Braniff Airways
N88709	31 Dec 48	"C" dropped from registration
	-- Oct 50	lsd to Northwest Airlines
	10 Jun 54	sub-lsd to Transocean Airlines
	10 Jun 54	dmgd on landing w/ Transocean at Keflavik Iceland
	-- --- 54	rtn to Braniff Airways
	-- Oct 54	lsd to Resort Airlines
	-- --- 55	rtn to Braniff Airways stored at Oakland
	-- Jun 61	pchs. by Channel Air Bridge

ATL-98-3

	-- Jun 61	trsf to Aviation Traders Ltd.
G-ARSF	17 Jul 61	reg Channel Air Bridge
	22 Jul 61	dlvr Stansted, conversion began
	31 Dec 61	nose dlvr SEN-STN
	01 Feb 62	reserved East Anglian Flying Svc (NTU) parent company of Channel Air ways placed order for 4 ships
	28 Jun 62	FRST FLT AS CARVAIR STN-SEN
	07 Jul 62	dlvr to Channel Air Bridge name: "PONT DE L'EUROPE"
	03 Aug 62	C of A A7367 issued, Inaugural flt to Geneva
	07 Aug 62	Inaugural flt to Calais
	28 Dec 62	crashed Rotterdam, Zestienhoven airport 250 m short of runway 4 crew, 14 psgrs. 1 fatality-crew
	05 Feb 63	British reg cxld
	23 Feb 63	wreckage removed to Southend
	-- Mar 63	wreckage moved to Stansted

C-54A-10-DC Douglas c/n 10338 (Built at Chicago)

42-72233	30 Jun 44	dlvr USAAF
NC65142	-- Nov 45	pchs by Braniff Airways
N65142	31 Dec 48	U.S. dropped "C" from registration
	-- Nov 54	pchs by Resort Airlines
	01 Jul 59	lsd to Slick Airways/Logair
	30 Jun 60	rtn to Resort Airlines
	-- Jun 61	pchs by Channel Air Bridge
	06 Jun 61	dlvr Channel Air Bridge
G-ARSH	17 Jun 61	reg Great Britain

ATL-98-4

	-- Jun 61	dlvr to Aviation Traders Ltd.
	27 Jul 61	arvd U.K. disassembly began Stansted
G-ARSH	09 Apr 62	reg cxld
	03 Jun 62	nose dlvrd. by road from Southend
G-41-2	05 Sep 62	FRST FLT AS CARVAIR STN-SEN
N9758F	12 Sep 62	reg. Intercontinental (USA) (F suffix for U.S. acft maintained abroad)
	13 Sep 62	Interocean flt crews begin training
	14 Sep 62	C of A
	20 Sep 62	pchs Intercontinental U.S. Inc. (USA) Acft left Southend for Frankfurt for FAA certification. Capt Morris in command
	15 Oct 62	Dlvr to Intercontinental
	29 Oct 62	rtn to Southend for maintenance (fitting of deicing boots)
	18 Nov 62	dprt Southend for Congo
LX-IOG	10 Dec 62	reg Interocean Airways (Luxembourg)
LX-IOH	-- Dec 62	reg Interocean Airways
	29 Dec 62	dlvr to Interocean Airways
	-- Jan 63	pchs by Interocean Airways (Luxembourg)
	26 Feb 63	rtn Southend for maintenance
	27 Feb 63	U.S. reg N9758F cxld
	06 Mar 63	dprt SEN for Congo
	-- Jul 63	UN Congo contract expired
	-- Mar 64	rtn Stansted modified from freighter to car-ferry standards, rear psgr windows added
	28 Apr 65	rtn to Stansted for repaint and maintenance. ramps removed
	-- May 65	pchs By CAT (Cie Air Transports
F-BMHU	19 Jun 65	dlvr. CAT (Cie Air Transport) (FRANCE) name: "CDT HENRI de MONTAL" ferry Stansted to Le Touquet
	04 Mar 67	landed Karachi with engine problem
	08 Mar 67	crashed after take-off Karachi, Pakistan en route from Paris to Tokyo 6 crew no psgrs. 4 fatalities-crew 7 fatalities on ground
	19 Dec 68	rmvd from French registration

C-54A-15-DC Douglas c/n 10365 (Built at Chicago)

42-72260	03 Aug 44	dlvr to USAAF
BU50843	03 Aug 44	trsf U.S. Navy as R5D-1
NC58003	10 Apr 46	pchs by Veterans Air Express
	19 May 47	Veterans Air Express bankrupt
	01 Nov 48	pchs by Matson Navigation Co
VP-CBD	20 Jan 49	pchs by Air Ceylon name: "LAXAPANA"
	03 Feb 49	dlvr Air Ceylon
CY-ACA	27 Sep 50	re-reg by Air Ceylon
VH-INY	16 Mar 51	pchs by Australian Nat'l Airlines Victoria Commonwealth Australia
	29 Mar 58	pchs by Twentieth Century Aircraft
	-- Mar 58	lsd to Seven Seas Airlines
N5520V	30 Apr 58	reg U.S.
D-ADAL	09 May 58	pchs by Trans-Avia Dusseldorf
	30 Sep 59	pchs by Lufttransport-Uternehman (LTU) name: "DUSSELDORF"
	07 Aug 60	pchs by British United Airways
	18 Aug 60	trsf Aviation Traders Ltd.
	04 Sep 62	German reg cxld
G-AREK	20 Sep 61	reg. Air Charter Ltd

ATL-98A-5

G-AREK	27 Dec 61	dlvr to Aviation Traders for conversion
	22 Jul 62	nose dlvr SEN-STN
	27 Jul 62	pchs by Winston Factors U.S.
N9757F	28 Aug 62	U.S. registration requested
	04 Sep 62	British reg cxld
	12 Sep 62	reg Intercontinental USA (F suffix for U.S. acft maintained abroad)
	25 Oct 62	frst engine runs
	02 Nov 62	FRST FLT AS CARVAIR over SEN and North Sea rtn STN
	13 Nov 62	C of A issued after 11 hr 20 min at SEN
	20 Nov 62	dlvr to Intercontinental U.S. Inc (USA) Ferried to Luxembourg by Capt Barron
LX-IOG	17 Dec 62	pchs by Interocean Airways (Luxembourg) (LX-IOG had been previously assigned to Carvair no. 4 and not picked up)
	27 Feb 63	U.S. reg N9757F cxld
	-- Apr 63	ONUC Congo
	-- Jul 63	UN Congo contract expired
	-- Sep 63	modified from freighter to car ferry standards, rear psgr windows added
	28 Apr 65	Ferry Flt to LUX-STN repaint CAT colors (maintenance work-ramps removed)
	30 May 65	Test Flt STN
	31 May 65	pchs CAT (Cie Air Transport) (France) name: "Cdt Max Geudt" Ferry Flt Stansted — Le Touquet
F-BMHV	30 May 65	Test Flt STN
	24 Jun 65	reg in France
	-- --- 70	pchs by TAR name: "Porky Pete"
	29 Apr 72	lsd by British Air Ferries
	18 Feb 73	pchs British air Ferries
G-41-1-73	18 Feb 73	re-reg British Air Ferries
G-AREK	23 Feb 73	re-reg British Air Ferries
	-- --- 73	repaint dark blue livery
	20 Jun 74	wet lsd by Paulings Middle East Ltd.
	08 Feb 76	rtn British Air Ferries
	-- Feb 76	wfu stored Southend Essex
	11 Apr 77	still in Pauling colors at Southend
	24 Jun 77	pchs by Uni Air (France)
	27 Jun 77	dlvr Uni Air
F-BYCL	06 Jul 77	reg in France — stored at Toulouse British reg cxld
	-- Apr 79	Falcon submitted letter of intent to pchs
N83FA	15 Jul 79	U.S. reg Requested
	20 Aug 79	pchs by Ruth A. May (USA) lsd to Falcon Airways
	30 Aug 79	U.S. registration granted
	15 Oct 79	registration discrepancy cleared
	-- Oct 80	wfu stored Meacham Field Fort Worth TX USA
	17 Feb 81	lsd to Gifford Aviation Inc to be operated by Kodiak Western subsidiary
	19 Jul 81	trsf to Kodiak Western Alaska
	-- Feb 82	rtn to Gifford Aviation Inc
	22 Feb 82	pchs by Gifford Aviation Inc
	15 Jul 82	Ownership certificate reg to Gifford Aviation
	08 Aug 85	repo by Mercantile Nat'l Bank of Dallas
	23 Aug 85	pchs Custom Air Service Robt. D. McSwiggan Griffin, GA (USA) (ad hoc charter work)
	04 Apr 97	CRASHED Griffin GA NTSB# ATL97FA057 2 crew-fatalities
	-- Jul 97	reg cxld — w/o

C-54A-5-DO Douglas c/n 7480 (Built at Santa Monica)

42-107461	11 Apr 44	dlvr USAAF
NC90431	10 Apr 46	pchs by American Airlines name: "Flagship Philadelphia"
N90431	01 Jan 49	"C" dropped from registration
	24 Sep 54	pchs by Airplane Enterprises
YV-C-AVH	-- Oct 54	pchs by AVENSA
N75298	-- Jan 55	pchs by Resort Airlines
	01 Jul 60	lsd by World Airways
	27 Jun 61	rtn to Resort Airlines (Logair titles)
	-- May 62	pchs Channel Air Bridge
G-ARZV	08 Jun 62	reg Channel Air Bridge

ATL-98-6

G-ARZV	18 Jun 62	dlvr Aviation Traders Ltd. Disassembly began
	09 Sep 62	nose dlvrd Stansted
	28 Oct 62	nose fitting completed
	21 Dec 62	FRST FLT AS CARVAIR STN-SEN (white top bare metal bottom no letters)
	21 Jan 63	British reg cxld
	05 Feb 63	C of A #317 issued
	23 Feb 63	Debut in Aer Lingus livery
EI-AMP	26 Feb 63	2nd flt Aer Lingus (1st flt full Aer full Aer Lingus livery)
	28 Feb 63	pchs Aer Lingus (Ireland) name: "St Albert" (left side) name: "Ailbhe" (rt side)
	01 Mar 63	compass check at Bascombe Down
	08 Mar 63	Irish C of A issued
	14 Mar 63	dlvr to Aer Lingus
	08 May 63	Dublin Liverpool service
	08 Jan 64	rtn to ATEL to raise ceiling under cockpit and install Rolamat floor. Nose wheel collapse on landing at Stansted
	15 Feb 64	rtn to svc with ceiling modification
	06 Oct 66	wfu stored Dublin
	30 Oct 66	last scheduled cargo flt Manchester — Dublin
CF-EPU	10 Jan 68	pchs Eastern Provincial Airwys (Canada) (reg not taken up)
	-- Feb 68	reg to Eastern Provincial Airwys (Canada)
	20 May 68	Ferry to Southend for cold weather mod
	29 May 68	Irish reg EI-AMP cxld
CF-EPX	01 Jul 68	First flt EPA frm Southend
	05 Jul 68	dprtd Southend for Gander
	24 Jul 68	Canadian C of A 14600 issued
	28 Sep 68	Crashed on approach, Twin Falls Labrador Canada. 3 crew, 33 psgrs. no fatalities
	31 Jan 69	Canadian reg cxld
	-- --- 69	parts salvaged — spares to Southend W/O

C-54A-1-DC Douglas c/n 10273 (Built at Chicago)

42-72168	06 Nov 43	dlvr USAAF
	03 May 46	trsf War Assets Administration
NC54373	03 May 46	pchs by California Eastern Airways
	15 May 46	reg California Eastern
N54373	31 Dec 48	"C" of registration NC dropped by U.S.
	29 Jul 49	lsd Arrow Airlines

	18 May 51	lsd to Twentieth Century Airlines
	15 Feb 53	pchs by Seaboard and Western Airlines name: "Wake Island Airtrader"
	-- --- 54	lsd by Transocean
	01 Dec 55	pchs by Air-World Leases Inc
	06 Oct 58	pchs by Seaboard and Western Airlines
	20 May 59	pchs by Marshall M. Landy
LX-BNG	07 Mar 61	pchs by Interocean Airways
	-- Mar 61	lsd to United Nations
	02 Aug 62	pchs by Aviation Traders Ltd.

ATL-98A-7

LX-BNG	17 Aug 62	dprt Congo for ATEL Stansted
	22 Aug 62	arvd STN conversion began
	-- Oct 62	airframe in jigs
	11 Nov 62	nose delivered Southend to Stansted
	-- Dec 62	nose placed in jigs
	21 Dec 62	nose fitted to DC-4 airframe
G-ASDC	11 Jan 63	reg Great Britain
	19 Mar 63	FRST FLT AS CARVAIR STN-SEN (Capt Cartlidge in command)
G-ASDC	26 Mar 63	rcvd C of A, pchs British United Air Ferries name: "Pont du Rhin" frst flt to Rotterdam
	13 Dec 63	frst Carvair to visit Lydd Kent
	21 Mar 64	inaug deep penetration routes from Lydd
	05 Dec 64	nose wheel collapse Rotterdam
	01 Oct 67	mgd British Air Ferries
	17 Feb 70	crash landing at Rotterdam
	04 Oct 72	name: "Big Louie"
	-- Jan 74	Stansted for maintenance and repaint
	08 Apr 74	rtn to svc BAF
	29 Jan 75	convert to all cargo at Southend
	11 Feb 75	debut in all cargo name chg: "Plain Jane"
	28 Oct 75	lsd to Invicta Intl Airlines
	08 Nov 75	rtn to British Air Ferries
	-- Jun 76	advertised for sale
	-- Jun 77	repaint yellow/black sash livery
	-- Dec 78	stored Southend
	-- Jan 79	advertised for sale
	04 Apr 79	last flt. British Air Ferries
N80FA	11 Apr 79	Falcon Airways applied for U.S. reg
	30 Apr 79	British reg G-ASDC cxld
	01 May 79	pchs Falcon Airways
	02 May 79	British reg removed from aircraft
	03 May 79	dprtd Southend for Dallas
	07 May 79	arvd U.S. as N80FA
	21 Mar 80	repo Mercantile Texas Capital
	07 Apr 80	trsf Mercantile Nat'l Bank of Dallas
	10 Oct 80	pchs Nasco Leasing Co
	22 Jan 81	reg Nasco Leasing Co
	-- Jun 81	stored Long Beach
	-- --- 81	rumored pchs. Calm Air Canada
	01 Dec 82	pchs James Blumenthal (Tucson)
	29 Dec 82	pchs Kemavia Inc. of Honolulu (Louis Khem)
	-- Jan 83	lsd Pacific Air Express
	08 Feb 83	reg trsf to Kemavia

	05 Mar 83	repaint in Pacific Air Express livery
	-- Mar 83	ferry LGB to HNL
	23 Mar 83	rqst to change registration to N103
N103	10 Apr 83	re-reg Pacific Air Express (USA)
	12 Jul 83	reg applied to acft
	-- Jul 86	reg Hondu Carib Cargo
	-- Oct 86	ferry HNL-TUS for storage
	08 Sep 87	pchs by Fred L. Mason (Great Southern Airways)
	27 Apr 88	ferry Tucson AZ to La Grange GA
	20 Dec 95	reg Yesterdays Wings Inc. (Frank Moss)
	17 Jan 96	maintenance at La Grange GA
	07 Jun 96	reg Great Arctic Airways
	12 Apr 96	Based at Fairbanks AK. (Hondu Carib)
	28 Jun 97	CRASHED Venetie AK. #2 eng fire Capt Frank Moss/3 crew no injuries NTSB report ANC07LA093

C-54B-1-DC Douglas c/n 10448 (Built at Chicago)

42-72343	15 Nov 44	dlvr USAAF
N88819	-- --- 50	Twentieth Century Aircraft
	-- --- 50	mgd acquisition by North American Airlines
	-- --- 55	pchs by Resort Airlines
	02 Jul 60	lsd to World Airlines (Logair titles)
	29 Jun 61	rtn to Resort Airlines
	-- --- 62	lsd by Slick Airways
	-- --- 62	rtn to Resort Airlines wearing Quicktrans titles
	23 Oct 62	pchs by Aviation Traders Ltd.

ATL-98-8

N88819	25 Oct 62	dlvr to Aviation Traders Ltd.
	18 Dec 62	nose transported SEN-STN
EI-AMR	05 Feb 63	issued Irish reg. #318
	30 Mar 63	Carvair nose/tail fitting complete engines mounted
	11 Apr 63	roll-out Stansted full Aer Lingus livery
	18 Apr 63	FIRST FLT AS CARVAIR
	19 Apr 63	ferry flt STN-SEN issue C of A
	29 Apr 63	dlvr Aer Lingus (Ireland) name: "St. Jarlath" (left side) name: "Larfhlaith" (right side)
	-- Dec 63	rtn to Southend for modification
	-- Feb 64	rtn Stansted for raise ceiling under cockpit and Rolamat floor
	09 Mar 64	work delayed at Stansted (ceiling mods)
	-- Jun 64	nose wheel collapse Dublin
	31 Oct 66	wfu stored Dublin IR.
	-- Sep 67	rtn to flying status for sale to Aerovias Nacionales Transcontiental Ecuador (ANTENA) of Ecuador
	07 Oct 67	recovered Viscount EI-AKK at Bristol
	10 Jan 68	pchs by Eastern Provincial Airways (Canada)
	16 Feb 68	rtn to Aviation Traders Southend for cold weather modifications
CF-EPV	20 May 68	roll-out in EPA colors
	21 May 68	frst flt after overhaul UK C of A for Export 7835
	24 May 68	reg Eastern Provincial Airways dprt SEN
	29 May 68	Irish reg cxld
	28 Jun 68	Canadian C of A 5799 issued
	-- Oct 69	wfu stored

	-- Sep 70	rtn to svc Eastern Provincial
	03 May 72	nose wheel collapse Gander Newfoundland
	23 Jul 73	pchs by R.C.S. Graham
	21 Sep 73	lsd British Air Ferries 3-6 mo (ntu) wfu stored Southend, Essex UK
LN-NAB	12 Jun 74	pchs Thor Tjontveit for Norwegian Overseas Airways (Norway) (option not exercised) Acft remained SEN in Eastern Provincial colors
	-- Aug 74	Graham lsd to Norwegian Overseas Airways
CF-EPV	09 Sep 74	Canadian reg CF-EPV cxld
LN-MDA	10 Sep 74	pchs by J. Jorgensen (Oslo Norway) acft remained at Southend
	17 Oct 74	lsd to Norwegian Overseas Airways— ntu
	-- Aug 75	impounded SEN for monies owed ATEL
	-- --- 76	stored Southend (no engines)
	13 Sep 78	b/o Southend, Essex UK, parts used on G-AOFW
	-- --- 80	cockpit pchs J Flanagan loaned to USAF 100th Bomb Group Museum at Thorpe Abbots, Norfolk
	-- --- 97	cockpit moved to USAF 8th Air Force Memorial Museum at Holton/ Halesworth, Suffolk UK
EI-AMR	-- May 05	cockpit remains on display 8th AF Museum Holton/Halesworth, Suffolk, UK

C-54B-20-DO Douglas c/n 27249 (Built at Santa Monica)

44-9023	11 Jan 45	dlvr USAAF
	15 Nov 45	trsf to Reconstruction Finance Corp
NC88816	23 Nov 45	U.S. registration
	01 Dec 45	pchs by Douglas Aircraft Corp DC-4 conversion no. 10 at El Segundo
	16 Oct 46	pchs by Western Air Lines
XA-MAA	18 Dec 56	pchs by Guest Aerovias Mexico
	19 Dec 56	U.S. reg cxld
HP-256	11 Sep 57	lsd by Aerovias Panama Airways
HP-268	-- --- 58	re-reg Aerovias Panama Airways
	-- Mar 60	rtn to Guest Aerovias Mexico
N9326R	22 Mar 60	pchs by The Babb Co. International Aircraft Sales
CP-682	01 Apr 60	lsd by Lloyd Aereo Bolivano
N9326R	17 Jan 62	pchs International Aircraft Services
	08 Feb 62	reg International Aircraft Services
	24 Oct 62	pchs Aviation Traders Ltd.
	-- --- 62	W. Remmert (prior to conversion details unknown, possibly Remmert-Werner acft brokers St Louis)

ATL-98-9

N9326R	25 Oct 62	maintenance work Orly field Paris
	29 Oct 62	ferry Orly-Stansted for disassembly
	27 Jan 63	nose assembly dlvr. SEN-STN
	13 Mar 63	U.S. reg cxld
G-ASHZ	06 May 63	rcvd British registration
	01 Jun 63	roll-out Stansted
	08 Jun 63	FIRST FLT AS CARVAIR
	14 Jun 63	C of A issued proving flt to Rotterdam
	15 Jun 63	pchs by British United Air Ferries name: "Maasbrug" at Rotterdam
	-- May 64	cvt to 3-car/55-psgr
	01 Oct 67	mgd British Air Ferries
	04 Oct 67	dlvr British Air Ferries
	17 Oct 67	repaint lettering to British Air Ferries

	18 Feb 68	overhaul at Southend repaint BAF sandstone blue
	01 Oct 70	new name: "Fat Annie"
	-- --- 73	repaint dark blue livery
	-- --- 76	repaint remove titles
	-- Jun 76	advertised for sale
	-- Nov 76	lsd: SOACO
	24 Feb 77	reg Trans World Leasing
	-- May 77	rtn BAF
	-- Jun 77	cvt to all cargo (2nd acft) repaint yw/bk sash c/s, name dropped
	06 Jul 77	roll-out all cargo
	-- Sep 77	nose door ripped off at Belfast in high winds.
	-- Jul 78	lsd to International Red Cross for relief work Middle and Far East
	07 Sep 78	rtn British Air Ferries
	24 Dec 78	stored Southend
	-- Jan 79	advertised for sale
N83FA	-- Apr 79	reg Falcon Airways (not exercised)
	14 May 79	British reg G-ASHZ cxld
N89FA	15 May 79	reg United States
	01 Jun 79	pchs by Falcon Airways (N83FA reg paint on acft in error)
	03 Jun 79	Falcon Arwys: departed Southend Dallas
	21 Mar 80	repo by Mercantile Nat'l Bank of Dallas
	07 Apr 80	reg to Mercantile Dallas Corp
	10 Oct 80	pchs by Nasco Leasing Corp
	07 Jan 81	repo by Mercantile Nat'l Bank of Dallas
	19 Feb 81	reg to Mercantile Nat'l Bank
	-- Jun 81	stored Tucson AZ
	-- --- 82	rumored to be pchs. by Calm Air Canada
	01 Dec 82	pchs by James R. Blumenthal
	-- Feb 83	reg to James R. Blumenthal remain stored
	14 Sep 86	pchs by NCBA Inc.(Steve Kehmeier)
	04 Mar 87	pchs by Robt D McSwiggan (Custom Air Svc)
	-- Oct 94	wfu Griffin Ga. USA
	01 Jan 96	used as engine test stand and parts support for N83FA Carvair #5
	-- Aug 97	heavy maintenance to bring back to flying status Griffin Ga.
	14 Apr 98	test flt from Griffin Ga. to Tara Field Hampton GA
	20 May 98	FAA certified airworthy after 3 hr check ride. Bruce McSwiggan in command TTL Hrs. 59,102
	01 Oct 98	acft at Arlington Municipal Airport, Arlington Texas
	-- Mar 00	nose wheel collapse Opa-Locka Florida
	09 Feb 01	pchs Avignon Inc. Cincinnati re-named: Fat Annie
	01 Nov 02	Owner changed to Gator Global Cincinnati based at Sherman/Dennison Texas

C-54A-15-DC Douglas c/n 10382 (Built at Chicago)

42-72277	24 Aug 44	dlvr USAAF
LN-HAU	13 Mar 47	pchs by Braathens SAFE name: "Norse Trader"/"Delhi Air Trader"
N1221V	17 Oct 50	pchs by Seaboard and Western Airlines name: "Oslo Air Trader"
	20 Jan 53	lsd by Trans Caribbean Airways
	-- --- 55	rtn to Seaboard and Western Airlines
	16 Sep 57	nose wheel collapse Brindisi Italy
	01 Sep 59	pchs by Lewman Corp
LX-BBP	07 Mar 61	pchs by Interocean Airways
	-- Mar 61	lsd by United Nations

	11 Nov 62	pchs by Aviation Traders Ltd.

ATL-98A-10

	12 Nov 62	dlvr to Aviation Traders Ltd.
	11 Dec 62	ready to enter assembly line Stansted
	24 Mar 63	nose dlvr by road SEN-STN fuselage still outside
G-ASKG	12 Jul 63	British registration assigned
	27 Jul 63	roll-out as Carvair
	29 Jul 63	FIRST FLT AS CARVAIR (flown by Capt Cartlidge)
	01 Aug 63	C of A issued
	06 Aug 63	dlvr to British United Air Ferries Southend, name: "Channel Bridge"
	07 Aug 63	British United titles painted out repaint Alisud
	08 Aug 63	lsd to Alisud — Cia Aerea Meridionale (short term lease to fly Naples-Palermo)
	15 Aug 63	frst Alisud Carvair flt Naples-Palermo
	25 Feb 64	rtn to British United Air Ferries ferry flt Palermo Southend
	25 Mar 64	rtn to svc. with new paint no lettering
	-- Dec 65	lsd by Air Ferry Ltd. (Rhodesia blockade fuel hauling Dar-es-Salaam Tanzania to Lusaka Zambia)
	28 Dec 65	dprt Southend for Dar-es-Salaam
	01 Jan 66	arvd Dar-es-Salaam
	11 Mar 66	rtn to Southend from Africa
	-- Mar 66	rtn to British United Air Ferries
	-- Nov 66	repaint BUA blue and sandstone
	25 Feb 67	flew last Southend Basle service
	04 Oct 67	mgd to British Air Ferries
	06 Jan 68	repaint BAF titles (remain sandstone blue stripe)
F-BRPT	03 Mar 69	reg Cie Air Transport (France) name: "Commandant Charles"
	17 Mar 69	rtn to British Air Ferries
	08 Jun 69	wfu parked Lydd Kent
	27 Jun 69	lsd by Transports Aeriens Reunis (TAR)
	-- Dec 69	pchs (CAT) Cie Air Transport (France)
	-- --- 70	moved to Nimes, parked
	-- Mar 70	pchs by TAR lsd to BAF name: "Big Joe"
	21 Apr 72	lsd by British Air Ferries
	26 Mar 73	pchs by British Air Ferries
	-- Mar 73	French reg cxld
	08 Apr 73	Debut in BAF livery (dark blue bottom)
G-ASKG	26 Apr 73	reg GB
TR-LUP	21 Feb 75	pchs Société Anoynme de Construction (Gabon)
	23 Feb 75	dlvr Société Anoynme de Construction (SOACO)
	25 Feb 75	British reg cxld
	-- Jun 76	overhauled and repaint at Southend
	-- --- 79	wfu
	-- Mar 79	confiscated by Empire of Central Africa Government
	-- Aug 79	Gabon reg cxld
	-- --- 80	stored Maya Maya Brazzaville Congo
TN-ADX	30 Oct 81	pchs by Aero Service (AFRIGO)
	16 Mar 83	wfu Maya Maya
	03 Feb 84	rtn to svc
	07 Oct 89	rmvd frm svc stored Brazzaville Congo
	-- --- 93	pchs by ECL Air Tabazaire
9Q-CTI	17 Mar 93	reg ECL Air (Zaire)
	19 Mar 93	test flt Brazzaville

	20 Mar 93	rtn to svc frst flt with ECL Air (flown by Mike Snow, Pierre Cote, and George de Mar)
	-- Apr 95	pchs Transair Congo
	-- Jul 95	wfu stored Kinshasa Africa (engines removed)
	05 Sep 95	destroyed Kinshasa Africa

C-54B-5-DO Douglas c/n 18333 (Built at Santa Monica)

43-17133	22 Jun 44	dlvr USAAF
	26 Dec 45	rtn to Douglas Aircraft Co. conversion # 34 to DC-4 at Santa Monica
N37477	27 Apr 46	pchs by Delta Air lines
	-- --- 54	pchs by North American Airlines
	-- --- 56	trsf to Trans American Airlines
	-- --- 57	trsf to Twentieth Century Airlines
D-ANET	13 May 58	re-reg. to Lufttransport-Unternehmen (LTU) (not exercised)
G-APNH	17 Jun 58	pchs by Independant Air Transport
	16 Mar 59	re-named Blue Air
	15 Oct 59	reg Blue Air
	-- Oct 59	Blue Air bankrupt
PH-EUR	-- Nov 59	re-reg. Euravia NV (not exercised)
	08 Mar 60	reg Astraeus Limited
G-APNH	25 Mar 60	pchs by Air Charter Ltd.
	01 Jul 60	mgd to British United Airways

ATL-98-11

G-APNH	01 Dec 62	dlvr to Aviation Traders Ltd. work began
	28 Feb 63	disassembly resumed at Southend
	02 May 63	new nose placed in position for joining
	-- Mar 64	name: "Pont de Crouch"
	07 Apr 64	roll-out of hangar in primer
	27 Aug 64	trsf to British United Air Ferries
	30 Dec 64	name: "Menai Bridge"
G-APNH	04 Jan 65	FIRST FLT AS CARVAIR over SEN
	09 Jan 65	C of A issued after 5hr 25min dlvr to British United Air Ferries
	-- --- 65	lsd to Aviation Traders Ltd. (equiped with 8 fuel tanks for long range work)
	11 Jan 65	dprt for Australia with rocket cargo
	16 Oct 65	World Demonstration Tour NZ and Australia
	-- Oct 65	sub-lsd to Straits Air Freight Express
	01 Dec 65	rtn to British Air Ferries
	03 Dec 65	dprt Lyneham for Lusaka Zambia
	-- Dec 65	lsd to Air Ferry Ltd.
	30 Dec 65	dprt Southend to Dar-es-Salaam in Air Ferry titles (Rhodesia blockade fuel hauling Dar-es-Salaam Tanzania to Lusaka Zambia)
	03 Jan 66	arvd Dar-es-Salaam
	02 Jun 66	rtn to British United Air Ferries
	-- Jun 66	cvt to 6 fuel tanks
	-- Nov 66	repaint BUA livery Jan 67 lease British Govt
	12 Apr 67	lsd to Shell in Trucial states. Dprtd SEN to Oman
	01 Oct 67	new corp name: British Air Ferries
	04 Oct 67	trsf from BUAF to BAF
	25 Oct 67	repaint BAF titles and overpaint Bird logo c/s
	07 Feb 68	recovery flt to Libya
	-- Jan 69	recovery flt to Milan

	27 Mar 69	recovery flt to Sudan
	18 Mar 71	Nose wheel collapsed on touchdown in cross wind at Le Touquet France. Acft slid to stop 1550 m from threshold. 7 crew 11 psgrs no fatalities. Damage beyond repair w/o
	02 Jun 71	British reg cxld

C-54A-15-DC Douglas c/n 10351 (Built at Chicago)

42-72246	15 Jul 44	dlvr to USAAF
	17 Jun 46	rtn to Douglas Aircraft Co. at El Segundo. conversion #25 to DC-4
NC88919	15 Mar 47	pchs by Pan American World Airways name: "Clipper Panama"
	-- --- --	re-name "Clipper Talisman"
	-- Oct 47	dmgd in landing accident in Alaska
	-- --- 48	pchs Twentieth Century Airlines
I-DALV	21 Apr 50	pchs by Alitalia (Italy) name: "Citta di Napoli"
N1346V	-- --- 54	pchs by California Eastern Airways
G-AOFW	09 Dec 55	pchs by Air Charter Ltd. name: "Jason"
	01 Jul 60	mgd to British United Airways

ATL-98-12

G-AOFW	23 Apr 63	dlvr to Aviation Traders Ltd
	09 Jun 63	nose dlvr by road SEN-STN
	11 Feb 64	FIRST FLT AS CARVAIR
	26 Mar 64	repaint AVIACO ferry to Southend
	01 Apr 64	C of A issued (GB)
	14 Apr 64	lsd by Iberia, Madrid
	15 Apr 64	Spanish C of A issued
EC-WVD	18 Apr 64	dlvr to Iberia (Spain) (reg for ferry flt)
EC-AVD	18 Apr 64	lsd to AVIACO (Spain)
EC-AVD	-- Apr 64	re-reg to AVIACO (Spain)
	01 May 64	began scheduled svc. AVIACO
	16 Nov 64	rtn to Aviation Traders stored at STN
	14 Jan 65	rmvd from storage
	25 Jan 65	issued British C of A
G-AOFW	08 Mar 65	pchs BUAF and re-reg GB
	26 Mar 65	arvd Southend for checks
	01 Apr 65	repaint BUAF dark stripe livery
	03 Apr 65	rtn to svc British United Air Ferries
	07 Feb 67	parked at Lydd Kent
	05 Mar 67	wfu stored at Lydd Kent
	01 Oct 67	company name chg British Air Ferries
	29 Apr 69	rmvd frm storage rtrn to svc in new two-tone blue BAF livery
	-- --- 72	name: "Big John"
	-- --- 73	repaint dark blue livery
	-- Aug 75	"Startrukin" logo for BTM rock group charters
	-- Jan 76	wfu Southend
	-- Jun 76	pchs by Secmafer (SFAir) not taken up
	01 Jan 77	last car ferry flt
	-- Feb 77	prkd Lydd Kent
	-- --- 77	mvd to Southend
	02 May 77	wfu stored at Southend (C of A expired)
	-- --- --	trsf to Panavia Air Cargo
	-- Jan 79	advertised for sale
	-- May 79	nose door damaged replaced w/door from salvage of CF-EPV

	-- Oct 81	static display Southend
	-- Jan 82	repainted displayed at Southend as a tribute to the community (last Carvair in BAF livery)
	-- Dec 83	b/o Southend, Essex. Some parts to Thameside Aviation Museum, Tilbury Essex
	30 May 85	British reg cxld

C-54-DO Douglas c/n 3058 (Built at Santa Monica)

41-37272	08 Feb 43	dlvr USAAF
NC79000	-- --- 46	pchs by Civil Aeronautics Administration
N79000	11 Apr 51	pchs by Eastern Air Lines name: "721"
	-- May 55	pchs Aero Leases (Company formed by California Eastern and North American)
	-- May 55	lsd to Grimley Engineering
	-- Jun 55	rtn to Aero Leases and sold to California Eastern
	-- Jun 55	pchs by Northwest Airlines
	-- --- 58	pchs by California Eastern Airways
	28 Feb 58	lsd to Transocean Air Lines (operated for one flight)
	-- Feb 58	rtrn California Eastern
	-- --- 59	pchs by World Airways
	-- --- 60	pchs by California Airmotive
	-- --- 60	lsd to Global Airways
	14 Oct 60	lsd to President Airlines
	-- --- 61	rtrn to California Airmotive
D-ADAM	-- May 61	pchs by Continentale Deutsche Leftreederei
	-- --- 61	repo Kunderkreditbank Essen
G-ASKN	22 Jul 63	pchs by Aviation Traders Ltd (oldest DC-4 converted to Carvair)

ATL-98-13

	25 Jul 63	ferry flt Hamburg to Stansted
	27 Jul 63	arvd Stansted
	28 Jul 63	nose arvd by road SEN-STN
	19 Dec 63	roll-out at Stansted
	08 Feb 64	FIRST FLT AS CARVAIR
	20 Feb 64	pchs. British United Air Ferries name: "Pont D'Avignon" Certificate of Airworthiness issued
	28 Feb 64	fully painted name not applied
	02 Mar 64	dlvr British United Air Ferries
	26 Mar 64	began schedule service to Ostend
	14 Dec 66	repaint in BUA livery
	14 Aug 67	lsd to Colin Beale operated as CB Flying Showcases (titles added)
	01 Oct 67	company name change to British Air Ferries
	04 Oct 67	dlvr British United Air Ferries
	-- --- 67	rtrn from CB lease Titles re-pain BUA
	30 Oct 67	ferry to Lydd for annual check change lettering BUA to BAF
	-- Dec 68	repaint BAF two-tone blue stripe c/s
	07 Dec 70	serious bird strike-nose door replaced with silver unit repaint one week later
	-- Nov 71	ATEL for maintenance and storage
	18 Dec 72	overhaul repaint dk-blue bottom for rtrn to svc. name "Big Bill"
	27 May 73	British Air Ferries name: "Big Bill"
	17 Jun 73	grounded due to engine shortage
	03 Jun 76	wfu parked Southend

TR-LWP	09 Jun 76	pchs Société Anonyme de Constr (SOACO) Libreville Gabon
	10 Jun 76	dlvr SOACO. British registration cxld
	08 Dec 76	Last flight
	20 Jun 78	Last eng run-up Brazzaville Congo Total time 60,983.47 hrs
	-- --- 79	confiscated by Empire of Central African Govt.
	-- --- 82	pchs by Aero Service (AFRIGO)
	-- May 86	b/o for spares Brazzaville Congo

C-54B-1-DC Douglas c/n 10458 (Built at Chicago)

42-72353	27 Nov 44	dlvr USAAF
	18 Jan 46	pchs by Douglas Aircraft Co. 17th cvrsn to DC-4 at El Segundo CA
NC88721	14 Sep 46	pchs by Western Air Lines
	12 Jan 56	pchs by SOBELAIR (totl time 27,467 hrs)
OO-SBO	15 Feb 56	reg to SOBELAIR
	06 Mar 56	dlvr Brussels name "Lomani"
	25 Mar 56	lsd by SABENA
	-- Jul 56	rtrn to SOBELAIR
F-BHVR	25 Nov 56	lsd by Union Aeromaritime de Transport
OO-SBO	15 Mar 57	rtrn to SOBELAIR
	15 Mar 57	lsd by SABENA
	31 Oct 59	rtrn to SOBELAIR
D-ABIB	08 Mar 60	pchs by Continentale Deutsche Luftreederei (totl time 33,992 hrs) Belgian reg cxld. German reg. NTU
D-ANEK	08 Apr 60	reg Continentale Deutsche
	18 Apr 60	pchs by Continentale Deutsche Luftreederei
	-- May 61	repo Marcard and Company. Grounded Hamburg
G-ASKD	09 Jul 63	pchs by Aviation Traders Ltd.

ATL-98A-14

G-ASKD	31 Jul 63	arvd at Stansted
	22 Sep 63	nose dlvr. by road SEN-STN
EI-ANJ	17 Apr 64	FIRST FLT AS CARVAIR
	22 Apr 64	C of A issued
	24 Apr 64	pchs by Aer Lingus name: "St. Seanan"/"Seanan" (modified for horse charters control runs in ceiling)
EI-ANJ	17 Jun 64	reg Ireland
	16 Feb 65	frst Aer Lingus Carvair to Scotland (race horse charter flight)
	-- Sep 66	wfu
	31 Oct 66	stored Dublin
	10 Jan 68	pchs by Eastern Provincial Airways (Canada)
	15 Feb 68	ferry to Southend for overhaul and cold weather mods Reg. Canada to Eastern Provincial Airways
EI-ANJ	29 May 68	Irish reg cxld
CF-EPW	04 Jun 68	frst flt after overhaul
	06 Jun 68	dprt Southend to Gander UK C of A for Export 7963
	04 Jul 68	Canadian C of A 5800 issued
	-- Oct 69	wfu
	-- Sep 70	rtrn to svc
	17 Apr 73	pchs by R.C.S Graham (Canada)
	18 Apr 73	lsd to Norwegian Overseas Airways for sub-lsd BAF
	20 Apr 73	nose wheel collapse Gander Newfoundland
	-- Jun 73	ferry Gander to Southend for service and repaint
	10 Jun 73	lsd to BAF

	03 Aug 73	Norwegian Overseas lease expired with R.C.S. Graham
G-ASKD	06 Aug 73	sub-lsd to BAF by Norwegian Overseas
	22 Aug 73	debut in BAF livery name: "Viking North Sea Express"
	02 Sep 73	lsd to Norwegian Overseas Airways
LN-NAA	17 Oct 74	pchs Norwegian Overseas Airways
	11 Nov 74	BAF lease expired with Norwegian Overseas
	02 Dec 74	trsf to Rorosfly Cargo (repaint Rorosfly lettering)
LN-NAA	19 Dec 74	reg to Thor Tjontveit Oslo Norway
	-- Jan 75	stored Athens
	24 Jan 75	rmvd frm storage for test flight/rtn to svc.
	28 Jan 75	Rorsfly take possession (sub-lsd to Rode Kors)
	02 Feb 75	ferry Southend to Oslo
	01 Apr 75	lsd to Intl' Red Cross
	08 May 75	engine change at Bahrain
	27 May 75	parked at Dubai (posbl Abu Dhabi)
	-- Jun 75	abandoned Bangkok
	09 Oct 75	lease expired Red Cross Intl
	-- Nov 75	wfu Bangkok Thailand
	31 Dec 75	listed as stored at Bangkok
	03 Jan 78	Graham rqst ferry auth BKK Terrace BC
	13 Nov 79	still intact at Bangkok
	01 Jun 81	Norwegian reg cxld
	-- --- 87	reported broken up Bangkok
	19 Jan 88	confirmed b/o Bangkok (exact date unknown)

C-54E-5-DO Douglas c/n 27311 (Built at Santa Monica)

44-9085	05 Apr 45	dlvr USAAF
NC88886	-- --- 45	trsf Reconstruction Finance Corp
	26 Sep 45	pchs by Douglas Aircraft Corp 2nd cvrsn to DC-4 at Santa Monica
	14 Nov 45	pchs Pan American World Airways name: "Clipper Mandarin" re-name "Clipper Frankfurt" re-name "Clipper Pegasus" re-name "Clipper Hannover"
OD-ADW	12 Jan 61	pchs Trans Mediterranean Airways-TMA
	-- Nov 63	pchs Aviation Traders Ltd.
	01 Nov 63	ferry Beirut-Basle-Stansted

ATL-98-15

OD-ADW	08 Nov 63	arvd Stansted
	-- Nov 63	disassembly Stansted
	10 Nov 63	nose dlvrd by road SEN-STN
	20 Dec 63	nose fitting completed
	-- Aug 64	completed at Stansted
	04 Oct 64	ready for flight testing
	-- Apr 65	stored less engines at Stansted up for sale
G-ATRV	07 Mar 66	British registration
	23 Mar 66	FIRST FLT AS CARVAIR
	25 Mar 66	pchs British United Air Ferries
	27 Mar 66	C of A issued ferry to Southend arvd with gray bottom
	01 Apr 66	began British United Air Ferries service Southend Rotterdam
	-- Apr 67	pchs (CAT) Cie Air Transport (France)
F-BOSU	06 May 67	frst flt with French registration in CAT colors
	07 May 67	trsf to (CAT) Cie Air Transport British reg cxld ferry Le Touquet-Nimes
	10 May 67	pchs by (CAT) Cie Air Transport name: "President Gamel"

	07 Jun 67	dlvr to CAT at Le Touquet
	-- --- 67	lsd to Shell Nigeria
	21 Nov 67	No.4 engine failed over Sahara desert
	-- Mar 70	French reg expired. Shell rtrn to CAT
	-- Aug 70	rtrn (CAT) Cie Air Transport wfu Stored Le Touquet
	-- Aug 70	pchs by TAR for spares/stored Le Touquet France
	-- Jan 72	French register list as still airworthy
	17 Apr 72	owner listed as Air Fret
	-- Nov 72	b/o Nimes-Garons, France
	-- Jun 74	insurance cxld
	-- --- 75	French reg cxld

C-54B-1-DC Douglas c/n 10485 (Built at Chicago)

42-72380	23 Dec 44	dlvr USAAF
NC90417	-- --- 46	pchs by American Airlines name: "Flagship Monterey"
EC-AEP	-- --- 49	pchs by Iberia Lineas Aereas Espanolas
	-- Apr 63	lsd by AVIACO
	15 Feb 64	trsf to Aviation Traders Ltd. ferry flt Barcelona to Stansted

ATL-98-16

	18 Feb 64	enter conversion Stansted
	02 Feb 64	nose dlvrd by road SEN-STN
EC-AXI	19 May 64	re-reg as Carvair (Spain)
	04 Jun 64	FIRST FLT AS CARVAIR
	17 Jun 64	C of A issued
	20 Jun 64	handed over to Iberia of Madrid
EC-WXI	24 Jun 64	dlvr AVIACO (Spain) (reg for ferry flight)
EC-AXI	-- Jun 64	re-reg AVIACO
	-- --- 68	Dominicana negotiate lease with option to pchs
HI-168	01 Feb 69	trsf Compañía Dominicana de Aviación (Dominicana) name: "President Malon"
	20 Feb 69	ferry Barcelona Santo Domingo
	25 Feb 69	rcvd Dominican C of A
	-- May 69	pchs Dominicana
	23 Jun 69	Crashed on takeoff Miami FL 3 crew 1 jump seat, 4 fatalities onboard 6 fatalities on ground
	01 Jan 70	Dominican registration cxld

C-54B-5-DO Douglas c/n 18342 (Built at Santa Monica)

43-17142	19 Jul 44	dlvr USAAF
NC30042	18 Oct 45	trsf Reconstruction Finance Corp
	07 Nov 45	pchs by Douglas Aircraft Corp 9th cvrsn to DC-4 at Santa Monica
	12 Feb 46	pchs by United Airlines name: "Cargoliner Rainier"
N30042	20 Mar 46	reg to United Airlines
	28 Dec 56	pchs Transocean Air Lines
	-- --- 58	lsd by Lufthansa
	-- --- 59	rtn to Transocean Air Lines
	21 Oct 59	pchs by United Airlines
	21 Oct 59	pchs by The Babb Co. Inc
LX-IOF	-- Mar 60	rqst reg Luxembourg
	21 Apr 60	pchs by U.S. Transport Corp

	-- Aug 60	lsd by Intercontinental Airways
	18 Apr 62	pchs by Interocean Airways
N30042	15 Jun 62	U.S. reg cxld

ATL-98A-17

LX-IOF	04 Feb 64	pchs by Aviation Traders Ltd. arvd Southend in error
	06 Feb 64	arvd Stansted
	-- Feb 64	attempt to interest AVIACO to cvt for Carvair 16
	10 Feb 64	entered line at Stansted
	22 Feb 64	nose dlvrd by SEN-STN
	02 Oct 64	completed Carvair stored at Stansted less engines and instruments
	-- Dec 68	prepared to remove from storage
G-AXAI	19 Feb 69	reg Great Britain
	17 Mar 69	work began to complete
	02 Apr 69	FIRST FLT AS CARVAIR STN-SEN
	04 Apr 69	C of A issued
	05 Apr 69	dlvr British Air Ferries no name
	-- --- 72	name: "Fat Albert" when painted dk-blue
F-BVEF	29 Dec 75	British reg cxld pending sale
G-AXAI	05 Jan 76	re-reg Britain after conversion to all cargo
	07 Jan 76	pchs by Secmafer dlvr to Nice
F-BVEF	27 Jan 76	reg France. British reg cxld
	01 Mar 76	SFAir created
	-- Mar 77	trsf SFAir
	-- --- 78	stored Nice France
	26 Nov 79	pchs by Airtime Corp. Portland OR (USA)
	07 Jan 80	French reg cxld
N55243	18 Feb 80	reg U.S.
	27 Feb 80	pchs by Aero Union
	14 Sep 80	remained in SF Air livery
	-- Apr 81	lsd to Kodiak Western
	-- --- 82	reg Pacific Air Express (not taken up)
	20 May 83	DC-8 roller floor sys installed
	27 Jul 83	pchs Pacific Air Express
	-- Aug 83	pchs Kemavia Inc. Honolulu
	-- Sep 83	lsd to Philippine Air Lines
	-- Jul 84	still in SFAir colors
	-- --- 84	pchs Pacific Air Express repaint Pacific Air Express livery
	-- --- 86	wfu Honolulu Hawaii
	24 Mar 88	ferry to Naples Florida
	26 Apr 88	pchs by Hondu Carib
	24 Jul 91	Naples placed lien on acft Owner Fred Lee Mason Great Southern Airways
	-- Aug 92	confiscated by City of Naples FL
	-- --- 92	pchs by Robt McSwiggan for salvage
	11 Feb 93	b/o at Naples FL Salvage parts removed to Griffin GA
	11 Jan 96	reg cxld.

C-54B-5-DO Douglas c/n 18340 (Built at Santa Monica)

43-17140	12 Jul 44	dlvr USAAF
NC90403	-- --- 45	pchs by American Airlines name: "Flagship Phoenix"
EC-AEO	-- --- 49	pchs by Iberia name "115"

ATL-98-18

EC-AEO	05 Nov 64	trsf to Aviation Traders Ltd.
	08 Nov 64	nose arvd by road SEN-STN
	10 Nov 64	acft arvd Stansted
	-- Feb 65	structure complete
EC-AZA	12 Mar 65	FIRST FLT AS CARVAIR
	15 Mar 65	C of A issued after 5.25 hrs of test
EC-AXY	-- --- 65	reg AVIACO (not taken up) (Spain)
	-- --- 65	AVIACO pchs from Iberia (Spain)
EC-WZA	-- Mar 65	reg for ferry flt to Barcelona
	26 Mar 65	dlvr over to AVIACO (Spain)
EC-AZA	-- Apr 65	re-reg Spain
	-- --- 68	wfu
HI-172	-- --- 68	Dominicana negotiate lease with option to pchs
	22 Mar 69	operated in Caribbean aid scheme
	-- May 69	pchs by Dominicana (Dominican Rep)
	12 Jan 74	wfu stored Santo Domingo
	30 Sep 76	new color scheme
	-- Jul 76	acft complete at Santo Domingo
	-- Nov 78	b/o Santo Domingo (MDSD)
	-- Jan 79	fuselage moved to El Embajador Hotel Santo Domingo DR. converted to restaurant
	-- Feb 84	cvt to night club.

DC-4-1009 Douglas c/n 42927 (Built at Santa Monica)

SE-BBD	17 May 46	dlvr to Svensk Interkontinental Luftrafik AB (SILA) name: "Monsun"
	-- --- 47	reg SAS (not taken up) name: "Styrbjorn Viking"
	01 Aug 48	mgd to SAS name: "Sigmund Viking"
JA6008	23 Feb 54	pchs by JAL name: "Zao"
	06 Aug 63	pchs By Ansett Airlines of New South Wales
	07 Aug 63	Japanese reg cxld VH-INJ Australian registration assigned
	09 Aug 63	dprt Tokyo
	12 Aug 63	arvd Melbourne/Essendon
	24 Feb 64	lsd to A.S.A.
	23 Mar 64	rtrn to Ansett

ATL-98-19

	26 Nov 64	nose arvd by road SEN-STN
VH-INJ	16 May 65	arvd Aviation Traders Ltd SEN
	17 May 65	ferry SEN-STN
	22 Jun 65	reg changed from DC-4 to Carvair
	14 Sep 65	FIRST FLT AS CARVAIR
	17 Sep 65	Australian C of A issued
	23 Sep 65	trsf STN-SEN
	24 Sep 65	Depart UK dlvr to Ansett ANA
	30 Sep 65	arvd Melbourne/Essendon
	25 Oct 65	First Carvair revenue flight
	01 Dec 68	Ansett renamed: Ansett Airlines of Australia
	01 Jul 69	ferry to Hong Kong for overhaul
	25 Nov 69	re-reg to Ansett Transport Ind
	02 Oct 70	struck by lghtng near Adelaide
	-- Apr 71	repaint in new "Alpha" c/s with Air Cargo titles

	01 Jun 71	struck by lghtng near Devonport
	09 Jun 72	last flt withdrawn frm svc. Melbourne/Tullamarine (Date Rprtd 01 FEB 72)
	-- Apr 73	pchs Southeast Asia Transport pending finance
	22 Jun 73	Ansett markings removed at Tullamarine AAS markings added.
	23 Jun 73	pchs Australian Aircraft Sales Sydney
	25 Jun 73	rqst for U.S. registration by Garfinkle
	29 Jun 73	sale cxld Southeast Asia Transport
	01 Jul 73	pchs by Jack M. Garfinkle (USA)
N33AC	04 Jul 73	U.S. reg applied to acft
	06 Jul 73	grnd by Australian Customs
	02 Aug 73	dprt Melbourne-Tullamarine
	02 Aug 73	lsd to South East Asia Transport Svc
	07 Aug 73	sub-lsd to Air Cambodge named: Barb
	26 Aug 74	lien filed by Robt Ferguson
	13 Apr 75	abandoned Pochentong Airport Phnom-Penh Cambodia (Khmer Rouge overran field)
	-- --- 76	presumed dismantled
	28 Apr 80	rqst de-reg U.S.
	31 Dec 80	rqst to cxl U.S. reg, acft destroyed
	13 Mar 81	U.S. reg cxld
	-- --- 82	verified at Phnom-Penh
	-- --- 97	fuselage reported intact Phnom-Penh (not verified)

DC-4-1009 Douglas c/n 42994 (Built at Santa Monica)

LN-IAE	24 Jun 46	pchs, dlvr DNL (Det Norske Luftfartsee Iskap Air Service)
	01 Aug 48	mgd with SAS name: "Olav Viking"
JA6012	27 Oct 56	pchs by Japan Airlines name: "Mikasa"
HL4003	29 Oct 62	lsd to Korean Airlines
JA6012	-- --- 63	rtrn to Japan Airlines
	08 Feb 64	Japanese reg cxld
VH-INK	21 Feb 64	pchs by Ansett reg Australia, overhauled at Hong Kong repaint in Ansett livery
	22 Apr 64	Ansett-ANA ferry Hong Kong, Melbourne
	20 Jun 65	trsf to Aviation Traders Ltd.

ATL-98A-20

VH-INK	25 Jun 65	arvd Aviation Traders Stansted
	19 Jul 65	Australian registration as DC-4 dropped
VH-INK	27 Oct 65	FIRST FLT AS CARVAIR
	28 Oct 65	reg chgd from DC-4 to Carvair recorded
	04 Nov 65	dprt UK for Melbourne/Essendon name: "Kasby I"
	11 Nov 65	arvd Ansett ANA Melbourne/Essendon
	01 Dec 68	Ansett renamed: Ansett Airlines of Australia
	12 Sep 69	repaint in new "Alpha" c/s with Air Cargo titles
	25 Nov 69	re-reg Ansett Transport Industries
	09 Mar 73	wfu stored Melbourne Tullamarine airport
	01 Jan 74	pchs (AAS) Australian Aircraft Sales Sydney (aka Air Australia Singapore)
	11 Feb 74	trsf Seulawah-Mandala (not taken up)
PK-	01 Jan 75	lsd to Seulawah-Mandala Air Services Jakarta
	04 Apr 75	repaint Seulawah Air titles
	22 Mar 75	prkd at Sydney

	-- May 75	Bau Air prkd at Melbourne
	03 Jul 75	test flt Melbourne
	19 Jul 75	dlvr Seulawah-Mandala Air Service Jakarta
	25 Aug 75	Australian registration cxld stored: Singapore-Seletar Airport
	-- Dec 75	Seulawah financing collapsed
	16 Jan 76	reverted to (AAS) Air Australia Singapore remained stored Singapore
	-- --- 76	pchs Air Express of Australia (not completed)
	-- Oct 76	Dwen Airmotive advertised for sale
	-- Aug 77	Australian Aircraft Sales advertised for sale.
N54598	17 Mar 78	pchs by James A. Cunningham (USA)
	-- Apr 78	overhaul by SINGAS at Singapore repaint Nationwide livery
ZK-EKY	17 Sep 78	reg to Nationwide Air (not taken up)
	18 Sep 78	planned ferry Brisbane, Auckland cxld
	28 Sep 78	pchs L. Leonard Lundy
ZK-NWA	01 Nov 78	reg Credit Finance Air Ltd. Auckland
	28 Nov 78	pchs by Nationwide Air Wellington New Zealand.
	-- Apr 79	Nationwide ceased operations
	-- Sep 79	wfu stored Nelson airport New Zealand.
	-- Oct 79	pchs James Air Ltd
	-- Feb 80	reg James Air New Zealand
	-- --- 80	Air Cargo Panama offer to pchs, sale not completed
N406JT	21 Dec 82	pchs by Turner Aviation Ltd. U.S. reg applied for
	-- Mar 83	repaint Pacific Aerolift livery (overhaul at Pacific Aerospace Hamilton)
	-- May 83	reg Pacific Aerolift (not taken up) (USA) name: "Ruth 1"
	-- Sep 83	test flt
	04 Nov 83	interest began accruing in NZ for unpaid balance on work performed
	-- --- 86	attempt to pchs by Golden Crown Aviation
N5459X	25 Jun 90	pchs by Hawaii Pacific Air rqst U.S. reg
ZK-NWA	02 Aug 90	still reg NZ
	-- Mar 92	dlvr Honolulu HI
N5459X	-- Apr 92	pchs by Air Cargo Hawaii (ptnrshp)
	13 Sep 93	reg Roberts Hawaii Inc
	-- Oct 93	wfu Honolulu
	-- Jun 95	remained wfu stored Honolulu HI
	29 Jul 96	Airline Marketing Consultants
	-- Aug 96	rtn to svc for dlvr to S. Africa
	-- Nov 96	dprt Honolulu ferry to Lakeland FL
	-- Feb 97	prkd Lakeland Fl due to engine failure
	-- Mar 97	Ferry to Griffin GA for maintenance
C-GAAH	-- Apr 97	pchs by HawkAir Terrace BC Canada
	14 Apr 97	U.S. reg cxld
	26 May 97	certification flt Griffin GA
	02 Jun 97	Ferried to Vancouver
	05 Jun 97	reg HawkAir
	30 Jun 97	nose wheel collapse Wrangell AK Nose wheel doors damaged
	-- Jul 97	Rolamat floor system removed due to unable to handle heavy cargo pallets
	-- Jun 99	prkd w/engines covered at Terrace British Columbia
	-- Oct 99	wfu Terrace BC. Up for sale
	03 Dec 02	pchs Brooks Fuel Ltd. Fairbanks AK
	27 Dec 02	de-reg Canada
	-- Dec 02	ferry to Fairbanks
	24 Feb 03	trsf Brooks Fuel Inc
N898AT	04 Mar 03	reg U.S. to Brooks Fuel Inc
	30 May 07	crashed McGrath Alaska NTSB #ANC07LAD40

C-54E-5-DO Douglas c/n 27314 (Built at Santa Monica)

44-9088	07 Apr 45	dlvr USAAF
NC88881	30 Aug 45	trsf to Reconstruction Finance Corp
	14 Nov 45	pchs by Douglas Aircraft Co 13th cvrsn to DC-4 at Santa Monica
	26 Dec 45	pchs Pan American World Airways name: "Clipper Kit Carson"
N88881	27 Dec 45	reg Pan American
	-- --- --	re-name: "Clipper Golden West"
	-- --- --	re-name: "Clipper Red Rover"
JA6015	17 Feb 58	pchs by Japan Air Lines name: "Amagi"
	27 Feb 58	U.S. reg cxld
	18 Mar 65	pchs by Ansett-ANA (Australia)
VH-INM	19 Mar 65	reg Australian National Airways
	23 Mar 65	depart Tokyo
	25 Mar 65	arvd Melbourne /Session
	29 Mar 65	Japanese reg cxld
	-- --- 65	lsd to MacRobertson Miller Airlines

ATL-98A-21

VH-INM	22 Feb 68	dprt Melbourne/Essendon for UK commanded by Capt J Coakley
	29 Feb 68	arvd Southend England
	13 Mar 68	DC-4 nose removed
	01 Mar 68	conversion begun at ATEL Southend
	16 Apr 68	Australian reg. as DC-4 dropped
	12 Jul 68	FIRST FLT AS CARVAIR
	16 Jul 68	reg re-instated as Carvair
	19 Jul 68	dlvr. to Ansett ANA (departed Southend) name: "Kasby II"
	26 Jul 68	arvd Melbourne Essondon Airport
	01 Aug 68	frst Carvair revenue flight AN861 Melbourne to Devonport
	01 Dec 68	Ansett renamed: Ansett Airlines of Australia
	25 Nov 69	re-reg Ansett Transport Industries Ltd.
	-- Jan 71	repaint in new "Alpha" c/s with Air Cargo titles
	01 Feb 73	wfu stored Melbourne Tullamarine airport
	01 Jan 74	pchs by Australian Aircraft Sales Sydney
	15 Jan 74	test flight Melbourne
	11 Feb 74	sale to Seulawah-Mandala collapsed
	04 Apr 75	repaint Seulawah Air titles
	-- May 75	attempt to pchs Bayu Air
	22 Jun 75	test flight Melbourne
PK-	04 Jul 75	repaint Seulawah Air livery
	07 Jul 75	dprt for Indonesia, cxld due to lease agreement not completed
	17 Jul 75	lsd by Seulawah-Mandala Air Services dprt Melbourne to Jakarta
	28 Aug 75	Australian reg cxld
	-- Dec 75	wfu collapse of Seulawah financing parked at Jakarta
	16 Jan 76	rtrn to Australian Aircraft Sales stored Singapore-Seletar
	-- --- 76	pchs Air Express of Australia (not completed)
	16 Aug 76	stored Singapore-Seletar
	-- Oct 76	Dwen Airmotive advertised for sale
N54596	17 Mar 78	pchs by James A. Cunningham (USA)
	-- Apr 78	overhaul by SINGAS at Singapore repaint in Nationwide livery
	-- Sep 78	planned ferry Brisbane, Auckland cxld
	14 Sep 78	pchs L. Leonard Lundy (Philadelphia)
ZK-EKZ	17 Sep 78	reg Nationwide Air (not taken up) (NZ)
N54596	29 Sep 78	reg to L. Lundy

	17 Oct 78	wing bob-wts removed, chgd from ATL-98A to ATL-98
	-- Nov 78	Credit Finance Air Ltd. Auckland
ZK-NWB	28 Nov 78	pchs by Nationwide Air (New Zealand) (Nationwide — merger between Akarana Air and Air North. Both div of Car Haulaways)
	29 Nov 78	de-reg U.S.
	-- Jun 79	failure No 4 engine at Christchurch three engine ferry to Hamilton
	16 Jul 79	Nationwide ceased operation acft prkd Nelson Airport NZ
	-- Sep 79	wfu stored Nelson
	-- Oct 79	pchs James Air Ltd.
	-- Feb 80	reg James Air of New Zealand
	-- --- 80	Air Cargo Panama offer to pchs, sale not completed
N407JT	21 Dec 82	reg Turner Aviation Ltd. (Hamilton NZ)
	21 Dec 82	reg Pacific Aerolift Cargo (not taken up) name: "Ruth II"
	-- --- 83	partial repaint Pacific Aerolift livery (overhaul at Pacific Aerospace Hamilton)
	-- --- --	stored Hamilton New Zealand
	-- --- 86	attempt to pchs by Golden Crown Aviation
	16 Feb 87	repaint two-tone orange cheat line
N5459M	-- Jun 90	pchs by Hawaii Pacific Air
	25 Aug 90	dlvr Hawaii Pacific Air
	17 Sep 90	de-reg New Zealand
	-- Mar 92	trsf Air Cargo Hawaii
	-- May 93	stored Honolulu
	13 Sep 93	trsf and reg. Roberts Hawaii
	-- Jun 95	wfu stored Honolulu Hi.
	-- Dec 95	advertised for sale
	29 Jul 96	pchs by Airline Marketing Consultants
	-- Aug 96	rtn to svc for trsf. to S. Africa
	27 Sep 96	reg Airline Marketing Consultants, Miami
	-- Oct 96	dprt Honolulu ferry to Africa
	31 Oct 96	prkd Griffin Georgia
	13 Nov 96	arvd South Africa Wonderboon airport lsd to Wonderair for Avia Air Charter
	-- Sep 97	rtn to Airline Marketing Consultants
	02 Nov 99	Triennial filed with FAA — acft destroyed
	15 Dec 99	Airline Marketing rqst U.S. de-register scrapped
	11 May 00	ferry Kinshasa to be broken up for spares
	05 Nov 00	ferry Pretoria — Johannesburg — Kinshasa upper fuselage painted white
	01 May 01	prkd Pietersburg S. Africa (for sale)
	-- Mar 02	pchs Phoebus Apollo, repaint
9J-PAA	-- Apr 02	reg Phoebus Apollo Johanesburg
	09 Apr 02	rqst to de-register U.S.
	30 Apr 02	U.S. reg cxld
	-- --- 05	restored to taxi status
	25 Jun 05	flew Johannesburg — Rand
	-- --- 07	stored Rand Airport South Africa without engines

Appendix 3: DC-4 Owners and Operators

Owners and Operators/ ATEL c/n Number

Aero Leases 13
Aerovias Panama Airways 9
Air Ceylon 5
Air Charter Limited 1, 11, 12
Air Charter Service Corp 1
Airplane Enterprise 6
Air World Leases 7
Alitalia 12
American Airlines 6, 16, 18
Ansett 19, 20, 21
Arrow Airlines 7
Astraeus 11
Australian National Airlines 5
AVENSA 6
AVIACO 16
Aviation Traders 14, 15
Blue Air 11
Braathens SAFE 10
Braniff Airways 1, 3, 4
British United Airways 1, 5, 12
California Airmotive 13
California Eastern Airways 2, 7, 12, 13
Channel Air Bridge 3, 4, 6
Civil Aeronautics Administration 13
Continentale Deutshce Leftreederei 13, 14
Delta Air Lines 11
Det Norske Luftfartsee Iskap (DNL) 20
Douglas Aircraft Corp. 0, 1, 9, 11, 12, 14, 15, 17, 21
Eastern Airlines 13
Federated Airlines 2
Global Aviation 13
Grimley Engineering 13
Guest Aerovias Mexico 9
Iberia 16, 18
Independent Air Transport 11
Intercontinental Airways 17
International Aircraft Sales 9

Interocean 7, 10, 17
Japan Airlines (JAL) 19, 20, 21
KLM Royal Dutch 0
Korean Airlines 20
Lewman Corp 10
Lloyd Aereo Bolivano 9
Lufthansa 17
Marshall Landy 7
Matson Navigation Company 5
North American Airlines 8, 11
Northwest Airlines 3, 13
Pan American World Airways 12, 15, 21
President Airlines 13
Remmert 9 ?
Resort Airlines 2, 3, 4, 6, 8
SABENA 14
Santa Fe Skyways 0
SAS 19, 20
Seaboard & Western 7, 10
Seven Seas Airways 5
Svensk Interkontinental Luftrafik AB (SILA) 19
Slick Airways 4, 8
SOBELAIR 14
The Babb Company 9, 17
Trans America Airlines 11
Trans Caribbean 10
Trans Mediterranean Airways 15
Transocean 2, 3, 7, 13, 17
Transavia Dusseldorf 5
Twentieth Century 5, 7, 8, 11, 12
Union Aeromaritime de Transport 14
United Airlines 17
United Nations 7, 10
USAAF 0, 1, 2, 3, 4, 5, 6, 7, 8, 9, 10, 11, 12, 13, 14, 15, 16, 17, 18, 21
US Navy 5
US Transport Corp 17
Veterans Air Express 5
Western Airlines 9, 14
World Airlines 1, 6, 8, 13

Appendix 4: ATL-98 Carvair Owners, Operators, Liveries

Owners, Operators, Liveries/ ATEL c/n Number

Aer Lingus 6, 8, 14
Aero Services 10, 11
Aero Union 17
Air Cambodge 19
Air Cargo Hawaii 20, 21
Air Ferry 10, 11
Air Fret 15
Airline Marketing Consultants 20, 21
Airtime Corp 17
Alisud 10
Ansett-ANA 19, 20, 21
Australian Aircraft Sales (AAS) 19, 20, 21
AVIACO 12, 16, 18
Aviation Traders 11, 13, 17
Avignon Inc. 9
British United Air Ferries (BUAF) 1, 2, 7, 9, 10, 11, 12, 13, 15,
British United Airways (BUA) 2, 7, 10, 11, 13
British Air Ferries (BAF) 5, 7, 8, 9, 10, 11, 12, 13, 14, 17
British Air Ferries (BAF Cargo) 7, 9
Brooks Fuel 20
Channel Air Bridge (CAB) 1, 2, 3
Colin Beale (CB Flying Showcases) 13
Compagnie [Cie] Air Transport (CAT) 4, 5, 10, 15
Custom Air Service 5, 9, 17
Dominicana 16, 18
Dwen Airmotive 20, 21
Eastern Provincial (EPA) 6, 8, 14
EclAir 10
Falcon Airways 5, 7, 9
Gator Global 9
Gifford Aviation 5, 17
Great Artic Airways 7
Great Southern Airways (Fred Mason) 7, 17
Hawaii Pacific Air (HPA) 20, 21
HawkAir 20
Hondu Carib (Frank Moss) 7, 17
Iberia (AVIACO) 12, 16, 18

Intercontinental 4, 5
International Red Cross 9, 14
Interocean 4, 5
Invicta 7
Jack M Garfinkle 19
James Air Limited 20, 21
James Blumenthal 7, 9
James Cunningham 20, 21
Kemavia 7, 17
Kodiak Western Alaska 5, 17
L Leonard Lundy 20, 21
Nasco Leasing 7, 9
Nationwide Air 20, 21
NCBA Inc 9
Norwegian Overseas Airways 8, 14
ONUC (United Nations) 4, 5
Pacific Aerolift (PAL) 20, 21
Pacific Air Express 7, 17
Panavia Air Cargo 12
Pauling Middle East 5
Philippine Air Lines 17
Phoebus Apollo 21
R C S Graham 8, 14
Roberts Hawaii 20, 21
Rode Kors 14
Rorosfly Cargo (RFN) 14
Ruth A May 5
Secmafer 17
Seulawah-Mandala 20, 21
SFAir 17
Shell Nigeria 15
Shell Oil 11
SOACO 9, 10, 13
Southeast Asia Air Transport (SEAAT) 19
Straits Air Freight Express (SAFE) 11
Transports Aeriens Reunis (TAR) 5, 10, 15
Transworld Leasing 9
Turner Aviation 20, 21
Uni Air 5
Wonderair 21
Yesterdays Wings 7

Appendix 5: DC-4 Registration

Registration/ATEL c/n Cross Reference

Civil

BOLIVIA

CP-682	9

CEYLON

CY-ACA	5

GERMANY

D-ABIB	4 (ntu)
D-ADAL	5
D-ADAM	13
D-ANEK	14
D-ANET	11 (ntu)

SPAIN

EC-AEO	18
EC-AEP	16

FRANCE

F-BHVR	14

GREAT BRITAIN

G-ANYB	1
G-AOFW	12
G-APNH	11
G-AREK	5
G-ARSH	4
G-ARZV	6
G-ASKD	14

SOUTH KOREA

HL-4003	20

PANAMA

HP-256	9
HP-258	9

ITALY

I-DALV	12

JAPAN

JA6008	19
JA6012	20
JA6015	21

NORWAY

LN-HAU	10
LN-IAE	20

LUXEMBOURG

LX-BBP	10
LX-BNG	7
LX-IOF	17

USA

N1221V	10
N1346V	12
N5520V	5
N9326R	9
N30042	17
N37477	11
N54373	7
N57670	2
N59952	1
N65142	4
N75298	6
N79000	13
N88709	3
N88819	8
N88881	21
N90431	6
NC30042	17
NC54373	7
NC57670	2
NC58003	5
NC59952	1
NC65142	4
NC88721	14
NC88723	1
N88816	9
NC88881	21
NC88886	15
NC88919	12
NC90403	18
NC90417	16
NC90431	6
NC90862	0
NX88881	21

LEBANON

OD-ADW	15

BELGIUM

OO-SBO	14

NETHERLANDS

PH-DBZ	0
PH-EUR	11 (ntu)
PH-TEZ	0

SWEDEN

SE-BBD	19

AUSTRALIA

VH-INJ	19
VH-INK	20
VH-INM	21
VH-INY	5

LEWARD ISLANDS

VP-CBD	5

VENEZUELA

YV-C-AVH	6

MEXICO

XA-MAA	9

Military

UNITED STATES

41–37272	13
42–7227	10
42–72168	7
42–72223	4
42–72246	12
42–72260	2
42–72343	8
42–72353	14
42–72375	0
42–72380	16
42–72423	1
42–107461	6
43–17133	11
43–17139	3
43–17140	18
43–17142	17
44–9023	9
44–9085	15
44–9088	21
BU50843	5

Appendix 6: Carvair Registration

Registration/ATEL c/n Cross Reference

CANADA

CF-EPU	6 (ntu)
CF-EPV	8
CF-EPW	14
CF-EPX	6
C-GAAH	20

SPAIN

EC-AEO	18
EC-AVD	12
EC-AXI	16
EC-AXY	18 (ntu)
EC-AZA	18
EC-WVD	12
EC-WXI	16
EC-WZA	18

IRELAND

EI-AMP	6
EI-AMR	8
EI-ANJ	14

FRANCE

F-BMHU	4
F-BMHV	5
F-BOSU	15
F-BRPT	10
F-BVEF	17
F-BYCL	5

GREAT BRITAIN

G-ANYB	1
G-AOFW	12
G-APNH	11
G-AREK	5
G-ARSD	2
G-ARSF	3
G-ATRV	15
G-ARZV	6
G-ASDC	7
G-ASHZ	9
G-ASKD	14
G-ASKG	10
G-ASKN	13
G-AXAI	17
G-41-1-73	5
G-41-2	4

DOMINICA

HI-168	16
HI-172	18

NORWAY

LN-NAA	14
LN-NAB	8
LN-MDA	8

LUXEMBOURG

LX-IOG	4
LX-IOG	5
LX-IOH	4

UNITED STATES

N33AC	19
N80FA	7
N83FA	5
N83FA	9 (ntu)
N89FA	9
N103	7
N406JT	20 (ntu)
N407JT	21 (ntu)
N898AT	20
N5459M	21
N5459X	20
N9757F	5
N9758F	4
N54596	21
N54598	20
N55243	17

INDONESIA

PK-	20 (ntu)
PK-	21 (ntu)

CONGO

TN-ADX	10

GABON

TR-LUP	10
TR-LWP	13

AUSTRALIA

VH-INJ	19
VH-INK	20
VH-INM	21

NEW ZEALAND

ZK-EKY	20 (ntu)
ZK-EKZ	21 (ntu)
ZK-NWA	20
ZK-NWB	21

ZAIRE

9Q-CTI	10

ZAMBIA

9J-PAA	21

Appendix 7: Aircraft Names

DC-4/ATEL c/n

Amagi	21
Atalanta	1
Cargoliner Rainier	17
Citta di Napoli	12
Clipper Frankfurt	15
Clipper Golden West	21
Clipper Hannover	15
Clipper Kit Carson	21
Clipper Mandarin	15
Clipper Panama	12
Clipper Pegasus	15
Clipper Red Rover	21
Clipper Talisman	12
Dusseldorf	5
Flagship Phoenix	18
Flagship Monterey	16
Flagship Philadelphia	6
Jason	12
Laxapana	5
Lomani	14
Mikasa	20
Monsun	19
Norse Trader	10
Olav Viking	20
Oslo Air Trader	10
Sigmund Viking	19
Sky Chief Apache	0
Styrbjorn Viking	19
Wake Island Trader	7
Zao	19
Zeeland	0
"115"	18
"721"	13

ATL-98 Carvair/ATEL c/n

Barb	19
Big Bill	13
Big Joe	10
Big John	12
Big Louie	7
Cdt Henry de Montal	4
Cdt Max Geudt	5
Channel Bridge	10
Chelsea Bridge	2
Commandant Charles	10
Crème Puff* (Ruth May — Falcon Airways)	5
Fat Albert	17
Fat Annie	9
Golden Gate Bridge	1
Kasby I	20
Kasby II	21
Maasbrug	9
Menai Bridge	11
Plain Jane	7
Pont de Avignon	13
Pont de Crouch* (During construction)	11
Pont De L'Europe	3
Pont du Rhin	7
Porky Pete	5
President Gamel	15
President Malon	16
Ruth I	20
Ruth II	21
St Albert (Ailbhe)	6
St Jarlath (Iarfhlaith)	8
St Seanan (Seanan)	14
Startrukin* (Summer of 1975)	12
Twiga* (During Rhodesian Blockade)	11
Viking North Sea Express	14

*Unofficial names

Glossary

AA   Automobile Association

AAM   Air America

AAS   Australian Aircraft Sales (AUS); Air Australia Singapore (SIN); American Aircraft Sales (USA)

ACL   Air Charter Limited

acft   aircraft

ADF   Automatic Direction Finder

AEA   Air Express Australia

AF   Air France

ANA   Australian National Airlines

AOG   Aircraft On Ground

ARB   Air Registration Board

arvd   arrived

ASA   Airline of Southern Australia

ATEL   Aviation Traders Engineering Limited

ANTENA   Aerovias Nacionales Transcontinental Ecuador

ATI   Ansett Transport Industries

ATL   Aviation Traders Limited

ATLB   Air Transport License Board

ATOG   Allowable Take-off Gross

aux   auxilary

AvGas   Aviation Gasoline

AVENSA   Aerovias Venezolanas Sociedad Anonima

AVIACO   Aviación y Comercio

BAF   British Air Ferries

BAS   British Air Service

BBC   British Broadcasting Corporation

BEA   British European Airways

b/o   broken-up

BOAC   British Overseas Airways

BST   British Standard Time

BTM   British Talent Management

BTU   British Thermal Unit

BUA   British United Airways

BUAF   British United Air Ferries

CAA   Civil Aeronautics Administration (US)

CAA   Civilian Aviation Authority (UK)

C of A   Certificate of Airworthiness

CAB   Civil Aeronautics Board (USA); Channel Air Bridge (UK)

CAF   Confederate Air Force

CASI   Continental Air Services

CAT   Compagnie [Cie] Air Transport

CG   Center of Gravity

c/n   Construction Number

Combi   Combination cargo-passenger

cu/ft   cubic feet

cvt   convert/converted

cvrsn   conversion

cxld   cancelled

CZUM   Churchill Falls (ICAO code)

dba   Doing business as

DCA   Australian Directorate of Civil Aviation

Dehmal   Flight Simulator (Invented by Richard Dehmal)

DHL   Dalsey, Hillblom, Lynn

DNL   Det Norske Luftfartsee Iskap Air Services (Norway)

dlvr   deliver/delivered

dmgd   damaged

DNKK   Kano Nigeria (ICAO code)

dprt   departed

EAL   Eastern Air Lines

EclAir   Entreprise Consolide pour des Livraisons par Air

EDDF   Frakfurt (ICAO code)

EGCC   Manchester (ICAO code)

EGJJ   Jersey (ICAO code)

EPA   Eastern Provincial Airlines

Eqpd   equiped

FAA   Federal Aviation Adminstration

FBO   Fix Based Operator

FCAA   Kinshasa Congo (ICAO code)

FCRF   Albertville Congo (ICAO code)

FKKD   Douala Cameron (ICAO code)

flt   flight

F/O   Flight Officer

FOD   Foreign object damage

frst   first

grnd   grounded

HAECO   Hong Kong Aircraft Engineering Co

Hg   Inches of Mercury

HNL   Honolulu (IATA code)

HPA   Hawaii Pacific Air

HSSS   Khartom Sudan (ICAO code)

IATA   International Air Transport Association

ICAO   International Civil Aviation Organization

ILS   Instrument Landing System

IPEC   Interstate Parcel Express Company

ISA   International Standard Atmosphere

Isud   Issued

JAL   Japan Airlines

KAPF   Naples Florida (ICAO code)

km   kilometers

LAMS   London Aero Motor Services

LGB   Long Beach (IATA code)

LIRP   Pisa Italy (ICAO code)

LMLI   Malta (ICAO code)

lndg   landing

lsd   leased

LTU   Luftransport/Uternehmen

MAA   McRobertson Miller Airlines

MAC   Mean Aerodynamic Chord

mb   Milibar(s) (1000 mb = 1 bar)

MATS   Military Air Transport Service

MDSD   Santo Domingo (ICAO code)

METAR   Aviation Routine Weather Report

METO   Maximum Except Takeoff

mgd   merged

MIT   Massachusets Institute of

mph   Miles per hour

MTOGW   Maximum Take-off Gross Weight

NOTAM   Notice to Airmen

NSI   No Stop Inbound

ntu   not taken up

ONUC   Organisation des Nations unies au Congo

P&W   Pratt & Whitney

PAE   Pacific Air Express

PAL   Pacific Aerolift

PanAm   American World Airways

pax   passengers

PNH   Phnom Penh

PPPPA   Pig Pilots of Phnom Penh Association

pchs   purchase

prkd   parked

psi   Pounds per square inch

QANTAS   Queenlands and Northern Territory Aerial Service

QC   Quick Change

QEC   Quick Engine Change Package

RAC   Royal Automobile Club

RATO   Rocket Assist Take-off

RCAF   Royal Canadian Air Force

RCMP   Royal Canadian Mounted Police

reg   registered

RFN   Rorosfly Cargo Norway

rmvd   removed

rpm   revolutions per minute

rtn   returned

SABENA   Société Anonyme Belge d'Exploitation de la Navigation Aerienne

SAFE   Straits Air Freight Express

SAGAT   Società Azionaria Gestione Aeroporto Torino

SAS   Scandinavian Airlines System

SCAL   Silver City Airlines (Societe Commerciale Aerienne du Littoral)

SEAAT   South East Asia Air Transport

SEN   Southend England (IATA code)

SILA   Svensk Interkontinental Luftrafik AB

SINGAS   Singapore General Aviation Services

SOACO   Société Anonyme de Construction

SOBELAIR   Société Belge de Transports par air

SST   Super Sonic Transport

Sta   Station

STC   Supplementary Type Certificate

STN   Stansted England (IATA code)

TAA   Trans-Australian Airlines

TAD   Trans-Avia Flug GMBH Duseldorf

TCA   Trans Caribbean Airways

TMA   Trans Mediterranean Airways

TMAC   Trans Meridian Airways (UK)

T/O   Takeoff

TPA   Trans Provincial Airlines (Canada)

trsf   transfered

TSO   Time Scheduled Overhaul

TWA   Trans World Airlines

TWL   TransWorld Leasing

UAE   United Arab Emirates

UAT   Union Aeromaritime de Transport (France)

UDI   Unilateral Declaration of Independence (Rhodesia)

USAF   United States Air Force

USAAF   United States Army Air Force

USD   United States Dollars

VARIG   Empresa de Viacão Aerea Rio Grandense

V1   Decision Speed (point on take-off that if an engine fails, the pilot decides to continue or abort take-off)

VHF   Very High Frequency

wfu   withdrawn from use

w/o   written-off

ZFW   Zero Fuel Weight

ZO   Horizontal Datum Line

Chapter Notes

Chapter 1

1. "From Berlin Airlift to Skytrain," *Take Off*, 1989, Vol. 7, Part 82, 2284–2291.
2. Ibid.
3. Peter Jackson, *The Sky Tramps* (London: Souvenir, 1965), Chapter 4, And Hay for the Generals Charger.
4. Air Charter Advertisement, *Flight*, 29 May 1963.
5. "From Berlin Airlift to Skytrain," 2288.

Chapter 2

1. Richard Goring, storekeeper, Aviation Traders, Southend, and co-editor, *Anglia Aeronews*, correspondence July 2006 — July 2007.
2. "New Routes New Aircraft for BAF," *Aviation News*, Spring 1973.
3. "Porky Pete and His Friends Take-off," *Evening Echo*, 30 January 1975.

Chapter 3

1. Richard Goring, correspondence, July 2006 — July 2007.
2. Ian Callier, co-founder, *Anglia Aeronews*, ATC Southend. Compiled list while staff for *Anglia Aeronews*, 1964, letter, 02 September 2002.
3. "Payload 100,000 Pounds, AF Evaluates XC-99 in Service," brochure prepared by Consolidated Vultee Aircraft Corporation, Fort Worth — San Diego.
4. Brian Kerry, chief aerodynamicist, ATEL; Cliff Berrett, chief draftsman ATEL.
5. Guy Craven, staff photographer, Aviation Traders, Southend. Personal account of early days at ATEL.
6. Ibid.
7. ATEL, Carvair Maintenance Manual. Elevator Control System, Chapter 15, p. 6, fig. 2. Rudder Control System, Chapter 15, p. 10, fig. 3.
8. Ian Callier, "The ATL-98 Carvair — Part 7, List of Potential Buyers," *Anglia Aeronews*, April 1970, 35.
9. Ian Howse, loadmaster, BUAF, letter, 07 October 2003.

Chapter 4

1. Paul Hawkins, co-founder, HawkAir, Terrace, British Columbia, correspondence, May — July 2007.
2. Goring, recognized subtle differences in paint schemes captured in his photos, July 2006 — July 2007.

3. Robert D. "Bob" McSwiggan, managing director, Custom Air Service, Griffin, Georgia, interview, 25 June 2007.

Chapter 5

1. Walter B. McCarthy, Interocean general manager, Congo Division, letter to UN requesting Congo identification for flight crews, 04 October 1962.
2. Robert Dedman, captain, Interocean Airways., correspondence, November 2006–June 2007.
3. Interocean memo to United Nations, 26 November 1962.
4. Bernard Goldberg, corporate officer, Intercontinental U.S., letter to FAA, 25 February 1963, and certificate of repossession, Mercantile Bank of Dallas, Texas, 22 February 1982.
5. G.R. Morris, Interocean director of operations, letter to Alan W. Cooper, United Nations procurement office, 30 April 1963.
6. Dedman, identified the carrier as Tran-State.
7. Michael O'Callaghan, captain, Aer Lingus Irish Airline, Dublin, Ireland, interview, comments on fond memories of flying the Carvair, 06 January 2003.
8. Ibid.
9. Ibid.
10. J.J. Sullivan, chief instructor, Aer Lingus Irish Airline, Dublin, Ireland, letter relating incident, 18 February 2003.
11. O'Callaghan, interview, 06 January 2003.
12. Joe Dible, captain, Aer Lingus Irish Airline, Dublin, Ireland, personal log book.
13. Fred Niven, Ansett Aviation historian, registration assignment research.
14. Lindsay Wise, Ansett engineer, Australia Department of Airworthiness surveyor, comments on Carvair conversion at Stansted.
15. Callier, *Anglia Aeronews*, April 1970, 35.
16. Kenny Smith, vice president, Cordova Airlines, "Airplanes in the Wrangells, 'Mergers,'" *St. Elias News*, Vol. 10, Issue 2, March-April 2001.
17. Louis Khem, president, Pacific Air Express and Kemavia, Honolulu, Hawaii, letter to FAA, 31 May 1983.

Chapter 7

1. Roger D. Launius, "Edward J. Daly, World Airways," in *The Airline Industry*, William M. Leary (New York: Facts on File, 1992), pp. 133–138.
2. Ronald Wilson, Key Publishing Limited Aviation Forum, 10 November 2006.

Chapter 8

1. "Across the Iron Curtain," *BUA Group News*, Autumn 1993, No 4.

Chapter 9

1. Frank J. Erhart, chief investigator, Netherlands Aviation Safety Board, Ministry of Aviation, Civil Aircraft Accident Report, (CAP 201) 30 October 1963 (London: Her Majesty's Stationery Office).
2. Ibid.
3. Ibid.
4. Ibid.
5. ATEL Carvair Operational Manual.
6. Ministry of Aviation, Civil Aircraft Accident Report, 30 October 1963.
7. Ibid.
8. Ibid.

Chapter 10

1. Dedman, personal recollections of Interocean Airways, November 2006 — June 2007.
2. Ibid.
3. Robert Dedman, pilot's logbook.
4. Dedman, personal recollections.
5. McCarthy, letter to chief of general services, United Nations, 04 October 1962.
6. Robert Dedman, "Congo Pilot, Flying Cargo During a Civil War," *Flight Journal*, April 2003, 60–64.
7. Dedman, 64.
8. Dedman, 63, and correspondence November 2006 — June 2007.
9. Dedman, correspondence November 2006 — June 2007.
10. *Flight*, 15 August 1963, and *Aviation Week and Space Technology*, 19 August 1963, 163.
11. Carvair Flight Manual, Operation Procedures, Weight Limitations, Engine Failure.

Chapter 11

1. FAA Registration/Airworthiness File, N58003 and N57777, CD-ROM.
2. William L. Worden. *Cargoes: Matson's First Century in the Pacific*. 1981, 102–105.
3. FAA Registration/Airworthiness File, N58003 and N57777, CD-ROM.
4. A.B. Eastwood and J. Roach, *Piston Airliner Production List* (Middlesex, England: Aviation Hobby Shop, 1991), 117, 124.
5. Ron Edwards, "Matson Airlines," *Ampersand*, Winter 1995–96, 17–19.
6. Ibid.
7. Earnest K. Gann, *Fate Is the Hunter* (New York: Simon & Schuster, 1961).
8. Ibid.
9. FAA Registration/Airworthiness File N9757F, Interocean, CD-ROM.
10. Paul Rakisits, captain, Interocean Airways, SEAAT, Pacific Air Express, interview, 20 May 2007, and correspondence, June 2007.
11. Brian Mees, engineer, British Air Ferries, taped interview and correspondence, August — November, 2005.

12. Ibid.
13. Ibid.
14. Ibid.
15. Ibid.
16. Ibid.
17. Ibid.
18. Ibid.
19. Ibid.
20. Ibid.
21. Ibid.
22. Ibid.
23. Ruth May, owner, N83FA, Falcon Airways, Dallas, Texas, interview and personal recollections, 03 April 1997, Honolulu; telephone interview, 25 May 1995.
24. Ibid.
25. Ibid.
26. Jean-Pierre Sauval, Uni-Air International, Toulouse, France, letter to Ruth May and spares list, 10 July 1979.
27. Goldberg, letter to FAA, 25 February 1963.
28. Sauval, letter of conformation, 17 August 1979.
29. Ruth May, delivery of N83FA at Toulouse, France.
30. Ibid.
31. Ibid.
32. Ruth May, interview, 03 April 1997.
33. L.A. Bassett, credit controller, BAF Engineering, letter to Ruth May.
34. Ruth May, ferry log N83FA, September 1979, 6.
35. Ruth May, personal account of ferry flight, interview, 03 April 1997.
36. Ibid.
37. Ibid.
38. Mercantile National Bank, loan documents for N83FA.
39. Ruth May, interview, 03 April 1997.
40. Lease agreement with Gifford Aviation.
41. John Gifford, president, Gifford Aviation, Anchorage, Alaska, letter to Ruth May, 19 May 1983.
42. Ruth May, personal notes, journal and files on N83FA.
43. Ruth May, handwritten note.
44. Ruth May, letters to potential buyers, 22 July 1983.
45. Gifford, letter, 20 October 1983.
46. Ruth May, memo from files.
47. Robert McSwiggan, interview, 20 May 1996.
48. Ibid.
49. Mark McSwiggan, Academy Airlines, Griffin, Georgia, interview, 20 May 1996.
50. Robert McSwiggan, interview, 20 May 1996.
51. NTSB ATL97FA057, incident report, crash of N83FA.
52. ATL-98 Operation Manual.
53. Patrick Dean, lanyard in all pre-crash cockpit photos tucked behind seat.
54. Mark McSwiggan, interview, 25 August 1997.
55. Ibid.
56. Robert McSwiggan, interview, 02 May 1998.
57. Ibid.

Chapter 12

1. O'Callaghan, interview, 06 January 2003.
2. Robert McSwiggan, interview, 02 May 1998.
3. O'Callaghan, interview, 06 January 2003, and letter, 03 June 2007.
4. Dible, letter, 29 December 2002.
5. "Tribute to Progress," *St. John's Telegram*, 25 October 1968.

6. Sullivan, letter, 18 June 2007.
7. Ibid.
8. Ibid.
9. Canadian Accident Investigation Board.
10. Ibid.
11. Ibid.

Chapter 13

1. FAA Registration/Airworthiness File N54373, CD-ROM.
2. Ibid.
3. Ibid.
4. Ibid.
5. John Erhart, Netherlands Air Accident Board. Ministry of Aviation Civil Aircraft Accident Report (CAP 201). London: Her Majesty's Stationery Office, 30 October 1963.
6. Graham Cowell, "Texas Bound with Plane Jane," *Propliner*, No. 3, Summer 1979, p. 29.
7. Ruth May, interview, 03 April 1997, and letter from Southend customs agent Richard Vandervord, 25 March 2006.
8. Cowell, "Texas Bound with Plane Jane," 29.
9. Ruth May, interview, 25 May 1999.
10. James Blumenthal, owner of N80FA and N89FA, Kingman, Arizona, interview, 02 November 1998.
11. A.P. Fairchild, Hawaii Pacific Air, Honolulu, Hawaii, interview, April 1990.
12. Khem, letter to FAA, 31 May 1983.
13. Frank Moss, owner, Honduras Caribbean (Hondu Carib) certificate, Tela, Honduras, interview, 20 July 2007.
14. Ibid.
15. Ibid.
16. Ibid.
17. Moss, telephone interview from Bronson Creek, Alaska, 20 July 1997.
18. Ibid.

Chapter 14

1. Sullivan, letter relating Irish Car-Ferry, 18 February 2003.
2. Dible, letter relating his experience with the Carvair, 28 December 2002.
3. Ibid.
4. O'Callaghan, interview, 06 January 2003.
5. Richard Gaudet, Dartmouth, Nova Scotia, letter relating his father's tenure as vice president of marketing for Eastern Provincial Airways, 05 February 2001.

Chapter 15

1. Blumenthal, interview, 02 November 1998.
2. Ruth May, interview, 03 April 1979.
3. Blumenthal, interview, 02 November 1998.
4. Ibid.
5. Steve Kehmeier, NCBA Inc., Denison, Texas, letter to Bob McSwiggan, 23 November 1986.
6. Robert McSwiggan, interview, 20 May 1996.
7. McSwiggan, interviews, 20 May 1996 and 25 June 2007.
8. George Dyess, owner, Ful-Air Aircraft Services Inc., Griffin, Georgia, interviews, 25 August 1997 and 20 May 1996.
9. Robert McSwiggan, interview, 25 June 2007.
10. Moss, interview, 20 July 2007.

Chapter 16

1. Mees, interviews, August — November 2005.
2. Michael Anciaux, SABENA, letter of 30 November 2000, describing condition of aircraft on 20 March 1993 and notations from aircraft logbook.

Chapter 17

1. Jackson, *The Sky Tramps*, 163.
2. Callier, letter, 27 September 2003.
3. Howse, letter, 07 October 2003.
4. Callier, letter, 07 September 2002.

Chapter 18

1. ATEL memo listing aircraft weights.
2. Mees, BAF engineer, letter, 01 November 2005.
3. Flight International, advertisements, June 1976 and May 1978.
4. "Carvair Hacked To Pieces," *Propliner*, No. 20, Winter 1983.
5. Keith Burton, Southend, England, witness to break-up of G-AOFW.

Chapter 19

1. Joseph E. Libby, "Stanley D. Weiss," in *The Airline Industry*, William M. Leary (New York: Facts on File, 1992), p. 497.
2. *Anglia Aeronews*, Southend. 10 September 1967.
3. Howse, letter, 07 October 2003.
4. T.D. Keegan, chairman, BAF, British Air Ferries internal memo, "All Engineers," 27 June, 1973.
5. Anciaux, logbook of TR-LWP.
6. Anciaux, interviewed Raymond Griesbaum, 25 May 1993.

Chapter 20

1. O'Callaghan, pilot's logbook.
2. O'Callaghan, correspondence, July 2007.
3. Dible, pilot's logbook.
4. O'Callaghan, correspondence regarding diversion to Gander, July 2007.
5. Mees, interview, August and November 2005.
6. Ibid.
7. Ibid.
8. Ibid.

Chapter 21

1. ATEL memo listing aircraft weights.

Chapter 22

1. Breve Historia del Avión Carvair, ATL98. Centro Documentación General, Iberia.
2. Ibid.
3. NTSB AAR-70-17, Aircraft Accident Report, Compañía Dominicana de Aviación, Miami, June 23, 1969, 7.

4. Breve Historia del Avión Carvair, Iberia.

5. Terry Johnson King, "Dominicans Are Amazed U.S. Let Plane Take Off," *The Miami News*, 25 June 1969, 1A.

6. NTSB AAR-70–17, Aircraft Accident Report, Miami, 7.

7. Ibid, 3.

8. Ibid, Appendix B, 1, 2.

9. Ibid, Appendix B, 1.

10. Ibid.

11. Ibid.

12. Ibid., Appendix B, 2.

13. Flight Manual, ATL-98 Carvair.

14. Fred Anderson and Frank Soler, "Only Fixed DC-4s Oil Valve, Moonlighting Mechanics Say," *The Miami Herald*, June 26, 1969, 1.

15. NTSB AAR-70–17, Aircraft Accident Report, Miami, 4.

16. Flight Manual, ATL-98 Carvair.

17. NTSB AAR-70–17, Aircraft Accident Report, Miami, 4.

18. "We're Going to Have a Crash," *The Miami News*, June 24, 1969, 7A.

19. NTSB AAR-70–17, Aircraft Accident Report, Miami, 24.

20. Ibid.

Chapter 23

1. FAA Registration/Airworthiness File N30042, CD-ROM.

2. Ibid.

3. Ibid.

4. Ibid.

5. Goring, letter, 22 January 2007.

6. FAA Registration/Airworthiness File N55243, CD-ROM.

7. Ruth May, handwritten note in personal papers and journal.

8. Ibid.

9. FAA Registration/Airworthiness File N55243, CD-ROM.

10. Ibid.

11. Moss, telephone interview, 14 July 2007.

12. FAA registration/Airworthiness File N55243, CD-ROM.

13. *Trade-A-Plane*, April 1992.

14. Robert McSwiggan, interview, 20 May 1996.

15. Dyess, interview, 20 May 1996.

Chapter 25

1. Fred Niven, Ansett Aviation historian, VH-INJ history.

2. Ibid.

3. Ibid.

4. Ibid.

5. FAA Registration/Airworthiness File N33AC.

6. Ibid.

7. Ibid.

8. Paul Howard, aviation writer-historian, Southeast Asia and Australia, Cairns, correspondence, June 2002 — July 2007.

9. Rakisits, interview, 20 May 2007.

10. Paul Howard, "Convairs of The Triple Nine Triangle," *Propliner* 62, Spring 1995, 38.

11. Rakisits, interview, 20 May 2007.

12. Ibid.

13. Ibid.

14. Ibid.

15. Ibid.

16. Ibid.

17. Ibid.

18. Howard, "Convairs of Triple Nine," 39.

19. Rakisits, interview, 20 May 2007.

20. Howard, "Convairs of Triple Nine," 41.

21. Robert M. Ferguson, SEAAT, letter to FAA, 26 August 1974.

22. Jack M. Garfinkle, president SEAAT, letter to FAA, 28 April 1980.

23. Garfinkle, FAA Form 8050–17, Deregistration of United States Civil

Aircraft, March 13, 1981.

24. Howard, letter, 30 December 2004.

25. Paul Howard and Bill Hickox, "Mercy Angels of Angkor," *Air Enthusiast*, November 2005, Issue 120, 28.

26. Rakisits, interview, 20 May 2007.

Chapter 26

1. Fred Niven, Ansett Aviation historian, VH-INK history.

2. Ibid.

3. Ibid.

4. Keith Gordon, engineering manager, SINGAS, Singapore, letter 21 August 2005.

5. Ibid.

6. Ibid.

7. Export Certificate E165071, FAA Registration/Airworthiness File N54598, CD-ROM.

8. Ibid.

9. Gordon, letter, 21 August 2005.

10. Aircraft Registration Application Form 8050–1, FAA Registration/Airworthiness File N54598, CD-ROM.

11. Murray Dryer, Pacific Aerospace, Hamilton, Condition Report ZK-NWA, compiled 1983.

12. Ibid.

13. Evaluation of Hawaii Inter-Island Air Cargo Service Using Two Carvair Aircraft, CAD Report 005, Auckland, New Zealand, December 1983, 27, 34, 35.

14. Aircraft Registration Application Form 8050–1, FAA Registration/Airworthiness File N5459X, CD-ROM.

15. FAA Registration/Airworthiness File N5459X, CD-ROM.

16. Ibid.

17. Ibid.

18. Ibid.

19. Ibid.

20. Robert McSwiggan, interview, 25 June 2007.

21. Transportation Safety Board of Canada, Report A96DO175, C-FGNI.

22. Robert McSwiggan, interview, 25 June 2007.

23. Hawkins, correspondence, May-July 2007.

24. Ibid.

25. Hawkins, letter, 12 June 2007.

26. Hawkins, letter, 01 June 2007.

27. Hawkins, letter, 11 February 2004.

28. Ibid.

29. Ibid.

30. Ibid.

31. Ibid.

32. Ibid.

33. Ibid.

34. Ibid.

35. Application for Airworthiness Certificate FAA Form 8130–6, N898AT, CD-ROM.

36. NTSB ANC07LAD40, Crash N898AT, 30 May 2007.

Chapter 27

1. Fred Niven, Ansett Aviation historian, VH-INM history.

2. Ibid.

3. Ibid.

4. Gordon, letter, 21 August 2005.

5. FAA Registration/Airworthiness File N54596, CD-ROM.

6. Ibid.

7. Aircraft Registration Application Form 8050–1, FAA Registration/Airworthiness File N54596, CD-ROM.

8. Evaluation of Hawaii Inter-Island Cargo Service, December 1983, Appendix C, 34.

9. Ruth May, personal notes, 07 June 1983.

10. Evaluation of Hawaii Inter-Island Cargo Service, December 1983, sec 2, 5.

11. Heavy Maintenance Report, Pacific Aerospace, Hamilton, Item 11, 2.

12. Evaluation of Hawaii Inter-Island Cargo Service, December 1983, Appendix C, 34.

13. Heavy Maintenance Report, Pacific Aerospace, Hamilton, Additional Modifications, Item 1.

14. Greg, Wiles, "Hawaii Pacific Air Taking Off," *Honolulu Advertiser*, February 1992, and FAA Registration File N5459M, CD-ROM.

15. FAA Registration/Airworthiness File N5459, CD-ROM.

16. Ibid.

17. Ibid.

18. Ibid.

19. Robert McSwiggan, interview, 25 June 2007.

20. Ibid.

21. FAA Registration File N5459M, CD-ROM.

22. Ibid.

23. Ibid.

24. Keith Gaskell, "Rising Star," *Propliner* 82, Spring 2000, 36.

25. Ibid.

26. Ibid.

27. "Smoked Out," Mailbag, *Airways*, July 2000, 63.

Bibliography

Books

Bilstein, Roger. "Thomas E. Braniff (Braniff Airways)," in *The Airline Industry*, William M. Leary, pp. 73–78.

_____ and Linda Bilstein. "Braniff Airways," in *The Airline Industry*, William M. Leary, p. 78.

Bishop, John C. "Earl F. Slick (Slick Airways)," in *The Airline Industry*, William M. Leary, pp. 433–434.

_____. "Seaboard World Airlines," in *The Airline Industry*, William M. Leary, p. 422.

Doyle, Paul. *Air Bridge 2*. Hertfordshire, England: Forward Airfield Research, 2000.

_____ and David Pugh. *Air Bridge 1*. Hertfordshire, England: Forward Airfield Research, 2005.

Eastwood, A.B., and J. Roach. *Piston Engine Airliner Production List*. Middlesex, England: Aviation Hobby Shop, 1991.

Finnis, Malcolm. *Twilight of the Pistons*. Eastbourne, U.K.: M.J. Finnis, 1997.

Francillon, Rene J. *McDonnell Douglas Aircraft Since 1920*, Vol. I. London: Putnam, 1988.

Gann, Earnest K. *Fate is the Hunter*. New York: Simon & Schuster, 1961.

Green, William, and Gerald Pollinger. *The Aircraft of the World*. Garden City, NY: Doubleday, 1965.

Hamlin, John F. *The Stansted Experience*. Peterborough, England: GMS Enterprises, 1997.

Jackson, A.J. *British Civil Aircraft Since 1919*, Vol. One. London: Putnum, 1974.

Jackson, Peter. *The Sky Tramps: The Story of Air Charter*. London: Souvenir Press, 1965.

Jones, Marsh. *The Little Airline that Could: Eastern Provincial Airlines*. St.Johns, Newfoundland: Creative, 1998.

Launius, Roger D. "Edward J. Daly (World Airways)," in *The Airline Industry*, William M. Leary, pp. 133–139.

_____. "Orvis M. Nelson," in *The Airline Industry*, William M. Leary, pp. 309–313.

_____. "Transocean Air Lines," in *The Airline Industry*, William M. Leary, p. 459.

Leary, William M. *The Airline Industry*. New York: Facts on File, 1992.

_____. "Trans Caribbean Airways," in *The Airline Industry*, William M. Leary, p. 457.

Libby, Joseph E. "North American Airlines," in *The Airline Industry*, William M. Leary, p. 315.

_____. "Stanley D. Weiss," in *The Airline Industry*, William M. Leary, pp. 496–497.

Pearcy, Arthur. *Douglas Propliners*. Shrewsbury, England: Airlife Publishing, 1995.

Porter, Malcolm. *Swingtail — The CL-44 Story*. Kent, England: Air Britain (Historians), 2004.

Smith, Hubert. *The Illustrated Guide to Aerodynamics*. Blue Ridge, PA: Tab, 1985.

Tambini, Anthony J. *Douglas Jumbos:The Globemaster*. Boston: Branden, 1999.

Taylor, W.R., and Kenneth Munson. *History of Aviation*. New York: Crown, 1972.

Tryckare, Tre. *The Lore of Flight*. Gothenberg, Sweden: Cagner, 1970.

Wegg, John. *Caravelle*. Sandpoint, ID: Airways, 2005.

Whybrow, Douglas. *Air Ferry: The Story of Silver City and Channel Air Bridge*. London: Tourism International, 1995.

Woods, John, and Maureen Woods. *DC-4*. Middlesex, England: Airline, 1980.

Worden, William L. *Cargoes: Matson's First Century in the Pacific*. Honolulu: University Press of Hawaii, 1981.

Wright, Alan J. *The British World Airlines Story*. Leicester, England: Midland, 1996.

Articles

Abt, Hanspeter. "Propellerjumbo der sechziger Jahre." *Jetstream*, August 1995, pp. 34–39.

"Across the Iron Curtain." *BUA Group News*, No. 4, Autumn 1963.

"Aer Lingus Gets the First Carvair." *BUA Group News*, No. 3, February 03, 1963.

"The Air Bridge's New Ferry." *Flight*, February 19, 1960, p. 255.

"Air Charter Limited." *Flight*, May 29, 1953.

"Air Ferries Expansion." *BUA Group News*, No. 4, Autumn 1963.

"Airline Liveries." *Scale Aircraft Modeling*, Vol. 19, No. 7, September 1993.

"Airplane Once Carried English Channel Trade." *Miami News*, June 24, 1969.

"Alisud Carvair Lease." *BUA Group News*, No. 4, Autumn 1963.

Allen, Margaret. "Trio Launch On-demand Air Freight Business." *Dallas Business Journal*, 24 May 2002.

"Aloha Sells Island Air Unit." Honolulu Starbulletin.com, December 6, 2003. http://starbulletin.com/2003/12/06/news/story1.html (17 March 2007).

"And 17 Got Out Alive." *Daily Sketch*, December 28, 1952.

Andersen, Fred. "DC4 Parts to Get Detailed Analysis." *Miami Herald*, June 27, 1969.

_____. "Plane Slams Into 36th St." *Miami Herald*, June 24 1969, p. 1A.

_____ and Frank Soler. "Only Fixed DC4's Oil Valve." *Miami Herald*, June 26, 1969.

"Arvid P. Fairchild," obituary. *Honolulu Advertiser*, November 26, 1995.

"The ATL-98 Carvair," parts 1–9. *Anglia Aeronews*, February 1969, May 1969, June 1969, January 1970, February 1970, March 1970, April 1970.

"Australian Air Fares." *Aircraft*, November 1965, p. 20.

"Automated Ansett." *Aeroplane and Commercial Aviation News (AeCAN)— Air Holdings Supplement*, September 23, 1965.

"The Aviation Traders Carvair." *Eagle, The Top Paper for Boys*, 30 June 1962, p. 20.

Bowers, Peter M. "Strato Freighter." *Airpower,* Volume 29, Number 4.

"Braniff International Airways 1930–1982." *AAHS,* Vol. 46 No. 3, p. 190.

"British Indepedent Airlines 1946–1976." *Propliner*, pp. 45–46, 84.

"British United Air Ferries." *Flight International*, 2 August 1962, p. 155.

Brouwer, Maarten. "Carvair, Heerser Van De Lucht Wordt Veerpont." *DDA Magazine*, December 1992, p. 14.

_____. "Dienstverband Voor Albion (DEEL 2)." *DDA Magazine*, February 1992, pp. 14–16.

_____. "Rotterdam EN DE Carvair." *DDA Magazine*, August 1992, pp. 14–16.

"BUAF Cuts Its Losses." *Flight International*, February 02, 1967.

Bulban, Erwin J. "Falcon Airways Doubling its DC-4 Fleet." *Aviation Week and Space Technology*, July 9, 1979, p. 33.

Burgi, Hansjorg. "What type of aircraft is that?" *Jetstream*, August 1995, pp. 40–43.

"CAB, Background to the Opening of Car Ferry Era 2." *Flight International*, January 04, 1962.

"Carvair Data Sheet." *The Aeroplane and Commercial Aviation News (AeCAN)— Air Holdings Supplement*, June 07, 1962.

"Carvair Ferry Proves Sturdy, Easy to Fly." *Aviation Week* September 23, 1963.

"Carvair Hacked to Pieces." *Propliner* 20, Winter 1983.

"Carvair in Service." *Flight*, February 22, 1962.

"Carvair — Ultimate In Space Utilization." *Flight International*, January 24, 1963.

"Carvair's Maiden Flight." *Flight*, 29 June 1961, p. 890.

"Carvairs for CAT." *The Aeroplane and Commercial Aviation News (AeCAN)— Air Holdings Supplement*, March 11, 1965.

"Carvairs for the Congo." *BUA Group News*, No. 2, August 02, 1962.

"Carvairs for East Anglian." *Flight International*, February 01, 1962.

"Carvairs Scheduled for U.K.–Swiss Routes." *Aviation Week and Space Technology*, September 4, 1961.

"Channel Air Bridge." *Flight*, April 13, 1961.

"Channel Air Bridge." *Flight*, August 24, 1961.

"Channel Air Bridge." *Flight International*, January 04, 1962.

"Channel Air Bridge, Background to the Opening of Car Ferry Era 2." *Flight International*, January 04, 1962, p. 10.

"Classic Aircraft Return to Hawaii." *Western-Pacific Intercom FAA,* Issue 92–07, 17 February 1992.

Clayton, Mike. "Carvair Carvair Carvair." *(Australian) Aircraft & Aerospace Magazine*, 1987, pp. 17–22.

Cooper, Martin. "The Aerial Dinosaur Flies On." *Propliner* 68, Autumn 1996, p. 28.

Cowell, Graham. "Texas Bound with Plain Jane." *Propliner* 3, Summer 1979.

Craven, Guy A. "Photography & The Carvair." *Model News*, Spring 2003, pp. 8–13.

Crump, Robert Earl. "Airbus Guppies & Belugas." *Airways* July/August 1996, p. 17; July 1999, p. 18.

Davies, Ed. "American Airlines and the Rocket DC-4." *Airliners* 50 March-April 1998, p. 42.

Davies, Victor. "Old Big Nose." *Aircraft Illustrated*, January 1995, p. 28.

"DC-4s For Hire, Matson Announces." *Honolulu Advertiser*, July 06, 1946.

Debus, Ralph. "Flying the Carvair." *Airborne Props and Jets*, Spring 1986, p. 12.

_____. "Prop News." *Propliner* 89, Winter 2001.

Dedman, Robert. "Congo Pilot." *Flight Journal*, April 2003, p. 60.

Dedwell, Don. "Pioneer DC4 Became Poor Man's Freighter." *Miami Herald*, June 24, 1969, p. 16A.

"Deeper Penetration for Channel Air Bridgehead." *Flight*, June 19, 1959, p. 824.

"Democratic Republic of Congo: A Republic with a Single-party System." http://infopls.com/spot/drcongo.html (13 January 1999).

"Dominicana de Aviación." www.answers.com/topic/dominicana-de-aviacion (18 June 2005).

Donohue, Ken. "HawkAir, from Reindeer to People." *Airways*, December 2003, p. 33.

_____. "LTU International Airways." *Airways*, March 2007, p. 32.

Edwards, Ron. "Matson Airlines." *Ampersand*, Winter 1995–1996, p. 17.

"Falcon Airways Introduces Carvair at Metroplex." *Dallas Trader Monthly Newsletter*, September 1979.

"Four-Engine Transports 1940–1945 Part VI." *AAHS Journal*, Vol. 46 No. 3, Fall 2001, p. 208.

"From Berlin Airlift to Skytrain." *Take Off* 82. Volume 7, p. 2284.

"From Broken Hill to BUAF." BUA Group News, No. 2, issue 92–07, February 17, 1992.

Gaskell, Keith. "European Super Guppy Operations." *Propliner* 63, Summer 1995, p. 26.

_____. "Phoebus Apollo's African Propliners." *Airways*, May 2007, p. 62.

_____. "Rising Star." *Propliner* 82, Spring 2000, p. 36.

Gates, Peter J. "The ATL.98 Carvair Down Under." *Airliners* 68, March/April 2007.

George, Hunter, and Bruce Giles. "Plane Cut $1 Million Path of Damage." *Miami Herald*, June 24, 1969, p. 1.

Goodall, Geoff. "Far Eastern Commandos." *Propliner* 9, Spring 1981, p. 31.

Goring, John. "The Carvair Story." *Aircraft Illustrated*, July 1971, p. 258.

Guillem, Jacques. "Air Bridge." *Airways*, March 2004, p. 30.

Gunson, Simon. "Aviation Traders Carvair, Nationwide Air History." www.geocities.com/anjapaul/nationwide.htm (06 December 2006).

Hayes, Karl, and Eamon Power. "Plain Jane Receives a Face-Lift." *Propliner* 67, Summer 1996, p. 10.

_____. "Roger Brooks and The Superlative Seven." *Propliner* 67, Summer 1996, p. 26.

Henderson, David Zeus. "North American Carvair Revival." *Propliner* 75, Summer 1998, p. 8.

Holland, Dick. "Pilot Might Have Been Confused." *Miami News*, June 24, 1969.

Howard, Paul. "Convairs of The Triple Nine Triangle." *Propliner* 62, Spring 1995, p. 38.

_____. "Oceanic Cargo Legend." *Propliner* 88, Autumn 2001, p. 16.

_____. "Supertramp." *Propliner* 77, Winter 1998, p. 38.

_____ and Bob Hickox. "Mercy Angels of Angkor, The Phnom Penh Airlift 1973 to 1975." *Air Enthusiast*, November/December 2005.

Hutchison, Iain. "The Leaping Lion, British World Airlines." *Airways*, November 2001, p. 31.

_____. "Royal Air Cambodge." *Airways*, May 1999, p. 31.

Jones, Tony Merton. "The Carvair in British Airline Service." *Propliner* 3, Summer 1979.

_____. "The Channel Air Bridge." *Propliner* 99, Summer 2004, p. 26.

_____. "The First European Airbus," Part Four. *Propliner* 81, Winter 99.

_____. "The First European Airbus," Part Five. *Propliner* 82, Spring 2000.

_____. "A Friendly Way to Fly," Part One. *Propliner* 87, Summer 2001, p. 26.

_____. "The Ocean Liner of The Air," Part Five. *Propliner* 76, Autumn 1998, p. 18.

_____. "Southend's Golden Wonder," Part Three. *Propliner* 53, Winter 1992, p. 14.

_____. "Sunshine Buccaneers," Part Two. *Propliner* 95, Summer 2003, p. 28.

Keating, Sean. "HawkAir Grows." *Propliner* 70, Spring 1997, p. 5.

Kinder, Steve. "On Safari." *Airways*, December 2001, p. 13.

King, Derek. "Commercial Failures, the Aviation Traders Accoutant." *Air Britain Aviation World*, Summer 2006, p. 58.

King, Terry Johnson. "Dominicans Are Amazed U.S. Let Plane Take Off." *Miami News*, June 25, 1969, p. 1A.

"Laker's Plump Partridge." *Propliner* 70, Spring 1997, p. 31.

"Lengthening the Bridge." *Flight*, June 14, 1962.

Leftley, A.C. "The Carvair Story." *Shell Aviation News*, 1966, No. 333, p. 2.

Linares, Jose Parejo. "Aviaco, 50 Anos." *Empuje*, Junio 1998, pp. 27–34.

Lo Bao, Phil. "The Carvair — Thirty Years On." *World Airline Fleet News*, October 1992.

Magnusson, Michael. "High Tech Hawkers." *Propliner* 80, Autumn 1999.

Mak, Chris J. "Antipodean Adventures." *Propliner* 73, Winter 1997, p. 39.

Marson, Peter J. "Air Bridge to Canada." *Propliner* 67, Summer 1996, p. 31.

Montgomerie, James. "The Rise and Fall of the Car Air Ferries." *Aviation News*, Vol. 22, No. 8, 10–23 September 1993, p. 359.

Murdoch, Simon. "Carvair — End of an Era." *Air Pictorial*, March 1982, p. 108.

_____. "Carvair, the Final Chapter." *Air Pictorial*, December 1997, p. 647.

_____. "The Carvair's Last Chapter." *Aircraft Illustrated*, October 1979, p. 487.

"A New Car-ferry." *The Aeroplane and Astronautics*, June 29, 1961, p. 749.

"New from Southend." *The Aeroplane and Astronautics*, June 29, 1961, p. 738.

"New Routes, New Aircraft for BAF." *Aviation News*, Spring 1973.

"News on Airways." *Airways*, July 2002, p. 6.

"Passengers, Crew Unhurt as EPA Plane Crashes." *(St. Johns Newfoundland) Evening Telegram*, September 30, 1968.

Payne, Richard. "Aviation Traders' Cargo Conversion." *Air International*, December 1965, p. 360.

Pearcy, Arthur. "ATL.98. Carvair." *Aviation News*, Vol. 19 No. 25, 26 April–9 May 1991.

Pettersen, Ralph. "Topflight Tanker Force." *Propliner* 86, Spring 2001, p. 40.

Peyer, Peter F. "The Con Bag Haulers Call It a Day." *Propliner* 79, Summer 1999, p. 7.

"Phoebus Apollo." *Airways*, July 2007, p. 63.

"Phoebus Apollo's African Propliners." *Airways*, May 2007, p. 62.

"Porky Pete and His Fat Friends Take Off." *Evening Echo*, January 30, 1975.

Power, Eamon. "Carvair, Five Years in Ireland." *Propliner* 15, Autumn 1982.

_____. "Carvair: 21 Years On." *Propliner* 15, Autumn 1982.

"Progress with the ATL-98." *Flight*, 24 June 1960, p. 875.

"Progress with the Carvair." *The Aeroplane*, February 8, 1962, p. 148.

Raith, Norbert, and Ralph Pettersen. "Carvair Tragedy." *Propliner* 70, Spring 1997, p. 7.

"Rethinking in Australia." *Aeroplane*, July 10, 1968.

"Second Generation Carvairs." *The Aeroplane and Commercial Aviation News (AeCAN) — Air Holdings Supplement*, April 08, 1965.

Seekings, John. "How It All Happened." *The Aeroplane*

and Commercial Aviation News (AeCAN)— Air Holdings Supplement, January 21, 1965.

Siegrist, Martin E. "A Jumbo Rejuvenated." *Propliner* 55, Summer 1993.

"Sir Freddie Laker." *Airways*, May/June 1995, p. 19.

"Sir Freddie Laker, 1922 — 2006." *Propliner* 105, Winter 2005, p. 19.

"Skymaster Story, the fourth Douglas Commercial." *Air Enthusiast*, Fifteen, June 1981, p. 38.

Smith, Kenny. "Airplanes in the Wrangells, Mergers." *St. Elias News*, Vol. 10, Issue 2, March-April 2001.

Smith, Kit. "Two Cargo Air Carriers Join Forces." *Honolulu Advertiser*, 1992.

"Smoked Out," Mailbag. *Airways*, July 2007, p. 13.

Soler, Frank, and Fred Andersen. "Two Colombians Anxious to Talk." *Miami Herald*, June 26, 1969, p. 4A.

Sosin, Milt. "Crashed Plane Just Repaired." *Miami News*, June 24, 1969, p. 1A.

Tenby, Henry. "HawkAir: BC's British Metal Act." *Airways*, March 1999, p. 53.

Tegler, Jan. "A over S: An Interview with Sir Freddie Laker." *Airways*, October 1999, p. 51.

Thiedeman, Roger. "Air Ceylon." *Airways*, August 1998, p. 55.

"Tightening the Belt." http://ourworld.compuserve.com/homepages/Rallport/belt.htm (12 January 1999).

"Tribute to Progress." *St. John's Telegram*, 25 October 1963.

"Turboprop Carvair." *Flight International*, 1964.

Van De Plas, Ferand. "The Douglas Swingtails." *Propliner* 76, Autumn 1998, p. 26.

"Vehicle Ferry Service." *Aersceala* (staff magazine of Aer Lingus), Vol XVI, No. 8, April 1963.

"We're Going to Have a Crash." *Miami News*, June 24, 1969, p. 7A.

Wiles, Greg, "Hawaii Pacific Air Taking Off." *Honolulu Advertiser*, February 1992.

Williams, Nicholas M. "Globemaster: The Douglas C-74." *AAHS Journal*, Vol. 25, No. 2, Summer 1980, p. 82.

Wood, J.R.T. "Rhodesian Insurgency." http://ourworld.compuserve.com/homepages/Rallport/wood1.htm (13 January 1999).

Wooley, David. "Diversity at Southend." *Flight International*, January 14, 1971, p. 60.

Wrixon, G.R. "Aviation Traders— A Combination of Two Philosophies." *The Aeroplane and Commercial Aviation News (AeCAN)— Air Holdings Supplement*, April 04, 1963.

Perodicals and Newspapers

Aeroplane and Aeronautics
Aeroplane and Commercial Aviation News
Air Britain Digest
Air Britain News
Air International
Air Letter
Air Pictorial (December 1962, January 1963, February 1963, April 1963, May 1963, July 1963, August 1963, February 1964, March 1964, April 1964, May 1964, June 1964, July 1964, August 1964, November 64, February 1965, April 1965, May 1965, August 1965, November 1965, December 1965, February 1966, June 1966).
Airways
Asia Plane
Aviation News
Evening Telegram, St. John's, Newfoundland, October 25, 1968.
Flight
Flight International
Propliner
Shell Aviation News
Southend Aeronews and Anglia Aeronews, various issues 1962–1979
Southend Standard, 1960–1965
Wingspan (ATEL newsletter), August 1967.

Advertisments, Pamphlets and Schedules

BAF Route Map & Souvenir.

Breve Historia del Avión Carvair ATL 98.

Silver City Air Ferry Schedule, 1954.

"The History of Southend Airport, 1914 to the Present." D. Hollander, ed. London Southend Airport Co. Ltd., Airport Marketing and Information Services, 2001.

Transmeridian-BAF Promotional Brochure.

Reports and Registers

Ansett Archive, Fred Nivens extracts.

British Aviation Register.

Carvair Production Data, ATEL Design Office.

Carvair General Data, Pacific Aerospace.

Complete Civil Aircraft Registers of The Netherlands since 1920, Herman Dekker.

Condition Report N406JT, N407JT. Harry Clarkson, October 1983.

Dail Eireann (Irish Parliament), extract from debate. Vol. 198, November 28, 1962, Traffic at Dublin Airport and Car Ferry Tourism; Vol. 207, February 25, 1964, Transport and Power debate; Vol. 216, June 09, 1965, Car Ferry and Carriage of Horses.

Evaluation of Hawaiian Inter-Island Air Cargo Service. CAD Report 005, Auckland, New Zealand, December 1983.

FAA Registration/Airworthiness File N33AC.

FAA Registration/Airworthiness File N83FA, CD-Rom.

FAA Registration/Airworthiness File N89FA, CD-Rom.

FAA Registration/Airworthiness File N103, CD-Rom.

FAA Registration/Airworthiness File N301JT, CD-Rom.

FAA Registration/Airworthiness File N898AT, CD-Rom.

FAA Registration/Airworthiness File N5459M, CD-Rom.

FAA Registration/Airworthiness File N55243, CD-Rom.

Heavy Maintenance Performed. Pacific Aerospace Report on ships 20 and 21, Nelson & Hamilton NZ. 1982.

Ministry of Civil Aviation. Civil Aircraft Accident Hermes G-ALDN, Mauritania (French West Africa). London: Her Majesty's Stationery Office, 26 May 1952.

Netherlands Air Accident Board. Ministry of Aviation, Civil Aircraft Accident Report (CAP 201). London: Her Majesty's Stationery Office, 30 October 1963.

NTSB AAR-70–17, Crash HI-168.

NTSB ATL97FAO57, Crash of N83FA.

NTSB ANC07LAO93, Crash of N103.

NTSB ANC07LAD40, Crash of N898AT.

Transportation Safety Board of Canada Report A96D0175, Crash of C-FGNI.

Manuals

Carvair Aircraft Operating Manual, Hawaii Pacific Air.

Carvair Maintenance Manual, Aviation Traders Ltd., July 1961.

Carvair Parts Manual, Aviation Traders Ltd.

Carvair Pilots Checklist, Aer Lingus.

Carvair Repair Manual, Aviation Traders Ltd.

Carvair Weight & Balance Manual, Hawaii Pacific Air.

DC-4 Maintenance Manual, Douglas Aircraft Division, December 15, 1945.

DC-4 Parts Manual, Power Plant, Douglas Aircraft.

Flight Manual for the ATL.98 Carvair CF-EPW, Eastern Provincial.

Logs

ATC Departure/Arrival Log, Southend, 18 September 1962.

Callier, Ian. Spotters Log, ATC Southend.

Dedman, Capt. Robert. Logbook.

Dible, Capt. Joe. Logbook.

Goring, Richarg. Spotters Log, ATEL Stores.

May, Ruth. Journal and personal notes.

Niven, Fred. Australian Air Log.

O'Callaghan, Capt. Michael. Logbook.

Rakisits, Capt. Paul. Logbook.

Sullivan, Capt. J.J. Logbook.

Websites

Aero Union Corporation. www.aerounion.com/index.htm (07 July 2001).

Air Charter — London. www.airlines.freeuk.com/a998.htm (02 August 2001).

Aircraft Traders Belgium. www.atbelgium.com/pages/ser_organization.html (29 June 2007).

Airwork. www.airlines2.freeuk.com/2a346.htm (01 August 2001).

Aviation Traders Ltd-ATL. Aviation Design Engineering Consultants. www.atl.aero/ (05 January 2007).

British United Air Ferries. www.airlines.freeuk.com/a994.htm (02 August 2001).

British United Airways. www.airlines.freeuk.com/a197.htm (02 August 2001).

Hunting-Clan Air Transport. www.arilines2.freeuk.com/2a345.htm (02 August 2001).

Phoebus Apollo's Carvair. www.phoebusapollo.co.za (02 March 2003).

Silver City Airways. www.airlines.freeuk.com/a992.htm (02 August 2001).

Interviews

Blumenthal, Jim. Owner, N80FA N89FA.

Callier, Ian. ATC Southend.

Craven, Guy. ATEL photographer.

Dedman, Robert. Interocean Airways 1962–63, captain.

Dible, Joe. Aer Lingus, captain.

Dyess, George. Full-Air, technician.

Fairchild, A.P. Interocean, PAL, PAE, HPA.

Gordon, Keith. Managing engineer, SINGAS.

Goring, Richard. Aviation Traders.

Hawkins Paul. Founder, HawkAir; owner, C-GAAH.

Jones, Kerry. FBO radio operator, Griffin Spalding Airport.

Kerry, Brian. ATEL chief aerodynamicist.

McSwiggan, Bruce. Academy Airlines.

McSwiggan, Mark. Academy Airlines.

McSwiggan, Robert D. "Bob," Owner, N83FA, N89FA, N55243; managing director, Custom Air Service.

May, Ruth. Owner, N83FA, Falcon Airways.

Mees, Brian. BAF inspection supervisor.

Moss, Frank. Owner, Hondu Carib and N103.

O'Callaghan, Michael. Aer Lingus, captain.

Rakisits, Paul. Interocean, SEAAT, PAE.

Sullivan, J.J. Aer Lingus, captain.

Tucker, Randy. PAE, captain.

Airlines, Museums, Historical Societies and Associations

Aer Lingus Irish International.

Air Britain (AB).

American Aviation Historical Society (AAHS).

Ansett Historical Archive.

British Air Ferries.

British World Airlines.

8th Airforce Museum, Holton, Halesworth, Suffolk, England.

Irish Aviation Archive, Meath Aero Museum.

Retired Irish Airline Pilots Association (RIALPA).

Southend Airport Authority.

The Airways Museum and Civil Aviation Historical Society (CAHS), Australia.

Index

Numbers in *bold italics* indicate pages with photographs or illustrations.